Lecture Notes in Mathematics   1847

Editors:
J.-M. Morel, Cachan
F. Takens, Groningen
B. Teissier, Paris

Tomasz R. Bielecki   Tomas Björk
Monique Jeanblanc   Marek Rutkowski
José A. Scheinkman   Wei Xiong

Paris-Princeton Lectures on Mathematical Finance 2003

Editorial Committee:

R. A. Carmona, E. Çinlar,
I. Ekeland, E. Jouini,
J. A. Scheinkman, N. Touzi

 Springer

Authors

Tomasz R. Bielecki
Department of Applied Mathematics
Illinois Institute of Technology
Chicago, IL 60616, USA
e-mail: bielecki@iit.edu

Tomas Björk
Department of Finance
Stockholm School of Economics
Box 6501
11383 Stockholm, Sweden
e-mail: tomas.bjork@hhs.se

Monique Jeanblanc
Equipe d'Analyse et Probabilités
Université d'Évry-Val d'Essonne
91025 Évry, France
e-mail:
Monique.Jeanblanc@maths.univ-
evry.fr

Marek Rutkowski
Faculty of Mathematics and
Information Science
Warsaw University of Technolgy
Pl. Politechniki 1
00-661 Warsaw, Poland
e-mail: markrut@mini.pw.edu.pl

José A. Scheinkman
Bendheim Center of Finance
Princeton University
Princeton NJ 08530, USA
e-mail: joses@princeton.edu

Wei Xiong
Bendheim Center of Finance
Princeton University
Princeton NJ 08530, USA
e-mail: wxiong@princeton.edu

[The addresses of the volume editors appear
on page IX]

Library of Congress Control Number:2004110085

Mathematics Subject Classification (2000): 92B24, 91B28, 91B44, 91B70, 60H30, 93E20

ISSN 0075-8434
ISBN 3-540-22266-9 Springer Berlin Heidelberg New York
DOI 10.1007/b98353

Typesetting: Camera-ready TeX output by the authors
41/3142-543210 - Printed on acid-free paper

Preface

This is the second volume of the Paris-Princeton Lectures in Mathematical Finance. The goal of this series is to publish cutting edge research in self-contained articles prepared by well known leaders in the field or promising young researchers invited by the editors. Particular attention is paid to the quality of the exposition, and the aim is at articles that can serve as an introductory reference for research in the field.

The series is a result of frequent exchanges between researchers in finance and financial mathematics in Paris and Princeton. Many of us felt that the field would benefit from timely exposés of topics in which there is important progress. René Carmona, Erhan Cinlar, Ivar Ekeland, Elyes Jouini, José Scheinkman and Nizar Touzi will serve in the first editorial board of the Paris-Princeton Lectures in Financial Mathematics. Although many of the chapters in future volumes will involve lectures given in Paris or Princeton, we will also invite other contributions. Given the current nature of the collaboration between the two poles, we expect to produce a volume per year. Springer Verlag kindly offered to host this enterprise under the umbrella of the Lecture Notes in Mathematics series, and we are thankful to Catriona Byrne for her encouragement and her help in the initial stage of the initiative.

This second volume contains three chapters. The first one is written by Tomasz Bielecki, Monique Jeanblanc and Marek Rutkowski. It reviews recent developments in the *reduced form* approach to credit risk and offers an exhaustive presentation of the hedging issues when contingent claims are subject to counterparty default. The second chapter is contributed by Tomas Bjork and is based on a short course given by him during the Spring of 2003 at Princeton University. It gives a detailed introduction to the geometric approach to mathematical models of fixed income markets. This contribution is a welcome addition to the long list of didactic surveys written by the author. Like the previous ones, it is bound to become a reference for the newcomers to mathematical finance interested in learning how and why the geometric point of view is so natural and so powerful as an analysis tool. The last chapter is due to José Scheinkman and Wei Xiong. It considers dynamic trading by agents with heterogeneous beliefs. Among other things, it uses short sale constraints and overconfidence of groups of agents to show that equilibrium prices can be consistent with speculative bubbles.

It is anticipated that the publication of this volume will coincide with the *Third World Congress* of the Bachelier Finance Society, to be held in Chicago (July 21-24, 2004).

The Editors
Paris / Princeton
June 04, 2004.

Contents

Heterogeneous Beliefs, Speculation and Trading in Financial Markets

Editors

René A. Carmona
Paul M. Wythes '55 Professor of Engineering and Finance
ORFE and Bendheim Center for Finance
Princeton University
Princeton NJ 08540, USA
email: rcarmona@princeton.edu

Erhan Çinlar
Norman J. Sollenberger Professor of Engineering
ORFE and Bendheim Center for Finance
Princeton University
Princeton NJ 08540, USA
email: cinlar@princeton.edu

Ivar Ekeland
Canada Research Chair in Mathematical Economics
Department of Mathematics, Annex 1210
University of British Columbia
1984 Mathematics Road
Vancouver, B.C., Canada V6T 1Z2
email: ekeland@math.ubc.ca

Elyes Jouini
CEREMADE, UFR Mathématiques de la Décision
Université Paris-Dauphine
Place du Maréchal de Lattre de Tassigny
75775 Paris Cedex 16, France
email: jouini@ceremade.dauphine.fr

José A. Scheinkman
Theodore Wells '29 Professor of Economics
Department of Economics and Bendheim Center for Finance
Princeton University
Princeton NJ 08540, USA
email: joses@princeton.edu

Nizar Touzi
Centre de Recherche en Economie et Statistique
15 Blvd Gabriel Péri
92241 Malakoff Cedex, France
email: touzi@ensae.fr

Hedging of Defaultable Claims

Tomasz R. Bielecki,[1] Monique Jeanblanc[2] and Marek Rutkowski[3]

[1] Department of Applied Mathematics
Illinois Institute of Technology
Chicago, USA
email: bielecki@iit.edu
[2] Equipe d'Analyse et Probabilités
Université d'Évry-Val d'Essonne
Évry, France
email: Monique.Jeanblanc@maths.univ-evry.fr
[3] Faculty of Mathematics and Information Science
Warsaw University of Technology
and
Institute of Mathematics of the Polish Academy of Sciences
Warszawa, Poland
email: markrut@mini.pw.edu.pl

Summary. The goal of this chapter is to present a survey of recent developments in the practically important and challenging area of hedging credit risk. In a companion work, Bielecki et al. (2004a), we presented techniques and results related to the valuation of defaultable claims. It should be emphasized that in most existing papers on credit risk, the risk-neutral valuation of defaultable claims is not supported by any other argument than the desire to produce an arbitrage-free model of default-free and defaultable assets. Here, we focus on the possibility of a perfect replication of defaultable claims and, if the latter is not feasible, on various approaches to hedging in an incomplete setting.

Key words: Defaultable claims, credit risk, perfect replication, incomplete markets, mean-variance hedging, expected utility maximization, indifference pricing.
MSC 2000 subject classification. 91B24, 91B28, 91B70, 60H30, 93E20

Acknowledgements: Tomasz R. Bielecki was supported in part by NSF Grant 0202851. Monique Jeanblanc thanks T.R.B. and M.R. for their hospitality during her visits to Chicago and Warsaw. Marek Rutkowski thanks M.J. for her hospitality during his visit to Evry. Marek Rutkowski was supported in part by KBN Grant PBZ-KBN-016/P03/1999.

Introduction

The present chapter is naturally divided into three different parts.

Part I is devoted to methods and results related to the replication of defaultable claims within the reduced-form approach (also known as the intensity-based approach). Let us mention that the replication of defaultable claims in the so-called structural approach, which was initiated by Merton (1973) and Black and Cox (1976), is entirely different (and rather standard), since the value of the firm is usually postulated to be a tradeable underlying asset. Since we work within the reduced-form framework, we focus on the possibility of an exact replication of a given defaultable claim through a trading strategy based on default-free and defaultable securities. First, we analyze (following, in particular, Vaillant (2001)) various classes of self-financing trading strategies based on default-free and defaultable primary assets. Subsequently, we present various applications of general results to financial models with default-free and defaultable primary assets are given. We develop a systematic approach to replication of a generic defaultable claim, and we provide closed-form expressions for prices and replicating strategies for several typical defaultable claims. Finally, we present a few examples of replicating strategies for particular credit derivatives. In the last section, we present, by means of an example, the PDE approach to the valuation and hedging of defaultable claims within the framework of a complete model.

In Part II, we formulate a new paradigm for pricing and hedging financial risks in incomplete markets, rooted in the classical Markowitz mean-variance portfolio selection principle and first examined within the context of credit risk by Bielecki and Jeanblanc (2003). We consider an investor who is interested in dynamic selection of her portfolio, so that the expected value of her wealth at the end of the pre-selected planning horizon is no less then some floor value, and so that the associated risk, as measured by the variance of the wealth at the end of the planning horizon, is minimized. If the perfect replication is not possible, then the determination of a price that the investor is willing to pay for the opportunity, will become subject to the investor's overall attitude towards trading. In case of our investor, the bid price and the corresponding hedging strategy is to be determined in accordance with the mean-variance paradigm.

The optimization techniques used in Part II are based on the mean-variance portfolio selection in continuous time. To the best of our knowledge, Zhou and Li (2000) were the first to use the embedding technique and linear-quadratic (LQ) optimal control theory to solve the continuous-time mean-variance problem with assets having deterministic diffusion coefficients. Their approach was subsequently developed in various directions by, among others, Li et al. (2001), Lim and Zhou (2002), Zhou and Yin (2002), and Bielecki et al. (2004b). For an excellent survey of most of these results, the interested reader is referred to Zhou (2003).

In the final part, we present a few alternative ways of pricing defaultable claims in the situation when perfect hedging is not possible. We study the indifference pricing approach, that was initiated by Hodges and Neuberger (1989). This method leads us to solving portfolio optimization problems in an incomplete market model, and we shall use the dynamic programming approach. In particular, we compare the indifference prices obtained using strategies adapted to the reference filtration to the indifference prices obtained using strategies based on the enlarged filtration, which encompasses also the observation of the default time. We also solve portfolio optimization problems for the case of the exponential utility; our method relies here on the ideas of Rouge and El Karoui (2000) and Musiela and Zariphopoulou (2004). Next, we study a particular indifference price based on the quadratic criterion; it will be referred to as the quadratic hedging price. In a default-free setting, a similar study was done by Kohlmann and Zhou (2000). Finally, we present a solution to a specific optimization problem, using the duality approach for exponential utilities.

Part I. Replication of Defaultable Claims

The goal of this part is the present some methods and results related to the replication of defaultable claims within the *reduced-form approach* (also known as the *intensity-based approach*). In contrast to some other related works, in which this issue was addressed by invoking a suitable version of the martingale representation theorem (see, for instance, Bélanger et al. (2001) or Blanchet-Scalliet and Jeanblanc (2004)), we analyze here the possibility of a perfect replication of a given defaultable claim through a trading strategy based on default-free and defaultable securities. Therefore, the important issue of the choice of primary assets that are used to replicate a defaultable claim (e.g., a vulnerable option or a credit derivative) is emphasized. Let us stress that replication of defaultable claims within the structural approach to credit risk is rather standard, since in this approach the default time is, typically, a predictable stopping time with respect to the filtration generated by prices of primary assets.

By contrast, in the intensity-based approach, the default time is not a stopping time with respect to the filtration generated by prices of default-free primary assets, and it is a totally inaccessible stopping time with respect to the enlarged filtration, that is, the filtration generated by the prices of primary assets and the jump process associated with the random moment of default.

Our research in this part was motivated, in particular, by the paper by Vaillant (2001). Other related works include: Wong (1998), Arvanitis and Laurent (1999), Greenfield (2000), Lukas (2001), Collin-Dufresne and Hugonnier (2002) and Jamshidian (2002).

For a more exhaustive presentation of the mathematical theory of credit risk, we refer to the monographs by Cossin and Pirotte (2000), Arvanitis and Gregory (2001), Bielecki and Rutkowski (2002), Duffie and Singleton (2003), or Schönbucher (2003).

This part is organized as follows. Section 1 is devoted to a brief description of the basic concepts that are used in what follows. In Section 2, we formally introduce the definition of a generic defaultable claim (X, Z, C, τ) and we examine the basic features of its ex-dividend price and pre-default value. The well-known valuation results for defaultable claims are also provided. In the next section, we analyze (following, in particular, Vaillant (2001)) various classes of self-financing trading strategies based on default-free and defaultable primary assets.

Section 4 deals with applications of results obtained in the preceding section to financial models with default-free and defaultable primary assets. We develop a systematic approach to replication of a generic defaultable claim, and we provide closed-form expressions for prices and replicating strategies for several typical defaultable claims. A few examples of replicating strategies for particular credit derivatives are presented.

Finally, in the last section, we examine the PDE approach to the valuation and hedging of defaultable claims.

1 Preliminaries

In this section, we introduce the basic notions that will be used in what follows. First, we introduce a default-free market model; for the sake of concreteness we focus on default-free zero-coupon bonds. Subsequently, we shall examine the concept of a random time associated with a prespecified hazard process.

1.1 Default-Free Market

Consider an economy in continuous time, with the time parameter $t \in \mathbb{R}_+$. We are given a filtered probability space $(\Omega, \mathbb{F}, \mathbb{P}^*)$ endowed with a d-dimensional standard Brownian motion W^*. It is convenient to assume that $\mathbb{F}$ is the $\mathbb{P}^*$-augmented and right-continuous version of the natural filtration generated by W^*. As we shall see in what follows, the filtration $\mathbb{F}$ will also play an important role of a *reference filtration* for the intensity of default event. Let us recall that any (local) martingale with respect to a Brownian filtration $\mathbb{F}$ is continuous; this well-known property will be of frequent use in what follows.

In the first step, we introduce an arbitrage-free default-free market. In this market, we have the following primary assets:

- A *money market account* B satisfying

$$dB_t = r_t B_t \, dt, \quad B_0 = 1,$$

  or, equivalently,

$$B_t = \exp\left(\int_0^t r_u \, du\right),$$

where r is an $\mathbb{F}$-progressively measurable stochastic process. Thus, B is an $\mathbb{F}$-adapted, continuous, and strictly positive process of finite variation.

• *Default-free zero-coupon bonds* with prices

$$B(t,T) = B_t \, \mathbb{E}_{\mathbb{P}^*}(B_T^{-1} \mid \mathcal{F}_t), \quad \forall\, t \leq T,$$

where T is the bond's maturity date. Since the filtration $\mathbb{F}$ is generated by a Brownian motion, for any maturity date $T > 0$ we have

$$dB(t,T) = B(t,T)\big(r_t \, dt + b(t,T) \, dW_t^*\big)$$

for some $\mathbb{F}$-predictable, $\mathbb{R}^d$-valued process $b(t,T)$, referred to as the *bond's volatility*.

For the sake of expositional simplicity, we shall postulate throughout that the default-free term structure model is complete. The probability $\mathbb{P}^*$ is thus the unique martingale measure for the default-free market model. This assumption is not essential, however. Notice that all price processes introduced above are continuous $\mathbb{F}$-semimartingales.

Remarks. The bond was chosen as a convenient and practically important example of a tradeable financial asset. We shall be illustrating our theoretical derivations with examples in which the bond market will play a prominent role. Most results can be easily applied to other classes of financial models, such as. models of equity markets, futures markets, or currency markets, as well as to models of LIBORs and swap rates.

1.2 Random Time

Let τ be a non-negative random variable on a probability space $(\Omega, \mathcal{G}, \mathbb{Q}^*)$, termed a *random time* (it will be later referred to as a *default time*). We introduce the jump process $H_t = \mathbb{1}_{\{\tau \leq t\}}$ and we denote by $\mathbb{H}$ the filtration generated by this process.

Hazard process. We now assume that some *reference filtration* $\mathbb{F}$ such that $\mathcal{F}_t \subseteq \mathcal{G}$ is given. We set $\mathbb{G} = \mathbb{F} \vee \mathbb{H}$ so that $\mathcal{G}_t = \mathcal{F}_t \vee \mathcal{H}_t = \sigma(\mathcal{F}_t, \mathcal{H}_t)$ for every $t \in \mathbb{R}_+$. The filtration $\mathbb{G}$ is referred to as to the *full filtration*. it includes the observations of default event. It is clear that τ is an $\mathbb{H}$-stopping time, as well as a $\mathbb{G}$-stopping time (but not necessarily an $\mathbb{F}$-stopping time). The concept of the hazard process of a random time τ is closely related to the process F_t which is defined as follows:

$$F_t = \mathbb{Q}^*\{\tau \leq t \mid \mathcal{F}_t\}, \quad \forall\, t \in \mathbb{R}_+.$$

Let us denote $G_t = 1 - F_t = \mathbb{Q}^*\{\tau > t \mid \mathcal{F}_t\}$ and let us assume that $G_t > 0$ for every $t \in \mathbb{R}_+$ (hence, we exclude the case where τ is an $\mathbb{F}$-stopping time). Then the process $\Gamma : \mathbb{R}_+ \to \mathbb{R}_+$, given by the formula

$$\Gamma_t = -\ln(1 - F_t) = -\ln G_t, \quad \forall t \in \mathbb{R}_+,$$

is termed the *hazard process* of a random time τ with respect to the reference filtration $\mathbb{F}$, or briefly the $\mathbb{F}$-*hazard process* of τ.

Notice that $\Gamma_\infty = \infty$ and Γ is an $\mathbb{F}$-submartingale, in general. We shall only consider the case when Γ is an increasing process (for a construction of a random time associated with a given hazard process Γ, see Section 1.2). This indeed is not a serious compromise to generality. We refer to Blanchet-Scalliet and Jeanblanc (2004) for a discussion regarding completeness of the underlying financial market and properties of the process Γ. They show that if the underlying financial market is complete then the so-called (H) hypothesis is satisfied and, as a consequence, the process Γ is indeed increasing.

Remarks. The simplifying assumption that $\mathbb{Q}^*\{\tau > t \,|\, \mathcal{F}_t\} > 0$ for every $t \in \mathbb{R}_+$ can be relaxed. First, if we fix a maturity date T, it suffices to postulate that $\mathbb{Q}^*\{\tau > T \,|\, \mathcal{F}_T\} > 0$. Second, if we have $\mathbb{Q}^*\{\tau \le T\} = 1$, so that the default time is bounded by some $U = \text{ess sup}\, \tau \le T$, then it suffices to postulate that $\mathbb{Q}^*\{\tau > t \,|\, \mathcal{F}_t\} > 0$ for every $t \in [0, U)$ and to examine separately the event $\{\tau = U\}$. For a general approach to hazard processes, the interested reader is referred to Bélanger et al. (2001).

Deterministic intensity. The study of a simple case when the reference filtration $\mathbb{F}$ is trivial (or when a random time τ is independent of the filtration $\mathbb{F}$, and thus the hazard process is deterministic) may be instructive. Assume that τ is such that the cumulative distribution function $F(t) = \mathbb{Q}^*\{\tau \le t\}$ is an absolutely continuous function, that is,

$$F(t) = \int_0^t f(u)\,du$$

for some density function $f : \mathbb{R}_+ \to \mathbb{R}_+$. Then we have

$$F(t) = 1 - e^{-\Gamma(t)} = 1 - e^{-\int_0^t \gamma(u)\,du}, \quad \forall t \in \mathbb{R}_+,$$

where (recall that we postulated that $G(t) = 1 - F(t) > 0$)

$$\gamma(t) = \frac{f(t)}{1 - F(t)}, \quad \forall t \in \mathbb{R}_+.$$

The function $\gamma : \mathbb{R}_+ \to \mathbb{R}$ is non-negative and satisfies $\int_0^\infty \gamma(u)\,du = \infty$. It is called the *intensity function* of τ (or the *hazard rate*). It can be checked by direct calculations that the process $H_t - \int_0^{t \wedge \tau} \gamma(u)\,du$ is an $\mathbb{H}$-martingale.

Stochastic intensity. Assume that the hazard process Γ is absolutely continuous with respect to the Lebesgue measure (and therefore an increasing process), so that there exists a process γ such that $\Gamma_t = \int_0^t \gamma_u\,du$ for every $t \in \mathbb{R}_+$. Then the $\mathbb{F}$-predictable version of γ is called the *stochastic intensity* of τ with respect to $\mathbb{F}$, or simply the $\mathbb{F}$-intensity of τ. In terms of the stochastic intensity, the conditional probability of the default event $\{t < \tau \leq T\}$, given the full information $\mathcal{G}_t$ available at time t, equals

$$\mathbb{Q}^*\{t < \tau \leq T \mid \mathcal{G}_t\} = \mathbb{1}_{\{\tau > t\}}\, \mathbb{E}_{\mathbb{Q}^*}\left(1 - e^{-\int_t^T \gamma_u\, du} \,\Big|\, \mathcal{F}_t\right).$$

Thus

$$\mathbb{Q}^*\{\tau > T \mid \mathcal{G}_t\} = \mathbb{1}_{\{\tau > t\}}\, \mathbb{E}_{\mathbb{Q}^*}\left(e^{-\int_t^T \gamma_u\, du} \,\Big|\, \mathcal{F}_t\right).$$

It can be shown (see, for instance, Jeanblanc and Rutkowski (2002) or Bielecki and Rutkowski (2004)) that the process

$$H_t - \Gamma_{\tau \wedge t} = H_t - \int_0^{\tau \wedge t} \gamma_u\, du = \int_0^t (1 - H_u)\gamma_u\, du, \quad \forall\, t \in \mathbb{R}_+,$$

is a (purely discontinuous) $\mathbb{G}$-martingale

Construction of a Random Time

We shall now briefly describe the most commonly used construction of a random time associated with a given hazard process Γ. It should be stressed that the random time obtained through this particular method – which will be called the *canonical construction* in what follows – has certain specific features that are not necessarily shared by all random times with a given $\mathbb{F}$-hazard process Γ. We start by assuming that we are given an $\mathbb{F}$-adapted, right-continuous, increasing process Γ defined on a filtered probability space $(\widetilde{\Omega}, \mathbb{P}^*, \mathbb{P}^{**})$. As usual, we assume that $\Gamma_0 = 0$ and $\Gamma_\infty = +\infty$. In many instances, the hazard process Γ is given by the equality

$$\Gamma_t = \int_0^t \gamma_u\, du, \quad \forall\, t \in \mathbb{R}_+,$$

for some non-negative, $\mathbb{F}$-predictable, stochastic intensity γ. To construct a random time τ such that Γ is the $\mathbb{F}$-hazard process of τ, we need to enlarge the underlying probability space $\widetilde{\Omega}$. This also means that Γ is not the $\mathbb{F}$-hazard process of τ under $\mathbb{P}^*$, but rather the $\mathbb{F}$-hazard process of τ under a suitable extension $\mathbb{Q}^*$ of the probability measure $\mathbb{P}^*$. Let ξ be a random variable defined on some probability space $(\widehat{\Omega}, \widehat{\mathcal{F}}, \widehat{\mathbb{Q}})$, uniformly distributed on the interval $[0, 1]$ under $\widehat{\mathbb{Q}}$. We consider the product space $\Omega = \widetilde{\Omega} \times \widehat{\Omega}$, endowed with the product σ-field $\mathcal{G} = \mathcal{F}_\infty \otimes \widehat{\mathcal{F}}$ and the product probability measure $\mathbb{Q}^* = \mathbb{P}^* \otimes \widehat{\mathbb{Q}}$. The latter equality means that for arbitrary events $A \in \mathcal{F}_\infty$ and $B \in \widehat{\mathcal{F}}$ we have $\mathbb{Q}^*\{A \times B\} = \mathbb{P}^*\{A\}\widehat{\mathbb{Q}}\{B\}$. We define the random time $\tau : \Omega \to \mathbb{R}_+$ by setting

$$\tau = \inf\{t \in \mathbb{R}_+ : e^{-\Gamma_t} \leq \xi\} = \inf\{t \in \mathbb{R}_+ : \Gamma_t \geq \eta\},$$

where the random variable $\eta = -\ln \xi$ has a unit exponential law under $\mathbb{Q}^*$. It is not difficult to find the process $F_t = \mathbb{Q}^*\{\tau \leq t \mid \mathcal{F}_t\}$. Indeed, since clearly $\{\tau > t\} = \{\xi < e^{-\Gamma_t}\}$ and the random variable Γ_t is $\mathcal{F}_\infty$-measurable, we obtain

$$\mathbb{Q}^*\{\tau > t \,|\, \mathcal{F}_\infty\} = \mathbb{Q}^*\{\xi < e^{-\Gamma_t} \,|\, \mathcal{F}_\infty\} = \widehat{\mathbb{Q}}\{\xi < e^{-x}\}_{x=\Gamma_t} = e^{-\Gamma_t}.$$

Consequently, we have

$$1 - F_t = \mathbb{Q}^*\{\tau > t \,|\, \mathcal{F}_t\} = \mathbb{E}_{\mathbb{Q}^*}\big(\mathbb{Q}^*\{\tau > t \,|\, \mathcal{F}_\infty\} \,|\, \mathcal{F}_t\big) = e^{-\Gamma_t},$$

and so F is an $\mathbb{F}$-adapted, right-continuous, increasing process. It is also clear that Γ is the $\mathbb{F}$-hazard process of τ under $\mathbb{Q}^*$. Finally, it can be checked that any $\mathbb{P}^*$-Brownian motion W^* with respect to $\mathbb{F}$ remains a Brownian motion under $\mathbb{Q}^*$ with respect to the enlarged filtration $\mathbb{G} = \mathbb{F} \vee \mathbb{H}$.

2 Defaultable Claims

A generic defaultable claim (X, C, Z, τ) with maturity date T consists of:

- The *default time* τ specifying the random time of default and thus also the default events $\{\tau \le t\}$ for every $t \in [0, T]$. It is always assumed that τ is strictly positive with probability 1.
- The *promised payoff* X, which represents the random payoff received by the owner of the claim at time T, if there was no default prior to or at time T. The actual payoff at time T associated with X thus equals $X \mathbb{1}_{\{\tau > T\}}$.
- The finite variation process C representing the *promised dividends* – that is, the stream of (continuous or discrete) random cash flows received by the owner of the claim prior to default or up to time T, whichever comes first. We assume that $C_T - C_{T-} = 0$.
- The *recovery process* Z, which specifies the recovery payoff Z_τ received by the owner of a claim at time of default, provided that the default occurs prior to or at maturity date T.

It is convenient to introduce the *dividend process* D, which represents all cash flows associated with a defaultable claim (X, C, Z, τ). Formally, the dividend process D is defined through the formula

$$D_t = X \mathbb{1}_{\{\tau > T\}} \mathbb{1}_{[T, \infty)}(t) + \int_{(0, t]} (1 - H_u) \, dC_u + \int_{(0, t]} Z_u \, dH_u,$$

where both integrals are (stochastic) Stieltjes integrals.

Definition 1. *The* ex-dividend price process U *of a defaultable claim of the form* (X, C, Z, τ) *which settles at time T is given as*

$$U_t = B_t \, \mathbb{E}_{\mathbb{Q}^*}\Big(\int_{(t, T]} B_u^{-1} \, dD_u \,\Big|\, \mathcal{G}_t \Big), \quad \forall t \in [0, T),$$

where $\mathbb{Q}^*$ is the spot martingale measure *and B is the savings account. In addition, at maturity date we set* $U_T = U_T(X) + U_T(Z) = X \mathbb{1}_{\{\tau > T\}} + Z_T \mathbb{1}_{\{\tau = T\}}$.

Observe that $U_t = U_t(X) + U_t(Z) + U_t(C)$, where the meaning of $U_t(X), U_t(Z)$ and $U_t(C)$ is clear. Recall also that the filtration $\mathbb{G}$ models the full information, that is, the observations of the default-free market and of the default event.

2.1 Default Time

We assume from now on that we are given an $\mathbb{F}$-adapted, right-continuous, increasing process Γ on $(\Omega, \mathbb{F}, \mathbb{P}^*)$ with $\Gamma_\infty = \infty$. The default time τ and the probability measure $\mathbb{Q}^*$ are constructed as in Section 1.2. The probability $\mathbb{Q}^*$ will play the role of the *martingale probability* for the defaultable market. It is essential to observe that:

- The Wiener process W^* is also a Wiener process with respect to $\mathbb{G}$ under the probability measure $\mathbb{Q}^*$.
- We have $\mathbb{Q}^*_{|\mathcal{F}_t} = \mathbb{P}^*_{|\mathcal{F}_t}$ for every $t \in [0, T]$.

If the hazard process Γ admits the integral representation $\Gamma_t = \int_0^t \gamma_u \, du$ then the process γ is called the (stochastic) *intensity of default* with respect to the reference filtration $\mathbb{F}$.

2.2 Risk-Neutral Valuation

We shall now present the well-known valuation formulae for defaultable claims within the reduced-form setup (see, e.g., Lando (1998), Schönbucher (1998), Bielecki and Rutkowski (2004) or Bielecki et al. (2004a)).

Terminal payoff. The valuation of the terminal payoff is based on the following classic result.

Lemma 1. *For any $\mathcal{G}$-measurable, integrable random variable X and any $t \leq T$ we have*

$$\mathbb{E}_{\mathbb{Q}^*}(\mathbb{1}_{\{\tau > T\}} X \mid \mathcal{G}_t) = \mathbb{1}_{\{\tau > t\}} \frac{\mathbb{E}_{\mathbb{Q}^*}(\mathbb{1}_{\{\tau > T\}} X \mid \mathcal{F}_t)}{\mathbb{Q}^*(\tau > t \mid \mathcal{F}_t)}.$$

If, in addition, X is $\mathcal{F}_T$-measurable then

$$\mathbb{E}_{\mathbb{Q}^*}(\mathbb{1}_{\{\tau > T\}} X \mid \mathcal{G}_t) = \mathbb{1}_{\{\tau > t\}} \mathbb{E}_{\mathbb{Q}^*}(e^{\Gamma_t - \Gamma_T} X \mid \mathcal{F}_t).$$

Let X be an $\mathcal{F}_T$-measurable random variable representing the promised payoff at maturity date T. We consider a defaultable claim of the form $\mathbb{1}_{\{\tau > T\}} X$ with zero recovery in case of default (i.e., we set $Z = C = 0$). Using the definition of the ex-dividend price of a defaultable claim, we get the following *risk-neutral valuation formula*

$$U_t(X) = B_t \, \mathbb{E}_{\mathbb{Q}^*}(B_T^{-1} \mathbb{1}_{\{\tau > T\}} X \,|\, \mathcal{G}_t)$$

which holds for any $t < T$. The next result is a straightforward consequence of Lemma 1.

Proposition 1. *The price of the promised payoff X satisfies for $t \in [0, T]$*

$$U_t(X) = B_t \, \mathbb{E}_{\mathbb{Q}^*}(B_T^{-1} X \mathbb{1}_{\{\tau > T\}} \,|\, \mathcal{G}_t) = \mathbb{1}_{\{\tau > t\}} \widetilde{U}_t(X), \tag{1}$$

where we define

$$\widetilde{U}_t(X) = B_t \, \mathbb{E}_{\mathbb{Q}^*}(B_T^{-1} e^{\Gamma_t - \Gamma_T} X \,|\, \mathcal{F}_t) = \widehat{B}_t \, \mathbb{E}_{\mathbb{Q}^*}(\widehat{B}_T^{-1} X \,|\, \mathcal{F}_t),$$

where the risk-adjusted savings account $\widehat{B}_t$ equals $\widehat{B}_t = B_t e^{\Gamma_t}$. If, in addition, the default time admits the intensity process γ then

$$\widehat{B}_t = \exp\left(\int_0^t (r_u + \gamma_u)\, du\right).$$

The process $\widetilde{U}_t(X)$ represents the *pre-default value* at time t of the promised payoff X. Notice that $\widetilde{U}_T(X) = X$ and the process $\widetilde{U}_t(X)/\widehat{B}_t$, $t \in [0, T]$, is a continuous $\mathbb{F}$-martingale (thus, the process $\widetilde{U}(X)$ is a continuous $\mathbb{F}$-semimartingale).

Remark. The valuation formula (1), as well as the concept of pre-default value, should be supported by replication arguments. To this end, we need first to construct a suitable model of a defaultable market. In fact, if we wish to use formula (1), we need to know the joint law of all random variables involved, and this appears to be a non-trivial issue, in general.

Recovery payoff. The following result appears to be useful in the valuation of the recovery payoff Z_τ which occurs at time τ. The process $\widetilde{U}(Z)$ introduced below represents the pre-default value of the recovery payoff.

For the proof of Proposition 2 we refer, for instance, to Bielecki and Rutkowski (2004) (see Propositions 5.1.1 and 8.2.1 therein).

Proposition 2. *Let the hazard process Γ be continuous, and let Z be an $\mathbb{F}$-predictable bounded process. Then for every $t \in [0, T]$ we have*

$$U_t(Z) = B_t \, \mathbb{E}_{\mathbb{Q}^*}(B_\tau^{-1} Z_\tau \mathbb{1}_{\{t < \tau \leq T\}} \,|\, \mathcal{G}_t)$$

$$= \mathbb{1}_{\{\tau > t\}} B_t \, \mathbb{E}_{\mathbb{Q}^*}\left(\int_t^T Z_u B_u^{-1} e^{\Gamma_t - \Gamma_u}\, d\Gamma_u \,\Big|\, \mathcal{F}_t\right) = \mathbb{1}_{\{\tau > t\}} \widetilde{U}_t(Z).$$

where we set

$$\widetilde{U}_t(Z) = \widehat{B}_t \, \mathbb{E}_{\mathbb{Q}^*}\left(\int_t^T Z_u \widehat{B}_u^{-1}\, d\Gamma_u \,\Big|\, \mathcal{F}_t\right), \quad \forall t \in [0, T].$$

If the default intensity γ with respect to $\mathbb{F}$ exists then we have

$$\widetilde{U}_t(Z) = \mathbb{E}_{\mathbb{Q}^*}\left(\int_t^T Z_u e^{-\int_t^u (r_v + \gamma_v)\, dv}\, \gamma_u \, du \,\Big|\, \mathcal{F}_t \right).$$

Remark. Notice that $\widetilde{U}_T(Z) = 0$ while, in general, $U_T(Z) = Z_T \mathbb{1}_{\{\tau = T\}}$ is non-zero. Note, however, that if the hazard process Γ is assumed to be continuous then we have $\mathbb{Q}^*\{\tau = T\} = 0$, and thus $\widetilde{U}_T(Z) = 0 = U_T(Z)$.

Promised dividends. To value the promised dividends C that are paid prior to default time τ we shall make use of the following result. Notice that at any date $t < T$ the process $\widetilde{U}(C)$ gives the pre-default value of future promised dividends.

Proposition 3. *Let the hazard process Γ be continuous, and let C be an $\mathbb{F}$-predictable, bounded process of finite variation. Then for every $t \in [0, T]$*

$$U_t(C) = B_t \, \mathbb{E}_{\mathbb{Q}^*}\left(\int_{(t,T]} B_u^{-1}(1 - H_u)\, dC_u \,\Big|\, \mathcal{G}_t \right)$$

$$= \mathbb{1}_{\{\tau > t\}} B_t \, \mathbb{E}_{\mathbb{Q}^*}\left(\int_{(t,T]} B_u^{-1} e^{\Gamma_t - \Gamma_u}\, dC_u \,\Big|\, \mathcal{F}_t \right) = \mathbb{1}_{\{\tau > t\}} \widetilde{U}_t(C),$$

where we define

$$\widetilde{U}_t(C) = \widehat{B}_t \, \mathbb{E}_{\mathbb{Q}^*}\left(\int_{(t,T]} \widehat{B}_u^{-1}\, dC_u \,\Big|\, \mathcal{F}_t \right), \quad \forall\, t \in [0, T].$$

If, in addition, the default time τ admits the intensity γ with respect to $\mathbb{F}$ then

$$\widetilde{U}_t(C) = \mathbb{E}_{\mathbb{Q}^*}\left(\int_{(t,T]} e^{-\int_t^u (r_v + \gamma_v)\, dv}\, dC_u \,\Big|\, \mathcal{F}_t \right).$$

2.3 Defaultable Term Structure

For a defaultable discount bond with zero recovery it is natural to adopt the following definition (the superscript 0 refers to the postulated zero recovery scheme) of the price

$$D^0(t, T) = B_t \, \mathbb{E}_{\mathbb{Q}^*}\left(B_T^{-1} \mathbb{1}_{\{\tau > T\}} \,|\, \mathcal{G}_t \right) = \mathbb{1}_{\{\tau > t\}} \widetilde{D}^0(t, T),$$

where $\widetilde{D}^0(t, T)$ stands for the *pre-default value* of the bond, which is given by the following equality:

$$\widetilde{D}^0(t, T) = \widehat{B}_t \, \mathbb{E}_{\mathbb{Q}^*}\left(\widehat{B}_T^{-1} \,|\, \mathcal{F}_t \right).$$

Since $\mathbb{F}$ is the Brownian filtration, the process $\widetilde{D}^0(t, T)/\widehat{B}_t$ is a continuous, strictly positive, $\mathbb{F}$-martingale. Therefore, the pre-default bond price $\widetilde{D}^0(t, T)$ is a continuous, strictly positive, $\mathbb{F}$-semimartingale. In the special case, when Γ is deterministic, we have $\widetilde{D}^0(t, T) = e^{\Gamma_t - \Gamma_T} B(t, T)$.

Remark. The case zero recovery is, of course, only a particular example. For more general recovery schemes and the corresponding bond valuation results, we refer, for instance, to Section 2.2.4 in Bielecki et al. (2004a).

Let $\mathbb{Q}_T$ stand for the *forward martingale measure*, given on $(\Omega, \mathcal{G}_T)$ (as well as on $(\Omega, \mathcal{F}_T)$) through the formula

$$\frac{d\mathbb{Q}_T}{d\mathbb{Q}^*} = \frac{1}{B_T B(0,T)}, \quad \mathbb{Q}^*\text{-a.s.,}$$

so that the process $W_t^T = W_t^* - \int_0^t b(u,T)\, du$ is a Brownian motion under $\mathbb{Q}_T$. Denote by $F(t,U,T) = B(t,U)(B(t,T))^{-1}$ the forward price of the U-maturity bond, so that

$$dF(t,U,T) = F(t,U,T)\big(b(t,U) - b(t,T)\big)dW_t^T.$$

Since the processes B_t and $B(t,T)$ are $\mathbb{F}$-adapted, it can be shown (see, e.g., Jamshidian (2002)) that Γ is also the $\mathbb{F}$-hazard process of τ under $\mathbb{Q}_T$, and thus

$$\mathbb{Q}_T\{t < \tau \le T \mid \mathcal{G}_t\} = \mathbb{1}_{\{\tau > t\}} \mathbb{E}_{\mathbb{Q}_T}(e^{\Gamma_t - \Gamma_T} \mid \mathcal{F}_t).$$

Let us define an auxiliary process $\Gamma(t,T) = \tilde{D}^0(t,T)(B(t,T))^{-1}$ (for a fixed $T > 0$). The next result examines the basic properties of the process $\Gamma(t,T)$.

Lemma 2. *Assume that the $\mathbb{F}$-hazard process Γ is continuous. The process $\Gamma(t,T)$, $t \in [0,T]$, is a continuous $\mathbb{F}$-submartingale and*

$$d\Gamma(t,T) = \Gamma(t,T)\big(d\Gamma_t + \beta(t,T)\, dW_t^T\big) \tag{2}$$

for some $\mathbb{F}$-predictable process $\beta(t,T)$. The process $\Gamma(t,T)$ is of finite variation if and only if the hazard process Γ is deterministic. In the latter case, we have $\Gamma(t,T) = e^{\Gamma_t - \Gamma_T}$.

Proof. Recall that $\widehat{B}_t = B_t e^{\Gamma_t}$ and notice that

$$\Gamma(t,T) = \frac{\tilde{D}^0(t,T)}{B(t,T)} = \frac{\widehat{B}_t\, \mathbb{E}_{\mathbb{Q}^*}(\widehat{B}_T^{-1} \mid \mathcal{F}_t)}{B_t\, \mathbb{E}_{\mathbb{Q}^*}(B_T^{-1} \mid \mathcal{F}_t)} = \mathbb{E}_{\mathbb{Q}_T}(e^{\Gamma_t - \Gamma_T} \mid \mathcal{F}_t) = e^{\Gamma_t} M_t,$$

where we set $M_t = \mathbb{E}_{\mathbb{Q}_T}(e^{-\Gamma_T} \mid \mathcal{F}_t)$. Recall that the filtration $\mathbb{F}$ is generated by a process W^*, which is a Wiener process with respect to $\mathbb{P}^*$ and $\mathbb{Q}^*$, and all martingales with respect to a Brownian filtration are continuous processes.

We conclude that $\Gamma(t,T)$ is the product of a strictly positive, increasing, right-continuous, $\mathbb{F}$-adapted process e^{Γ_t}, and a strictly positive, continuous, $\mathbb{F}$-martingale M. Furthermore, there exists an $\mathbb{F}$-predictable process $\widehat{\beta}(t,T)$ such that M satisfies

$$dM_t = M_t \widehat{\beta}(t,T)\, dW_t^T$$

with the initial condition $M_0 = \mathbb{E}_{\mathbb{Q}_T}(e^{-\Gamma_T})$. Formula (2) follows by an application of Itô's formula, by setting $\beta(t, T) = e^{-\Gamma_t}\widehat{\beta}(t, T)$. To complete the proof, it suffices to recall that a continuous martingale is never of finite variation, unless it is a constant process. $\square$

Suppose that $\Gamma_t = \int_0^t \gamma_u \, du$. Then (2) yields

$$d\Gamma(t, T) = \Gamma(t, T)\big(\gamma_t \, dt + \beta(t, T) \, dW_t^T\big).$$

Consequently, the pre-default price $\widetilde{D}^0(t, T) = \Gamma(t, T)B(t, T)$ is governed by

$$d\widetilde{D}^0(t, T) = \widetilde{D}^0(t, T)\Big(\big(r_t + \gamma_t + \beta(t, T)b(t, T)\big) \, dt + \widetilde{b}(t, T) \, dW_t^*\Big), \quad (3)$$

where the volatility process equals $\widetilde{b}(t, T) = \beta(t, T) + b(t, T)$.

3 Properties of Trading Strategies

In this section, we shall examine the most basic properties of the wealth process of a self-financing trading strategy. First, we concentrate on trading in default-free assets. In the next step, we also include defaultable assets in our portfolio.

3.1 Default-Free Primary Assets

Our goal in this section is to present some auxiliary results related to the concept of a self-financing trading strategy for a market model involving default-free and defaultable securities. For the sake of the reader's convenience, we shall first discuss briefly the classic concepts of self-financing cash and futures strategies in the context of default-free market model. It appears that in case of defaultable securities only minor adjustments of definitions and results are needed (see, Vaillant (2001) or Blanchet-Scalliet and Jeanblanc (2004)).

Cash Strategies

Let Y_t^1 and Y_t^2 stand for the cash prices at time $t \in [0, T]$ of two tradeable assets. We postulate that Y^1 and Y^2 are continuous semimartingales. We assume, in addition, that the process Y^1 is strictly positive, so that it can be used as a numeraire.

Remark. We chose the convention that price processes of default-free securities are continuous semimartingales. Results of this section can be extended to the case of general semimartingales (for instance, jump diffusions). Our choice was motivated by the desire of providing relatively simple closed-form expressions.

Let $\phi = (\phi^1, \phi^2)$ be a trading strategy for default-free market so that, in particular, processes ϕ^1 and ϕ^2 are predictable with respect to the reference filtration $\mathbb{F}$ (the same measurability assumption will be valid for components $\phi^1, \ldots, \phi^k$ of a k-dimensional trading strategy). The component ϕ_t^i represents the number of units of the ith asset held in the portfolio at time t.

Let $V_t(\phi)$ denote the wealth of the cash strategy $\phi = (\phi^1, \phi^2)$ at time t, so that

$$V_t(\phi) = \phi_t^1 Y_t^1 + \phi_t^2 Y_t^2, \quad \forall t \in [0, T].$$

We say that the cash strategy ϕ is *self-financing* if

$$V_t(\phi) = V_0(\phi) + \int_0^t \phi_u^1 \, dY_u^1 + \int_0^t \phi_u^2 \, dY_u^2, \quad \forall t \in [0, T],$$

that is,

$$dV_t(\phi) = \phi_t^1 \, dY_t^1 + \phi_t^2 \, dY_t^2.$$

This yields

$$dV_t(\phi) = (V_t(\phi) - \phi_t^2 Y_t^2)(Y_t^1)^{-1} \, dY_t^1 + \phi_t^2 \, dY_t^2.$$

Let us introduce the relative values:

$$V_t^1(\phi) = V_t(\phi)(Y_t^1)^{-1}, \quad Y_t^{2,1} = Y_t^2 (Y_t^1)^{-1}.$$

A simple application of Itô's formula yields

$$V_t^1(\phi) = V_0^1(\phi) + \int_0^t \phi_u^2 \, dY_u^{2,1}.$$

It is well known that a similar result holds for any finite number of cash assets. Let $Y_t^1, Y_t^2, \ldots, Y_t^k$ represent that cash values at time t of k assets. We postulate that $Y^1, Y^2, \ldots, Y^k$ are continuous semimartingales. Then the wealth $V_t(\phi)$ of a trading strategy $\phi = (\phi^1, \phi^2, \ldots, \phi^k)$ equals

$$V_t(\phi) = \sum_{i=1}^k \phi_t^i Y_t^i, \quad \forall t \in [0, T], \tag{4}$$

and ϕ is said to be a *self-financing cash strategy* if

$$V_t(\phi) = V_0(\phi) + \sum_{i=1}^k \int_0^t \phi_u^i \, dY_u^i, \quad \forall t \in [0, T]. \tag{5}$$

Suppose that the process Y^1 is strictly positive. Then by combining the last two formulae, we obtain

$$dV_t(\phi) = \left(V_t(\phi) - \sum_{i=2}^k \phi_t^i Y_t^i \right)(Y_t^1)^{-1} \, dY_t^1 + \sum_{i=2}^k \phi_t^i \, dY_t^i.$$

The latter representation shows that the wealth process depends only on $k - 1$ components of ϕ. Choosing Y^1 as a numeraire asset, and denoting $V_t^1(\phi) = V_t(\phi)(Y_t^1)^{-1}$, $Y_t^{i,1} = Y_t^i (Y_t^1)^{-1}$, we get the following well-known result.

Lemma 3. *Let $\phi = (\phi^1, \phi^2, \ldots, \phi^k)$ be a self-financing cash strategy. Then we have*

$$V_t^1(\phi) = V_0^1(\phi) + \sum_{i=2}^{k} \int_0^t \phi_u^i \, dY_u^{i,1}, \quad \forall t \in [0, T].$$

Cash-Futures Strategies

Let us first consider the special case of two assets. Assume that Y_t^1 and Y_t^2 represent the cash and futures prices at time $t \in [0, T]$ of some assets, respectively. As before, we postulate that Y^1 and Y^2 are continuous semimartingales. Moreover, Y^1 is assumed to be a strictly positive process. In view of specific features of a futures contract, it is natural to postulate that the wealth $V_t(\phi)$ satisfies

$$V_t(\phi) = \phi_t^1 Y_t^1 + \phi_t^2 0 = \phi_t^1 Y_t^1, \quad \forall t \in [0, T].$$

The cash-futures strategy $\phi = (\phi^1, \phi^2)$ is self-financing if

$$dV_t(\phi) = \phi_t^1 \, dY_t^1 + \phi_t^2 \, dY_t^2, \tag{6}$$

which yields, provided that Y^1 is strictly positive,

$$dV_t(\phi) = V_t(\phi)(Y_t^1)^{-1} \, dY_t^1 + \phi_t^2 \, dY_t^2.$$

Remark. Let us recall that the futures price Y_t^2 (that is, the quotation of a futures contract at time t) has different features than the cash price of an asset. Specifically, we make the standard assumption that it is possible to enter a futures contract at no initial cost. The gains or losses from futures contracts are associated with *marking to market* (see, for instance, Duffie (2003) or Musiela and Rutkowski (1997)). Note that the 0 in the formula defining $V_t(\phi)$ is aimed to represent the value of a futures contract at time t, as opposed to the futures price Y_t^2 at this date.

Lemma 4. *Let $\phi = (\phi^1, \phi^2)$ be a self-financing cash-futures strategy. Suppose that the processes Y^1 and Y^2 are strictly positive. Then the relative wealth process $V_t^1(\phi) = V_t(\phi)(Y_t^1)^{-1}$ satisfies*

$$V_t^1(\phi) = V_0^1(\phi) + \int_0^t \widehat{\phi}_u^{2,1} \, d\widehat{Y}_u^{2,1}, \quad \forall t \in [0, T],$$

where $\widehat{\phi}_t^{2,1} = \phi_t^2 (Y_t^1)^{-1} e^{\alpha_t^{2,1}}$, $\widehat{Y}_t^{2,1} = Y_t^2 e^{-\alpha_t^{2,1}}$ and

$$\alpha_t^{2,1} = \langle \ln Y^2, \ln Y^1 \rangle_t = \int_0^t (Y_u^2)^{-1} (Y_u^1)^{-1} \, d\langle Y^2, Y^1 \rangle_u.$$

Proof. For brevity, we write $V_t = V_t(\phi)$ and $V_t^1 = V_t^1(\phi)$. The Itô formula, combined with (6), yields

$$
\begin{aligned}
dV_t^1 &= (Y_t^1)^{-1} dV_t + V_t \, d(Y_t^1)^{-1} + d\langle (Y^1)^{-1}, V \rangle_t \\
&= \phi_t^1 (Y_t^1)^{-1} \, dY_t^1 + \phi_t^2 (Y_t^1)^{-1} \, dY_t^2 + \phi_t^1 Y_t^1 \, d(Y_t^1)^{-1} \\
&\quad - \phi_t^1 (Y_t^1)^{-2} \, d\langle Y^1, Y^1 \rangle_t - \phi_t^2 (Y_t^1)^{-2} \, d\langle Y^1, Y^2 \rangle_t \\
&= \phi_t^2 (Y_t^1)^{-1} dY_t^2 - \phi_t^2 (Y_t^1)^{-2} \, d\langle Y^1, Y^2 \rangle_t \\
&= \phi_t^2 e^{\alpha_t^{2,1}} (Y_t^1)^{-1} \big(e^{-\alpha_t^{2,1}} dY_t^2 - Y_t^2 e^{-\alpha_t^{2,1}} d\alpha_t^{2,1} \big) = \widehat{\phi}_t^{2,1} \, d\widehat{Y}_t^{2,1}
\end{aligned}
$$

and the result follows.   □

Let $Y^1, \ldots, Y^l$ be the cash prices of l assets, and let $Y^{l+1}, \ldots, Y^k$ represent the futures prices of $k - l$ assets. Then the wealth process of a trading strategy $\phi = (\phi^1, \phi^2, \ldots, \phi^k)$ is given by the formula

$$
V_t(\phi) = \sum_{i=1}^{l} \phi_t^i Y_t^i, \quad \forall t \in [0, T], \tag{7}
$$

and ϕ is a *self-financing cash-futures strategy* whenever

$$
V_t(\phi) = V_0(\phi) + \sum_{i=1}^{k} \int_0^t \phi_u^i \, dY_u^i, \quad \forall t \in [0, T].
$$

The proof of the next result relies on the similar calculations as the proofs of Lemmas 3 and 4.

Lemma 5. *Let $\phi = (\phi^1, \phi^2, \ldots, \phi^k)$ be a self-financing cash-futures strategy. Suppose that the processes Y^1 and $Y^{l+1}, \ldots, Y^k$ are strictly positive. Then the relative wealth process $V_t^1(\phi) = V_t(\phi)(Y_t^1)^{-1}$ satisfies, for every $t \in [0, T]$,*

$$
V_t^1(\phi) = V_0^1(\phi) + \sum_{i=2}^{l} \int_0^t \phi_u^i \, dY_u^{i,1} + \sum_{i=l+1}^{k} \int_0^t \widehat{\phi}_u^{i,1} \, d\widehat{Y}_u^{i,1},
$$

where we denote $Y_t^{i,1} = Y_t^i (Y_t^1)^{-1}$, $\widehat{\phi}_t^{i,1} = \phi_t^i (Y_t^1)^{-1} e^{\alpha_t^{i1}}$, $\widehat{Y}_t^{i,1} = Y_t^i e^{-\alpha_t^{i1}}$, and

$$
\alpha_t^{i1} = \langle \ln Y^i, \ln Y^1 \rangle_t = \int_0^t (Y_u^i)^{-1} (Y_u^1)^{-1} \, d\langle Y^i, Y^1 \rangle_u.
$$

Constrained Cash Strategies

We continue the analysis of cash strategies for some $k \geq 3$. Price processes $Y^1, Y^2, \ldots, Y^k$ are assumed to be continuous semimartingales. We postulate, in addition, that Y^1 and $Y^{l+1}, \ldots, Y^k$ are strictly positive processes, where $1 < l+1 \leq k$.

Let $\phi = (\phi^1, \phi^2, \ldots, \phi^k)$ be a self-financing trading strategy, so that the wealth process $V(\phi)$ satisfies (4)-(5). We shall consider three particular cases of increasing generality.

Strategies with zero net investment in $Y^{l+1}, \ldots, Y^k$. Assume first that at any time t there is zero net investment in assets $Y^{l+1}, \ldots, Y^k$. Specifically, we postulate that the strategy is subject to the following constraint:

$$\sum_{i=l+1}^{k} \phi_t^i Y_t^i = 0, \quad \forall t \in [0, T], \tag{8}$$

so that the wealth process $V_t(\phi)$ is given by (7). Equivalently, we have $\phi_t^k = -\sum_{i=l+1}^{k-1} \phi_t^i Y_t^i (Y_t^k)^{-1}$. Combining the last equality with (5), we obtain

$$dV_t(\phi) = \left(V_t(\phi) - \sum_{i=2}^{l} \phi_t^i Y_t^i \right) (Y_t^1)^{-1} dY_t^1$$

$$+ \sum_{i=2}^{l} \phi_t^i \, dY_t^i + \sum_{i=l+1}^{k-1} \phi_t^i \left(dY_t^i - Y_t^i (Y_t^k)^{-1} dY_t^k \right).$$

It is thus clear that the wealth process $V(\phi)$ depends only on $k - 2$ components $\phi^2, \ldots, \phi^{k-1}$ of the k-dimensional trading strategy ϕ. The following result, which can be seen as an extension of Lemma 4, provides a more convenient representation for the (relative) wealth process.

Lemma 6. *Let $\phi = (\phi^1, \phi^2, \ldots, \phi^k)$ be a self-financing cash strategy such that (8) holds. Assume that the processes $Y^1, Y^{l+1}, \ldots, Y^k$ are strictly positive. Then the relative wealth process $V_t^1(\phi) = V_t(\phi)(Y_t^1)^{-1}$ satisfies*

$$V_t^1(\phi) = V_0^1(\phi) + \sum_{i=2}^{l} \int_0^t \phi_u^i \, dY_u^{i,1} + \sum_{i=l+1}^{k-1} \int_0^t \widehat{\phi}_u^{i,k,1} \, d\widehat{Y}_u^{i,k,1}, \quad \forall t \in [0, T],$$

where we denote

$$\widehat{\phi}_t^{i,k,1} = \phi_t^i (Y_t^{1,k})^{-1} e^{\alpha_t^{i,k,1}}, \quad \widehat{Y}_t^{i,k,1} = Y_t^{i,k} e^{-\alpha_t^{i,k,1}}, \tag{9}$$

with $Y_t^{i,k} = Y_t^i (Y_t^k)^{-1}$ and

$$\alpha_t^{i,k,1} = \langle \ln Y^{i,k}, \ln Y^{1,k} \rangle_t = \int_0^t (Y_u^{i,k})^{-1} (Y_u^{1,k})^{-1} \, d\langle Y^{i,k}, Y^{1,k} \rangle_u. \tag{10}$$

Proof. Let us consider the relative values of all processes, with the price Y^k chosen as a numeraire, and let us consider the process

$$V_t^k(\phi) := V_t(\phi)(Y_t^k)^{-1} = \sum_{i=1}^{k} \phi_t^i Y_t^{i,k}.$$

In view of the constraint (8) we have that $V_t^k(\phi) = \sum_{i=1}^{l} \phi_t^i Y_t^{i,k}$. In addition, similarly as in Lemma 3, we obtain

$$dV_t^k(\phi) = \sum_{i=1}^{k-1} \phi_t^i \, dY_t^{i,k}.$$

Since

$$Y_t^{i,k}(Y_t^{1,k})^{-1} = Y_t^{i,1}, \quad V_t^1(\phi) = V_t^k(\phi)(Y_t^{1,k})^{-1},$$

using argument analogous as in proof of Lemma 4, we obtain

$$V_t^1(\phi) = V_0^1(\phi) + \sum_{i=2}^{l} \int_0^t \phi_u^i \, dY_u^{i,1} + \sum_{i=l+1}^{k-1} \int_0^t \widehat{\phi}_u^{i,k,1} \, d\widehat{Y}_u^{i,k,1}, \quad \forall t \in [0,T],$$

where the processes $\widehat{\phi}_t^{i,k,1}$, $\widehat{Y}_t^{i,k,1}$ and $\alpha_t^{i,k,1}$ are given by (9)-(10). $\square$

Strategies with a pre-specified net investment Z in $Y^{l+1}, \ldots, Y^k$. We shall now postulate that the strategy ϕ is such that

$$\sum_{i=l+1}^{k} \phi_t^i Y_t^i = Z_t, \quad \forall t \in [0,T], \tag{11}$$

for a pre-specified, $\mathbb{F}$-progressively measurable, process Z. The following result is a rather straightforward extension of Lemma 6.

Lemma 7. *Let $\phi = (\phi^1, \phi^2, \ldots, \phi^k)$ be a self-financing cash strategy such that (11) holds. Assume that the processes $Y^1, Y^{l+1}, \ldots, Y^k$ are strictly positive. Then the relative wealth process $V_t^1(\phi) = V_t(\phi)(Y_t^1)^{-1}$ satisfies*

$$V_t^1(\phi) = V_0^1(\phi) + \sum_{i=2}^{l} \int_0^t \phi_u^i \, dY_u^{i,1} + \sum_{i=l+1}^{k-1} \int_0^t \widehat{\phi}_u^{i,k,1} \, d\widehat{Y}_u^{i,k,1}$$

$$+ \int_0^t Z_u (Y_u^k)^{-1} \, d(Y_u^{1,k})^{-1},$$

where $\widehat{\phi}_t^{i,k,1}$, $\widehat{Y}_t^{i,k,1}$ and $\alpha_t^{i,k,1}$ are given by (9)-(10).

Proof. Let us sketch the proof of the lemma for $k = 3$. Then $l = 2$ and $\phi_t^2 Y_t^2 + \phi_t^3 Y_t^3 = Z_t$ for every $t \in [0,T]$. Consequently, for the process $V^3(\phi) = V(\phi)(Y^3)^{-1}$ we get

$$V_t^3(\phi) = \sum_{i=1}^{3} \phi_t^i Y_t^i (Y_t^3)^{-1} = \phi_t^1 Y_t^{1,3} + Z_t (Y_t^3)^{-1}, \quad \forall t \in [0, T].$$

Furthermore, the self-financing condition yields

$$dV_t^3(\phi) = \phi_t^1 \, dY_t^{1,3} + \phi_t^2 \, dY_t^{2,3}.$$

Proceeding in an analogous way as in the proof of Lemma 4, we obtain for $V_t^1(\phi) = V_t^3(\phi)(Y_t^{1,3})^{-1}$

$$\begin{aligned}
dV_t^1(\phi) &= \phi_t^2 e^{\alpha_t^{2,3,1}} (Y_t^{1,3})^{-1} \left(e^{-\alpha_t^{2,3,1}} dY_t^{2,3} - Y_t^{2,3} e^{-\alpha_t^{2,3,1}} \, d\alpha_t^{2,3,1} \right) \\
&\quad + Z_t (Y_t^3)^{-1} d(Y_t^{1,3})^{-1} \\
&= \widehat{\phi}_u^{2,3,1} \, d\widehat{Y}_u^{2,3,1} + Z_t (Y_t^3)^{-1} d(Y_t^{1,3})^{-1},
\end{aligned}$$

where $\widehat{\phi}_t^{2,3,1} = \phi_t^2 (Y_t^{1,3})^{-1} e^{\alpha_t^{2,3,1}}$, $\widehat{Y}_t^{2,3,1} = Y_t^{2,3} e^{-\alpha_t^{2,3,1}}$ and

$$\alpha_t^{2,3,1} = \langle \ln Y^{2,3}, \ln Y^{1,3} \rangle_t = \int_0^t (Y_u^{2,3})^{-1} (Y_u^{1,3})^{-1} \, d\langle Y^{2,3}, Y^{1,3} \rangle_u.$$

The proof for the general case is based on similar calculations.   □

Strategies with consumption and a pre-specified net investment Z in $Y^{l+1}, \ldots,$ Y^k. Let consumption be given by an $\mathbb{F}$-adapted process A of finite variation, with $A_0 = 0$. We consider a self-financing cash strategy ϕ with *consumption process A*, so that the wealth process $V(\phi)$ satisfies:

$$V_t(\phi) - \sum_{i=1}^{k} \psi_t^i Y_t^i = \sum_{i=1}^{l} \phi_t^i Y_t^i + Z_t, \quad \forall t \in [0, T],$$

and

$$V_t(\phi) = V_0(\phi) + \sum_{i=1}^{k} \int_0^t \phi_u^i \, dY_u^i + A_t, \quad \forall t \in [0, T].$$

Then it suffices to modify the formula established in Lemma 7 by adding a term associated with the consumption process A. Specifically, for the relative wealth process $V_t^1(\phi) = V_t(\phi)(Y_t^1)^{-1}$ we obtain the following integral representation, which is valid for every $t \in [0, T]$

$$\begin{aligned}
V_t^1(\phi) &= V_0^1(\phi) + \sum_{i=2}^{l} \int_0^t \phi_u^i \, dY_u^{i,1} + \sum_{i=l+1}^{k-1} \int_0^t \widehat{\phi}_u^{i,k,1} \, d\widehat{Y}_u^{i,k,1} \\
&\quad + \int_0^t Z_u (Y_u^k)^{-1} \, d(Y_u^{1,k})^{-1} + \int_0^t (Y_u^1)^{-1} \, dA_u.
\end{aligned}$$

Remark. We use here a generic term 'consumption' to reflect the impact of A on the wealth. The financial interpretation of A depends on particular circumstances. For instance, an increasing process A represents the inflows of cash, rather than the outflows of cash (the latter case is commonly referred to as consumption in the financial literature).

3.2 Defaultable and Default-Free Primary Assets

Let $Y^1, \ldots, Y^m$ be prices of m defaultable assets, and let $Y^{m+1}, \ldots, Y^k$ represent prices of $k - m$ default-free assets. Processes $Y^{m+1}, \ldots, Y^k$ are assumed to be continuous semimartingales. We make here an essential assumption that τ is the default time for each defaultable asset Y^i, $i = 1, \ldots, m$. Of course, in the case of defaultable assets with different default times (e.g., when dealing with the first-to-default claim), some definitions should be modified in a natural way. A special case of first-to-default claims is examined in Section 4.4.

Self-Financing Trading Strategies

The following definition is a rather obvious extension of conditions (4)-(5). We postulate here that the processes $\phi^1, \ldots, \phi^k$ are $\mathbb{G}$-predictable processes, in general.

Definition 2. *The wealth $V_t(\phi)$ of a trading strategy $\phi = (\phi^1, \phi^2, \ldots, \phi^k)$ equals $V_t(\phi) = \sum_{i=1}^k \phi_t^i Y_t^i$ for every $t \in [0, T]$. A strategy ϕ is said to be self-financing if for every $t \in [0, T]$*

$$V_t(\phi) = V_0(\phi) + \sum_{i=1}^m \int_0^t \phi_{u-}^i \, dY_u^i + \sum_{i=m+1}^k \int_0^t \phi_u^i \, dY_u^i.$$

Although Definition 2 is formulated in a general setup, it can be simplified for our further purposes. Indeed, since we shall deal only with defaultable claims with default time τ, we shall only examine a particular trading strategy ϕ prior to and at default time τ or, more precisely, on the stochastic interval $[\![0, \tau \wedge T]\!]$, where $[\![0, \tau \wedge T]\!] = \{(t, \omega) \in \mathbb{R}_+ \times \Omega : 0 \le t \le \tau(\omega) \wedge T\}$.

In fact, we shall examine separately the following issues: (i) the behavior of the wealth process $V(\phi)$ on the random interval $[\![0, \tau \wedge T[\![= \{(t, \omega) \in \mathbb{R}_+ \times \Omega : 0 \le t < \tau(\omega) \wedge T\}$ and (ii) the size of its jump at the random time moment $\tau \wedge T$ or, equivalently, the value of $V_{\tau \wedge T}$. Such a study is, of course, sufficient in our setup, since we only consider the case where a recovery payment (if any) is made at the default time (and not after this date). Consequently, since we never deal with a trading strategy after the random time $\tau \wedge T$, we may and do assume from now on that all components $\phi^1, \phi^2, \ldots, \phi^k$ of a portfolio ϕ are $\mathbb{F}$-predictable, rather than $\mathbb{G}$-predictable processes.

It is worthwhile to mention, that in the next two parts we will examine the importance of the measurability property of an admissible trading strategy within the framework of optimization problems in incomplete market.

Remark. It can be formally shown that for any $\mathbb{R}^k$-valued $\mathbb{G}$-predictable process ϕ there exists a unique $\mathbb{F}$-predictable process ψ such that the equality $\mathbb{1}_{\{\tau \ge t\}} \phi_t =$

$\mathbb{1}_{\{\tau \geq t\}} \psi_t$ holds for every $t \in [0, T]$. In addition, we find it convenient to postulate, by convention, that the price processes $Y^{m+1}, \ldots, Y^k$ are also stopped at the random time $\tau \wedge T$.

We have the following definition of a trading strategy.

Definition 3. *By a trading strategy* $\phi = (\phi^1, \phi^2, \ldots, \phi^k)$ *we mean a family* ϕ^1, ϕ^2, $\ldots$, ϕ^k *of* $\mathbb{F}$*-predictable stochastic processes.*

Let us stress that if a trading strategy considered in this section is self-financing on $[\![0, \tau \wedge T[\![$ then it is also self-financing on $[\![0, \tau \wedge T]\!]$. At the intuitive level, the portfolio is not rebalanced at time $\tau \wedge T$, but it is rather sold out in order to cover liabilities. Let $\widetilde{Y}_t^i$ stands for the pre-default value of the ith defaultable asset at time t. We postulate throughout that processes $\widetilde{Y}^i$, $i = 1, \ldots, m$ are continuous $\mathbb{F}$-semimartingales.

Definition 4. *We define the pre-default wealth process* $\widetilde{V}(\phi)$ *of a trading strategy* $\phi = (\phi^1, \phi^2, \ldots, \phi^k)$ *by setting for every* $t \in [0, T]$,

$$\widetilde{V}_t(\phi) = \sum_{i=1}^{m} \phi_t^i \widetilde{Y}_t^i + \sum_{i=m+1}^{k} \phi_t^i Y_t^i.$$

A strategy ϕ is said to be self-financing prior to default if for every $t \in [0, T]$

$$\widetilde{V}_t(\phi) = \widetilde{V}_0(\phi) + \sum_{i=1}^{m} \int_0^t \phi_u^i \, d\widetilde{Y}_u^i + \sum_{i=m+1}^{k} \int_0^t \phi_u^i \, dY_u^i.$$

Note that $\widetilde{V}_0(\phi) = V_0(\phi)$, since $\mathbb{P}^* \{\tau > 0\} = 1$. Let us stress that if a trading strategy ϕ is self-financing prior to default then ϕ is also self-financing on $[0, T]$. Indeed, we always postulate that trading ceases at time of default, and the terminal wealth at time $\tau \wedge T$ equals

$$V_{\tau \wedge T}(\phi) = \sum_{i=1}^{k} \phi_{\tau \wedge T}^i Y_{\tau \wedge T}^i.$$

Of course, on the event $\{\tau > T\}$ we also have

$$V_{\tau \wedge T}(\phi) = V_T(\phi) = \widetilde{V}_T(\phi) = \sum_{i=1}^{m} \phi_T^i \widetilde{Y}_T^i + \sum_{i=m+1}^{k} \phi_T^i Y_T^i.$$

Hence, we shall not distinguish in what follows between the concept of a self-financing trading strategy and a trading strategy self-financing prior to default.

Zero Recovery for Defaultable Assets

The following assumption corresponds to the simplest situation of zero recovery for all defaultable primary assets that are used for replication. Manifestly, this assumption is not practical, and thus it will be later relaxed.

Assumption (A). The defaultable primary assets $Y^1, \ldots, Y^m$ are all subject to the zero recovery scheme, and they have a common default time τ.

By virtue of Assumption (A), the prices $Y^1, \ldots, Y^m$ vanish at default time τ, and thus also after this date. Consequently, for every $i = 1, \ldots, m$ we have $Y_t^i = \mathbb{1}_{\{\tau > t\}} \widetilde{Y}_t^i$ for every $t \in [0, T]$ for some $\mathbb{F}$-predictable processes $\widetilde{Y}^1, \ldots, \widetilde{Y}^m$. In other words, for any $i = 1, \ldots, m$ the price Y^i jumps from $\widetilde{Y}_{\tau-}^i$ to $\widetilde{Y}_\tau^i = 0$ at the time of default. We make a technical assumption that the pre-default values $\widetilde{Y}^1, \ldots, \widetilde{Y}^m$ are continuous $\mathbb{F}$-semimartingales.

In order to be able to use the price Y^1 as a numeraire prior to default, we assume that the pre-default price $\widetilde{Y}^1$ is a strictly positive continuous $\mathbb{F}$-semimartingale. Notice that $\widetilde{Y}_0^1 = Y_0^1$.

Assume first zero recovery for the defaultable contingent claim we wish to replicate. Thus, at time τ the wealth process of any strategy that is capable to replicate the claim $\mathbb{1}_{\{\tau > T\}} X$ should necessarily jump to zero, provided that $\tau \leq T$. We can achieve this by considering only self-financing strategies $\phi = (\phi^1, \phi^2, \ldots, \phi^k)$ such that at any time the net investment in default-free assets $Y^{m+1}, \ldots, Y^k$ equals zero, so that we have

$$\sum_{i=m+1}^{k} \phi_t^i Y_t^i = 0, \quad \forall t \in [0, T]. \tag{12}$$

In the general case, that is, when Z is a pre-specified non-zero recovery process for a defaultable claim under consideration, it suffices to consider self-financing strategies $\phi = (\phi^1, \phi^2, \ldots, \phi^k)$ such that

$$\sum_{i=m+1}^{k} \phi_t^i Y_t^i = Z_t, \quad \forall t \in [0, T]. \tag{13}$$

Notice that prior to default time (that is, on the event $\{\tau > t\}$) we have $V_t(\phi) = \sum_{i=1}^{m} \phi_t^i \widetilde{Y}_t^i + Z_t$, and the self-financing property of ϕ prior to default time τ takes the following form

$$dV_t(\phi) = \sum_{i=1}^{m} \phi_t^i \, d\widetilde{Y}_t^i + \sum_{i=m+1}^{k} \phi_t^i \, dY_t^i. \tag{14}$$

At default time τ, we have $V_\tau(\phi) = Z_\tau$ on the set $\{\tau \leq T\}$.

The next goal is to examine the existence of ϕ with the properties described above. To this end, we denote $\widetilde{Y}_t^{i,1} = \widetilde{Y}_t^i (\widetilde{Y}_t^1)^{-1}$ for $i = 2, \ldots, m$ and $\widetilde{Y}_t^{1,k} = \widetilde{Y}_t^1 (Y_t^k)^{-1}$. As before, we write $Y_t^{i,k} = Y_t^i(Y_t^k)^{-1}$. Using Lemma 7, we obtain the following auxiliary result that will be later used to establish the existence of a replicating strategy for a defaultable claim.

Proposition 4. (i) *Let $\phi = (\phi^1, \phi^2, \ldots, \phi^k)$ be a self-financing strategy such that (13) holds. Assume that the processes $\widetilde{Y}^1, Y^{m+1}, \ldots, Y^k$ are strictly positive. Then the pre-default wealth process $\widetilde{V}(\phi)$ satisfies for every $t \in [0, T]$*

$$\widetilde{V}_t(\phi) = \widetilde{Y}_t^1 \left(\widetilde{V}_0^1(\phi) + \sum_{i=2}^m \int_0^t \phi_u^i \, d\widetilde{Y}_u^{i,1} + \sum_{i=m+1}^{k-1} \int_0^t \widetilde{\phi}_u^{i,k,1} \, d\widehat{Y}_u^{i,k,1} \right.$$
$$\left. + \int_0^t Z_u(Y_u^k)^{-1} \, d(\widetilde{Y}_u^{1,k})^{-1} \right),$$

where we denote

$$\widetilde{\phi}_t^{i,k,1} = \phi_t^i(\widetilde{Y}_t^{1,k})^{-1} e^{\widetilde{\alpha}_t^{i,k,1}}, \quad \widehat{Y}_t^{i,k,1} = Y_t^{i,k} e^{-\widetilde{\alpha}_t^{i,k,1}},$$

and

$$\widetilde{\alpha}_t^{i,k,1} = \langle \ln Y^{i,k}, \ln \widetilde{Y}^{1,k} \rangle_t = \int_0^t (Y_u^{i,k})^{-1}(\widetilde{Y}_u^{1,k})^{-1} \, d\langle Y^{i,k}, \widetilde{Y}^{1,k} \rangle_u.$$

In addition, at default time the wealth of ϕ equals $V_\tau(\phi) = Z_\tau$ on the event $\{\tau \le T\}$.
(ii) *Suppose that the $\mathbb{F}$-predictable processes ψ^i, $i = 2, \ldots, m$ and $\widetilde{\psi}^{i,k,1}$, $i = m + 1, \ldots, k - 1$ are given. For an arbitrary constant $c \in \mathbb{R}$, we define the process V by setting, for $t \in [0, T]$,*

$$\widetilde{V}_t = c + \sum_{i=2}^m \int_0^t \psi_u^i \, d\widetilde{Y}_u^{i,1} + \sum_{i=m+1}^{k-1} \int_0^t \widetilde{\psi}_u^{i,k,1} \, d\widehat{Y}_u^{i,k,1} + \int_0^t Z_u(Y_u^k)^{-1} \, d(\widetilde{Y}_u^{1,k})^{-1}.$$

Then there exists a self-financing trading strategy $\phi = (\phi^1, \phi^2, \ldots, \phi^k)$ such that:
(a) *$\phi_t^i = \psi_t^i$ for $i = 2, \ldots, m$ and $\phi_t^i = \widetilde{\psi}_t^{i,k,1} \widetilde{Y}_t^{1,k} e^{-\widetilde{\alpha}_t^{i,k,1}}$ for $i = m + 1, \ldots, k - 1$,*
(b) *ϕ satisfies (13), so that $\sum_{i=m+1}^k \phi_t^i Y_t^i = Z_t$ for every $t \in [0, T]$,*
(c) *the pre-default wealth $\widetilde{V}(\phi)$ of ϕ equals $\widetilde{V}$,*
(d) *at default time the wealth of ϕ equals $V_\tau(\phi) = Z_\tau$ on the event $\{\tau \le T\}$.*

Proof. Part (i) is an almost immediate consequence of Lemma 7. Therefore, we shall focus on the second part. The idea of the proof of part (ii) is also rather clear. First, let ϕ^i, $i = 2, \ldots, m$ and ϕ^i, $i = m + 1, \ldots, k - 1$ be defined from processes ψ^i and $\widetilde{\psi}_t^{i,k,1}$ as in (a). Given the processes ϕ^i for $i = m + 1, \ldots, k - 1$, we observe that the component ϕ^k is uniquely specified by condition (13). Thus, it remains to check that there exists a (unique) component ϕ^1 such that the resulting k-dimensional trading strategy is self-financing prior to default, in the sense of Definition 4. Let us set

$$\phi_t^1 = \left(\widetilde{V}_t - \sum_{i=2}^{m} \phi_t^i Y_t^i - Z_t\right)(\widetilde{Y}_t^1)^{-1} = \left(\widetilde{V}_t - \sum_{i=2}^{k} \phi_t^i Y_t^i\right)(\widetilde{Y}_t^1)^{-1}.$$

It is clear that $\widetilde{V}_t(\phi) = \widetilde{V}_t$ for every $t \in [0, T]$. To show that the strategy $(\phi^1, \phi^2, \ldots, \phi^k)$ described above is self-financing prior to default, it suffices to show that for the discounted pre-default wealth

$$\widetilde{V}_t^1(\phi) = \sum_{i=1}^{m} \phi_t^i \widetilde{Y}_t^{i,1} + \sum_{i=m+1}^{k} \phi_t^i Y_t^{i,1}$$

we have for every $t \in [0, T]$

$$\widetilde{V}_t^1(\phi) = \widetilde{V}_0^1(\phi) + \sum_{i=2}^{m} \int_0^t \phi_u^i d\widetilde{Y}_u^{i,1} + \sum_{i=m+1}^{k} \int_0^t \phi_u^i dY_u^{i,1}.$$

Towards this end, it is enough observe that $\widetilde{V}_t^1(\phi) = (\widetilde{Y}_t^1)^{-1}\widetilde{V}_t = \widetilde{V}_t^1$, and then to verify that

$$\widetilde{V}_t^1 = \widetilde{V}_0^1 + \sum_{i=2}^{m} \int_0^t \phi_u^i d\widetilde{Y}_u^{i,1} + \sum_{i=m+1}^{k} \int_0^t \phi_u^i dY_u^{i,1}$$

for every $t \in [0, T]$. To establish that last equality, it suffices to use the definition of the process $\widetilde{V}^1$ and to observe that

$$\sum_{i=m+1}^{k-1} \widetilde{\psi}_t^{i,k,1} d\widehat{Y}_t^{i,k,1} = \sum_{i=m+1}^{k} \phi_t^i dY_t^{i,1},$$

which follows by direct calculations, using the definitions of ϕ^i, $i = m+1, \ldots, k$. It is easy to see that the strategy ϕ satisfies conditions (a)-(d).   □

Remarks. Let us observe that the equality established in Proposition 4 is in fact valid on the random interval $[\![0, \tau[\![$ on the event $\{\tau \leq T\}$ and on the interval $[0, T]$ on the event $\{\tau > T\}$. It is also important to notice that the assumption of zero recovery for $Y^1, \ldots, Y^m$ is not essential for the validity of the statements in the last result, except for the last part, that is, the equality $V_\tau(\phi) = Z_\tau$. Indeed, the proof of Proposition 4 relies on conditions (13) and (14). Therefore, if defaultable primary assets $Y^1, \ldots, Y^m$ are subject to non-zero recovery, it will be possible to modify Proposition 4 accordingly (see Section 3.2 below).

When dealing with defaultable claims with no recovery, that is, claims for which the recovery process Z vanishes, it will be convenient to use directly the following corollary to Proposition 4.

Corollary 1. Let $\phi = (\phi^1, \phi^2, \ldots, \phi^k)$ be a self-financing strategy such that condition (12) holds.

(i) *Assume that the processes* $\widetilde{Y}^1, Y^{m+1}, \ldots, Y^k$ *are strictly positive. Then the wealth process* $V(\phi)$ *satisfies for every* $t \in [0, T]$

$$V_t(\phi) = Y_t^1 \left(V_0^1(\phi) + \sum_{i=2}^{m} \int_0^t \phi_u^i \, d\widetilde{Y}_u^{i,1} + \sum_{i=m+1}^{k-1} \int_0^t \widetilde{\phi}_u^{i,k,1} \, d\widehat{Y}_u^{i,k,1} \right).$$

(ii) *Assume that all primary assets are defaultable, that is,* $m = k$, *and the pre-default value* $\widetilde{Y}^1$ *is a strictly positive process. Then the wealth process* $V(\phi)$ *satisfies for every* $t \in [0, T]$

$$V_t(\phi) = Y_t^1 \left(V_0^1(\phi) + \sum_{i=2}^{m} \int_0^t \phi_u^i \, d\widetilde{Y}_u^{i,1} \right).$$

Of course, the counterparts of part (ii) in Proposition 4 are also valid and they will be used in what follows, although they are not explicitly formulated here.

Remark. Consider the special case of two primary assets, defaultable and default-free, with prices $Y_t^1 = \mathbb{1}_{\{\tau > t\}} \widetilde{Y}_t^1$ and Y_t^2, respectively, where $\widetilde{Y}^1$ and Y^2 are strictly positive, continuous, $\mathbb{F}$-semimartingales. Suppose we wish to replicate a defaultable claim with zero recovery. We have

$$V_t(\phi) = \phi_t^1 Y_t^1 + \phi_t^2 Y_t^2 = \phi_t^1 \mathbb{1}_{\{\tau > t\}} \widetilde{Y}_t^1 + \phi_t^2 Y_t^2$$

and

$$dV_t(\phi) = \left(V_{t-}(\phi) - \phi_t^2 Y_t^2 \right) (\widetilde{Y}_t^1)^{-1} d\widetilde{Y}_t^1 + \phi_t^2 \, dY_t^2.$$

It is rather clear that the equality $V_t(\phi) = 0$ on $\{\tau \leq t\}$ implies that $\phi_t^2 = 0$ for every $t \in [0, T]$. Therefore,

$$dV_t(\phi) = V_{t-}(\phi)(\widetilde{Y}_t^1)^{-1} dY_t^1$$

and the existence of replicating strategy for a defaultable claim with zero-recovery is unlikely within the present setup (except for some trivial cases).

Non-Zero Recovery for Defaultable Assets

In this section, the assumption of zero recovery for defaultable primary assets $Y^1, \ldots, Y^m$ is relaxed. To be more specific, Assumption (A) is replaced by the following weaker restriction.

Assumption (B). We assume that the defaultable assets $Y^1, \ldots, Y^m$ are subject to an arbitrary recovery scheme, and they have a common default time τ.

Under Assumption (B), condition (13) no longer implies that $V_\tau(\phi) = Z_\tau$ on the set $\{\tau \leq T\}$. We can achieve this requirement by substituting (13) with the following constraint

$$\sum_{i=1}^{m} \phi_t^i \bar{Y}_t^i + \sum_{i=m+1}^{k} \phi_t^i Y_t^i = Z_t, \quad \forall t \in [0, T], \tag{15}$$

where $\bar{Y}^i$ represents the recovery payoff of the defaultable asset Y^i, so that $Y_\tau^i = \bar{Y}_\tau^i$ for $i = 1, 2, \ldots, m$. In this general setup, condition (15) does not seem to be sufficiently restrictive for more explicit calculations. It is plausible, however, that it can be used to derive a replicating strategy in several non-trivial and practically interesting cases.

It is not difficult to see that Proposition 4 can be extended to the case of non-zero recovery for defaultable assets, provided, of course, that we are in a position to find a priori the wealth invested in non-defaultable assets, that is, if the process $\beta_t := \sum_{i=m+1}^{k} \phi_t^i Y_t^i$ is known beforehand. By arguing as in Proposition 4, we then obtain for every $t \in [0, T]$

$$\widetilde{V}_t(\phi) = \widetilde{Y}_t^1 \left(\widetilde{V}_0^1(\phi) + \sum_{i=2}^{m} \int_0^t \phi_u^i \, d\widetilde{Y}_u^{i,1} + \sum_{i=m+1}^{k-1} \int_0^t \widetilde{\phi}_u^{i,k,1} \, d\widehat{Y}_u^{i,k,1} \right.$$

$$\left. + \int_0^t \beta_u (Y_u^k)^{-1} \, d(\widetilde{Y}_u^{1,k})^{-1} \right).$$

In view of (15), we also have that

$$\bar{\alpha}_t := \sum_{i=1}^{m} \phi_t^i \bar{Y}_t^i = Z_t - \beta_t, \quad \forall t \in [0, T], \tag{16}$$

thereby imposing an additional constraint on the wealth invested in defaultable assets. Condition (16) is not directly accounted for in the last formula for $\widetilde{V}(\phi)$, however, and thus the problem at hand is not completely solved. For further considerations related to non-zero recovery of defaultable primary assets, see Section 4.1 and 4.2.

Fractional recovery of market value. As an example of a non-zero recovery scheme, we consider the so-called fractional recovery of (pre-default) market value (FRMV) scheme with constant recovery rates $\delta_i \neq 1$ (typically, $0 \leq \delta_i < 1$). Then we have $\bar{Y}_t^i = \delta_i \widetilde{Y}_t^i$ for every $i = 1, 2, \ldots, m$, and thus (15) becomes

$$\sum_{i=1}^{m} \phi_t^i \delta_i \widetilde{Y}_t^i + \sum_{i=m+1}^{k} \phi_t^i Y_t^i = Z_t, \quad \forall t \in [0, T]. \tag{17}$$

Let us mention that in the case of a defaultable zero-coupon bond, the FRMV scheme results in the following expression for the pre-default value of a defaultable bond with unit face value (see, for instance, Section 2.2.4 in Bielecki et al. (2004a))

$$\widetilde{D}_M^\delta(t, T) = \mathbb{E}_{\mathbb{Q}^*} \left(e^{-\int_t^T (r_u + (1-\delta)\gamma_u) du} \,\Big|\, \mathcal{F}_t \right),$$

where the recovery rate δ may depend on the bond's maturity T, in general. In particular, if the default intensity γ is deterministic then we have

$$\widetilde{D}_M^\delta(t, T) = e^{-\int_t^T (1-\delta)\gamma(u)\, du}\, B(t, T).$$

Manifestly, we always have $D_M^\delta(\tau, T) = \delta D_M^\delta(\tau-, T)$ on the set $\{\tau \le T\}$ under the FRMV scheme.

4 Replication of Defaultable Claims

We are in a position to examine the issue of an exact replication of a generic defaultable claim. By a *replicating strategy* we mean here a self-financing trading strategy ϕ such that the wealth process $V(\phi)$ matches exactly the pre-default value of the claim at any time prior to default (and prior to the maturity date), as well as coincides with the claim's payoff at default time or at maturity date, whichever comes first. Using our notation introduced in Section 2, this can be formalized as follows.

Definition 5. *A self-financing trading strategy ϕ is a replicating strategy for a defaultable claim $(X, 0, Z, \tau)$ if and only if the following hold:*
(i) $V_t(\phi) = \widetilde{U}_t(X) + \widetilde{U}_t(Z)$ on the random interval $[\![0, \tau \wedge T[\![$,
(ii) $V_\tau(\phi) = Z_\tau$ on the set $\{\tau \le T\}$,
(iii) $V_T(\phi) = X$ on the set $\{\tau > T\}$.
We say that a defaultable claim is attainable if it admits at least one replicating strategy

The last definition is suitable only in the case of a defaultable claim with no promised dividends. Some comments regarding replication of promised dividends are given in Section 4.3.

4.1 Replication of a Promised Payoff

We shall first examine the possibility of an exact replication of a defaultable contingent claim of the form $(X, 0, 0, \tau)$, that is, a defaultable claim with zero recovery and with no promised dividends. Our approach will be based on Proposition 4. Thus, we assume that processes $Y^1, \ldots, Y^m$ represent prices of defaultable primary assets and $Y^{m+1}, \ldots, Y^k$ are prices of default-free primary assets. Processes $\widetilde{Y}^1, \ldots, \widetilde{Y}^m, Y^{m+1}, \ldots, Y^k$ are assumed to be continuous $\mathbb{F}$-semimartingales, and processes $\widetilde{Y}^1, Y^{m+1}, \ldots, Y^k$ are strictly positive.

Zero Recovery for Defaultable Primary Assets

Unless explicitly stated otherwise, we postulate that Assumption (A) is valid. Recall that $\widetilde{U}_t(X)$ stands for the pre-default value at time $t \in [0, T]$ of a defaultable claim $(X, 0, 0, \tau)$. In the statement of following result we preserve the notation of Proposition 4.

Proposition 5. *Suppose that there exist a constant $\widetilde{V}_0^1$, and $\mathbb{F}$-predictable processes ψ^i, $i = 2, \ldots, m$ and $\widetilde{\psi}^{i,k,1}$, $i = m + 1, \ldots, k - 1$ such that*

$$\widetilde{Y}_T^1 \left(\widetilde{V}_0^1 + \sum_{i=2}^{m} \int_0^T \psi_u^i \, d\widetilde{Y}_u^{i,1} + \sum_{i=m+1}^{k-1} \int_0^T \widetilde{\psi}_u^{i,k,1} \, d\widehat{Y}_u^{i,k,1} \right) = X. \qquad (18)$$

Let $\widetilde{V}_t = \widetilde{Y}_t^1 \widetilde{V}_t^1$, where the process $\widetilde{V}_t^1$ is defined as, for every $t \in [0, T]$,

$$\widetilde{V}_t^1 = \widetilde{V}_0^1 + \sum_{i=2}^{m} \int_0^t \psi_u^i \, d\widetilde{Y}_u^{i,1} + \sum_{i=m+1}^{k-1} \int_0^t \widetilde{\psi}_u^{i,k,1} \, d\widehat{Y}_u^{i,k,1}.$$

Then the trading strategy $\phi = (\phi^1, \phi^2, \ldots, \phi^k)$ defined by

$$\phi_t^1 = \left(\widetilde{V}_t - \sum_{i=2}^{m} \psi_t^i Y_t^i \right) (\widetilde{Y}_t^1)^{-1},$$

$$\phi_t^i = \psi_t^i, \quad i = 2, \ldots, m,$$

$$\phi_t^i = \widetilde{\psi}_t^{i,k,1} \widetilde{Y}_t^{1,k} e^{-\widetilde{\alpha}_t^{i,k,1}}, \quad i = m + 1, \ldots, k - 1,$$

$$\phi_t^k = - \sum_{i=m+1}^{k-1} \psi_t^i Y_t^i (Y_t^k)^{-1},$$

is self-financing and it replicates the claim $(X, 0, 0, \tau)$. In particular, we have $\widetilde{V}_t(\phi) = \widetilde{V}_t = \widetilde{U}_t(X)$, so that $\widetilde{V}$ represents the pre-default value of $(X, 0, 0, \tau)$.

Proof. The statement is an almost immediate consequence of part (ii) of Proposition 4 (see also Corollary 1). The strategy $(\phi^1, \phi^2, \ldots, \phi^k)$ introduced in the statement of the proposition is self-financing, and at the default time τ the wealth $V(\phi)$ jumps to zero. Finally, $V_T(\phi) = \widetilde{V}_T(\phi) = X$ on the event $\{\tau > T\}$. We conclude that ϕ is self-financing and it replicates $(X, 0, 0, \tau)$. $\square$

The following corollary to Proposition 5 provides the risk-neutral characterization of the process $\widetilde{U}_t(X)$, and thereby it furnishes a convenient method for the valuation of a promised payoff.

Corollary 2. *Assume that a defaultable claim $(X, 0, 0, \tau)$ is attainable. Suppose that there exists a probability measure $\widetilde{\mathbb{Q}}$ such that the processes $\widetilde{Y}^{i,1}$, $i = 2, \ldots, m - 1$ and processes $\widehat{Y}^{i,k,1}$, $i = m + 1, \ldots, k - 1$ are $\mathbb{F}$-martingales under $\widetilde{\mathbb{Q}}$. If all stochastic integrals in (18) are $\widetilde{\mathbb{Q}}$-martingales, rather than $\widetilde{\mathbb{Q}}$-local martingales, then the pre-default value of $(X, 0, 0, \tau)$ equals, for every $t \in [0, T]$,*

$$\widetilde{U}_t(X) = \widetilde{Y}_t^1 \, \mathbb{E}_{\widetilde{\mathbb{Q}}} \left(X (\widetilde{Y}_T^1)^{-1} \mid \mathcal{F}_t \right).$$

Defaultable asset and two default-free assets. In the case when $m = 1$ and $k = 2$, Proposition 5 reduces to the following result. Recall that we denote

$$\tilde{\alpha}_t^{2,3,1} = \langle \ln Y^{2,3}, \ln \tilde{Y}^{1,3} \rangle_t = \int_0^t (Y_u^{2,3})^{-1} (\tilde{Y}_u^{1,3})^{-1} \, d\langle Y^{2,3}, \tilde{Y}^{1,3} \rangle_u,$$

where in turn $\tilde{Y}_t^{1,3} = \tilde{Y}_t^1 (Y_t^3)^{-1}$ and $Y_t^{2,3} = Y_t^2 (Y_t^3)^{-1}$. Moreover, $\hat{Y}_t^{2,3,1} = Y_t^{2,3} e^{-\tilde{\alpha}_t^{2,3,1}}$. We postulate that the processes $\tilde{Y}^1, Y^2$ and Y^3 are strictly positive.

Corollary 3. *Suppose that there exists a constant $\tilde{V}_0^1$ and an $\mathbb{F}$-predictable process $\tilde{\psi}^{2,3,1}$ such that*

$$\tilde{Y}_T^1 \left(\tilde{V}_0^1 + \int_0^T \tilde{\psi}_u^{2,3,1} \, d\hat{Y}_u^{2,3,1} \right) = X. \tag{19}$$

Let us set $\tilde{V}_t = \tilde{Y}_t^1 \tilde{V}_t^1$, where for every $t \in [0,T]$ the process $\tilde{V}_t^1$ is given by

$$\tilde{V}_t^1 = \tilde{V}_0^1 + \int_0^t \tilde{\psi}_u^{2,3,1} \, d\hat{Y}_u^{2,3,1}. \tag{20}$$

Then the trading strategy $\phi = (\phi^1, \phi^2, \phi^3)$, given by the expressions

$$\phi_t^1 = \tilde{V}_t (\tilde{Y}_t^1)^{-1}, \quad \phi_t^2 = \tilde{\psi}_t^{2,3,1} \tilde{Y}_t^{1,3} e^{-\tilde{\alpha}_t^{2,3,1}}, \quad \phi_t^3 = -\phi_t^2 Y_t^2 (Y_t^3)^{-1},$$

is self-financing prior to default and it replicates a claim $(X, 0, 0, \tau)$.

Assume that a claim $(X, 0, 0, \tau)$ is attainable, and let $\tilde{\mathbb{Q}}$ be a probability measure such that $\hat{Y}^{2,3,1}$ is an $\mathbb{F}$-martingale under $\mathbb{Q}$. Then the pre-default value of $(X, 0, 0, \tau)$ equals, for every $t \in [0,T]$,

$$U_t(X) = \tilde{Y}_t^1 \, \mathbb{E}_{\tilde{\mathbb{Q}}} \left(X (\tilde{Y}_T^1)^{-1} \mid \mathcal{F}_t \right), \tag{21}$$

provided that the integral in (20) is also a $\tilde{\mathbb{Q}}$-martingale.

Example 1. Assume that

$$d\tilde{Y}_t^1 = Y_t^1 (\mu_t \, dt + \sigma_t^1 \, dW_t)$$

and

$$dY_t^i = Y_t^i (r_t \, dt + \sigma_t^i \, dW_t^*)$$

for $i = 2, 3$, where W^* is a one-dimensional standard Brownian motion with respect to the filtration $\mathbb{F} = \mathbb{F}^{W^*}$ under the martingale measure $\mathbb{Q}^*$. Then for the processes $\tilde{Y}_t^{1,3} = \tilde{Y}_t^1 (Y_t^3)^{-1}$ and $Y_t^{2,3} = Y_t^2 (Y_t^3)^{-1}$ we get

$$d\tilde{Y}_t^{1,3} = \tilde{Y}_t^{1,3} \left((\mu_t - r_t + \sigma_t^3(\sigma_t^3 - \sigma_t^1)) \, dt + (\sigma_t^1 - \sigma_t^3) \, dW_t^* \right),$$

$$dY_t^{2,3} = Y_t^{2,3} \left(\sigma_t^3(\sigma_t^3 - \sigma_t^2) \, dt + (\sigma_t^2 - \sigma_t^3) \, dW_t^* \right),$$

and thus

$$\tilde{\alpha}_t^{2,3,1} = \int_0^t (\sigma_u^3 - \sigma_u^1)(\sigma_u^3 - \sigma_u^2)\, du.$$

Hence, the process $\hat{Y}_t^{2,3,1} = Y_t^{2,3} e^{-\tilde{\alpha}_t^{2,3,1}}$ satisfies

$$d\hat{Y}_t^{2,3,1} = \hat{Y}_t^{2,3,1}\left(\sigma_t^1(\sigma_t^3 - \sigma_t^2)\, dt + (\sigma_t^2 - \sigma_t^3)dW_t^*\right).$$

If $\sigma^2 \neq \sigma^3$ then, under mild technical assumptions, there exists a probability measure $\mathbb{Q}$ such that $\hat{Y}^{2,3,1}$ is a martingale. To conclude, it suffices to use the fact that an $\mathcal{F}_T$-measurable random variable $X(\tilde{Y}_T^1)^{-1}$ can be represented (by virtue of the predictable representation theorem) as follows

$$X(\tilde{Y}_T^1)^{-1} = \tilde{U}_0(X) + \int_0^T \hat{\phi}_u^{2,3,1}\, d\hat{Y}_u^{2,3,1}$$

for some $\mathbb{F}$-predictable process $\hat{\phi}^{2,3,1}$. It is natural to conjecture that within the present setup all defaultable claims with zero recovery and no promised dividends will be attainable, provided that the underlying default-free market is assumed to be complete, and provided we can use in our hedging portfolio a defaultable asset that is sensitive to the same default risk as the defaultable claim that we want to hedge.

Two defaultable assets. Let us examine the case when $m = k = 2$. We thus consider two defaultable primary assets Y^1 and Y^2 with zero recovery at default.

Corollary 4. *Suppose that there exists a constant $\tilde{V}_0^1$ and an $\mathbb{F}$-predictable process ψ^2 such that*

$$\tilde{Y}_T^1\left(\tilde{V}_0^1 + \int_0^T \psi_u^2\, d\tilde{Y}_u^{2,1}\right) = X, \tag{22}$$

where $\tilde{Y}_t^{2,1} = \tilde{Y}_t^2(\tilde{Y}_t^1)^{-1}$. Let us set $\tilde{V}_t = \tilde{Y}_t^1\tilde{V}_t^1$, where for every $t \in [0,T]$ the process $\tilde{V}_t^1$ is given by

$$\tilde{V}_t^1 = \tilde{V}_0^1 + \int_0^t \psi_u^2\, d\tilde{Y}_u^{2,1}. \tag{23}$$

Then the trading strategy $\phi = (\phi^1, \phi^2)$ where, for every $t \in [0,T]$,

$$\phi_t^1 = (\tilde{V}_t^1 - \psi_t^2\tilde{Y}_t^2)(\tilde{Y}_t^1)^{-1}, \quad \phi_t^2 = \psi_t^2,$$

is self-financing and it replicates a defaultable claim $(X, 0, 0, \tau)$.

Suppose that $(X, 0, 0, \tau)$ is an attainable claim. Let $\tilde{\mathbb{Q}}$ be a probability measure such that $\tilde{Y}^{2,1}$ is an $\mathbb{F}$-martingale under $\tilde{\mathbb{Q}}$. If the stochastic integral in (23) is a $\tilde{\mathbb{Q}}$-martingale, then the pre-default value of $(X, 0, 0, \tau)$ satisfies, for every $t \in [0,T]$,

$$\tilde{U}_t(X) = \tilde{Y}_0^1\, \mathbb{E}_{\tilde{\mathbb{Q}}}\left(X(\tilde{Y}_T^1)^{-1}\,\big|\, \mathcal{F}_t\right). \tag{24}$$

Remark. Under the assumptions of Corollary 4, a defaultable claim $(X, 0, 0, \tau)$ is attainable since the associated promised payoff X can be achieved by trading in the pre-default values $\widetilde{Y}^1$ and $\widetilde{Y}^2$. If we introduce, in addition, some default-free assets, a replicating strategy for an arbitrary defaultable claim $(X, 0, 0, \tau)$ will typically have a zero net investment in default-free assets. Therefore, default-free assets are not relevant if we restrict our attention to defaultable claims of the form $(X, 0, 0, \tau)$.

Non-Zero Recovery for Defaultable Primary Assets

We relax Assumption (A), and we postulate instead that Assumption (B) is valid. Specifically, let us consider m defaultable primary assets with a common default time τ that are subject to a fractional recovery of market value (see Section 3.2) with $\delta_i = \delta \neq 1$ for $i = 1, 2, \ldots, m$. Let us denote

$$\widetilde{\alpha}_t = \sum_{i=1}^{m} \phi_t^i \widetilde{Y}_t^i, \quad \beta_t = \sum_{i=m+1}^{k} \phi_t^i Y_t^i.$$

so that $\widetilde{\alpha}_t + \beta_t$ represents the pre-default wealth of ϕ. As usual, $\widetilde{U}_t(X)$ stands for the pre-default value at time t of the promised payoff X. It is rather clear that the processes $\widetilde{\alpha}_t$ and β_t should be chosen in such a way that $\widetilde{\alpha}_t + \beta_t = \widetilde{U}_t(X)$ and $\bar{\alpha}_t + \beta_t = \delta \widetilde{\alpha}_t + \beta_t = 0$ for every $t \in [0, T]$ (for the latter equality, see (16) and (17)). By solving these equations, we obtain, for every $t \in [0, T]$,

$$\alpha_t = (1 - \delta)^{-1} \widetilde{U}_t(X), \quad \beta_t = (\delta - 1)^{-1} \delta \widetilde{U}_t(X)$$

We end up with the following equation

$$\widetilde{Y}_T^1 \left(\widetilde{U}_0(X) + \sum_{i=2}^{m} \int_0^T \phi_u^i \, d\widetilde{Y}_u^{i,1} + \sum_{i=m+1}^{k-1} \int_0^T \widetilde{\phi}_u^{i,k,1} \, d\widehat{Y}_u^{i,k,1} \right.$$
$$\left. + \int_0^T \beta_u (Y_u^k)^{-1} \, d(\widetilde{Y}_u^{1,k})^{-1} \right) = X.$$

Using the latter equation, one may try to establish a suitable extension of Proposition 5. Notice that the process β depends explicitly on the pre-default value $\widetilde{U}(X)$. In addition, we need to take care of the constraint $\widetilde{\alpha}_t = (1 - \delta)^{-1} \widetilde{U}_t(X)$ for every $t \in [0, T]$. Thus, the problem of replication of a promised payoff under non-zero recovery for defaultable primary assets seems to be rather difficult to solve, in general.

4.2 Replication of a Recovery Payoff

Let us now focus on the recovery payoff Z at time of default. As before, we write $\widetilde{U}_t(Z)$ to denote the pre-default value at time $t \in [0, T]$ of the claim $(0, 0, Z, \tau)$. Recall that $\widetilde{U}_T(Z) = 0$ (and $U_T(Z) = 0$ on the event $\{\tau > T\}$.

Zero Recovery for Defaultable Primary Assets

In order to examine the replicating strategy, we shall once again make use of Proposition 4. As already explained, in this case we need to assume that condition (11) is imposed on a strategy ϕ we are looking for, that is, we necessarily have $\sum_{i=m+1}^{k} \phi_t^i Y_t^i = Z_t$ for every $t \in [0, T]$.

Proposition 6. *Suppose that there exist a constant $\widetilde{V}_0^1$, and $\mathbb{F}$-predictable processes ψ^i, $i = 2, \ldots, m$ and $\widetilde{\psi}^{i,k,1}$, $i = m + 1, \ldots, k - 1$ such that*

$$\widetilde{Y}_T^1 \Big(\widetilde{V}_0^1 + \sum_{i=2}^{m} \int_0^T \psi_u^i \, d\widetilde{Y}_u^{i,1} + \sum_{i=m+1}^{k-1} \int_0^T \widetilde{\psi}_u^{i,k,1} \, d\widetilde{Y}_u^{i,k,1} $$

$$+ \int_0^T Z_u (Y_u^k)^{-1} \, d(\widetilde{Y}_u^{1,k})^{-1} \Big) = 0. \tag{25}$$

Let $\widetilde{V}_t = \widetilde{Y}_t^1 \widetilde{V}_t^1$, where the process $\widetilde{V}_t^1$ is defined as

$$\widetilde{V}_t^1 = \widetilde{V}_0^1 + \sum_{i=2}^{m} \int_0^t \psi_u^i \, d\widetilde{Y}_u^{i,1} + \sum_{i=m+1}^{k-1} \int_0^t \widetilde{\psi}_u^{i,k,1} \, d\widetilde{Y}_u^{i,k,1} $$

$$+ \int_0^t Z_u (Y_u^k)^{-1} \, d(\widetilde{Y}_u^{1,k})^{-1}. $$

Then the replicating strategy $\phi = (\phi^1, \phi^2, \ldots, \phi^k)$ for $(0, 0, Z, \tau)$ is given by

$$\phi_t^1 = \Big(\widetilde{V}_t - Z_t - \sum_{i=2}^{m} \phi_t^i Y_t^i \Big) (\widetilde{Y}_t^1)^{-1},$$

$$\phi_t^i = \psi_t^i, \quad \forall i = 2, \ldots, m,$$

$$\phi_t^i = \widetilde{\psi}_t^{i,k,1} \widetilde{Y}_t^{1,k} e^{-\widetilde{\alpha}_t^{i,k,1}}, \quad \forall i = m + 1, \ldots, k - 1,$$

$$\phi_t^k = \Big(Z_t - \sum_{i=m+1}^{k-1} \phi_t^i Y_t^i \Big) (Y_t^k)^{-1}.$$

Proof. The proof is based on an application of part (ii) of Proposition 4. First, notice that by virtue of the specification of the strategy ϕ we have $\widetilde{V}_t(\phi) = \widetilde{V}_t$ for every $t \in [0, T]$. Moreover, $V_\tau(\phi) = Z_\tau$ on the set $\{\tau \leq T\}$. Finally, $V_T(\phi) = \widetilde{V}_T(\phi) = 0$ on the event $\{\tau > T\}$. □

Defaultable asset and two default-free assets. For the ease of reference, we consider here a special case of Proposition 6. We take $m = 1$ and $k = 3$, and we postulate that the processes $\widetilde{Y}^1, Y^2$ and Y^3 are strictly positive. Recall that the recovery process Z, and thus also its pre-default value process $\widetilde{U}(Z)$, are prespecified.

Corollary 5. *Suppose that there exists a constant $\widetilde{V}_0^1$ and an $\mathbb{F}$-predictable process $\widehat{\psi}^{2,3,1}$ such that*

$$\widetilde{Y}_T^1 \left(\widetilde{V}_0^1 + \int_0^T \widehat{\psi}_u^{2,3,1} \, d\widehat{Y}_u^{2,3,1} + \int_0^T Z_u (Y_u^3)^{-1} \, d(\widetilde{Y}_u^{1,3})^{-1} \right) = 0. \quad (26)$$

Let $\widetilde{V}_t = \widetilde{Y}_t^1 \widetilde{V}_t^1$, where the process $\widetilde{V}_t^1$ is defined as

$$\widetilde{V}_t = \widetilde{V}_0^1 + \int_0^t \widehat{\psi}_u^{2,3,1} \, d\widehat{Y}_u^{2,3,1} + \int_0^t Z_u (Y_u^3)^{-1} \, d(\widetilde{Y}_u^{1,3})^{-1}.$$

Then the replicating strategy for the claim $(0, 0, Z, \tau)$ equals

$$\phi_t^1 = (\widetilde{V}_t - Z_t)(\widetilde{Y}_t^1)^{-1}, \quad \phi_t^2 = \widehat{\psi}_t^{2,3,1} \widetilde{Y}_t^{1,3} e^{-\widetilde{\alpha}_t^{2,3,1}}, \quad \phi_t^3 = (Z_t - \phi_t^2 Y_t^2)(Y_t^3)^{-1}.$$

The existence of $\widehat{\psi}^{2,3,1}$, as well as the possibility of deriving a closed-form expression for ϕ are not obvious. One needs to impose more specific assumptions on the price processes of primary assets and the recovery process in order to obtain results that would be more practical.

If there exists a probability $\mathbb{Q}^*$ such that $\widehat{Y}^{2,3,1}$ is an $\mathbb{F}$-martingale, then the (ex-dividend) value of Z^0 equals

$$U_t(Z) = Y_t^1 \, \mathbb{E}_{\mathbb{Q}^*} \left(\int_t^T Z_u (Y_u^3)^{-1} \, d(\widetilde{Y}_u^{1,3})^{-1} \,\Big|\, \mathcal{F}_t \right).$$

Two defaultable assets. Of course, if both defaultable primary assets are subject to the zero recovery scheme, and no other asset is available for trade, no replicating strategy exists in the case of a non-zero recovery process Z. Thus, we need to postulate a more general recovery scheme for defaultable assets if we wish to have a positive result.

Non-Zero Recovery for Defaultable Primary Assets

Suppose now that Assumption (B) is valid and $Y^1, \ldots, Y^m$ are defaultable primary assets with a fractional recovery of market value. We assume that $\delta_i = \delta \neq 1$ for $i = 1, 2, \ldots, m$, and we proceed along the similar lines as in Section 4.1. Recall that we denote

$$\widetilde{\alpha}_t = \sum_{i=1}^m \phi_t^i \widetilde{Y}_t^i, \quad \beta_t = \sum_{i=m+1}^k \phi_t^i Y_t^i.$$

We now postulate that $\widetilde{\alpha}_t + \beta_t = \widetilde{U}_t(Z)$ and $\bar{\alpha}_t + \beta_t = \delta \widetilde{\alpha}_t + \beta_t = Z_t$ for every $t \in [0, T]$, where $\widetilde{U}_t(Z)$ is the pre-default value of $(0, 0, Z, \tau)$. Consequently, for every $t \in [0, T]$ we have

$$\widetilde{\alpha}_t = (\delta - 1)^{-1}(Z_t - \widetilde{U}_t(Z)), \quad \beta_t = (\delta - 1)^{-1}(\delta \widetilde{U}_t(Z) - Z_t).$$

To find a replicating strategy for a defaultable claim $(0, 0, Z, \tau)$, we need, in particular, to find $\mathbb{F}$-predictable processes ψ^i and $\widehat{\psi}^{i,k,1}$ such that the equality

$$\widetilde{U}_t(Z) = \widetilde{Y}_t^1 \Big(U_0(Z) + \sum_{i=2}^{m} \int_0^t \psi_u^i \, d\widetilde{Y}_u^{i,1} + \sum_{i=m+1}^{k-1} \int_0^t \widehat{\psi}_u^{i,k,1} \, d\widehat{Y}_u^{i,k,1}$$
$$+ \int_0^t \beta_u (Y_u^k)^{-1} \, d(\widetilde{Y}_u^{1,k})^{-1} \Big)$$

is satisfied for every $t \in [0, T]$. Similarly as in Section 4.2, we conclude that the considered problem is non-trivial, in general.

4.3 Replication of Promised Dividends

We return to the case of zero recovery for defaultable primary assets, and we consider a defaultable claim $(0, C, 0, \tau)$. In principle, replication of the stream of promised dividends can reduced to previously considered cases (that's why it was possible to postulate in Definition 5 that $C = 0$). Specifically, it suffices to introduce the recovery process Z^C generated by C by setting, for every $t \in [0, T]$,

$$Z_t^C = \int_{(0,t)} B^{-1}(u, t) \, dC_u,$$

and to combine it with the terminal payoff $\mathbb{1}_{\{\tau > T\}} X^C$, where the promised payoff X^C associated with C equals

$$X^C = \int_{(0,T]} B^{-1}(u, T) \, dC_u.$$

It should be stressed, however, that the pre-default price of an "equivalent" defaultable claim $(X^C, 0, Z^C, \tau)$ introduced above does not coincide with the pre-default price of the original claim $(0, C, 0, \tau)$, that is, processes $\widetilde{U}(C)$ and $\widetilde{U}(Z^C) + \widetilde{U}(X^C)$ are not identical. But, clearly, the equality $U_0(C) = U_0(Z^C) + U_0(X^C)$ is satisfied, and thus at time 0 the replicating strategies for both claims coincide.

Remark. It is apparent that the concept of the (ex-dividend) pre-default price $\widetilde{U}(C)$ does not fit well into study of replication of promised dividends if one only considers non-dividend paying primary assets. It would be much more convenient to use in the case of dividend-paying (default-free or defaultable) primary assets. For instance, it is sometimes legitimate to postulate the existence of a default-free version of the defaultable claim $(0, C, 0, \tau)$, that is, a default-free asset with the dividend stream C.

If we insist on working directly with the process $\widetilde{U}(C)$, then we derive the following set of necessary conditions for a self-financing trading strategy ϕ with the consumption process $A = -C$

$$\sum_{i=m+1}^{k} \phi_t^i Y_t^i = 0, \quad V_t(\phi) = \sum_{i=1}^{m} \phi_t^i \widetilde{Y}_t^i = \widetilde{U}_t(C), \quad (27)$$

and

$$dV_t(\phi) - \sum_{i=1}^{m} \phi_t^i \, d\widetilde{Y}_t^i + \sum_{i=m+1}^{k} \phi_t^i \, dY_t^i - dC_t = d\widetilde{U}_t(C). \quad (28)$$

The existence of a strategy $\phi = (\phi^1, \phi^2, \ldots, \phi^k)$ with consumption process $A = -C$, which satisfies (27)-(28) is not evident, however.

Example 2. Let us take, for instance, $m = 1$ and $k = 3$. Then conditions (27)-(28) become:

$$\phi_t^1 \widetilde{Y}_t^1 = \widetilde{U}_t(C), \quad \phi_t^2 Y_t^2 + \phi_t^3 Y_t^3 = 0,$$

and

$$\phi_t^1 \, d\widetilde{Y}_t^1 + \phi_t^2 \, dY_t^2 + \phi_t^3 dY_t^3 = d\widetilde{U}_t(C) + dC_t.$$

Assume that under $\mathbb{Q}^*$ we have

$$d\widetilde{Y}^1 = \mu_t \, dt + \sigma_t^1 \, dW_t^*,$$
$$dY_t^i = r_t \, dt + \sigma_t^i \, dW_t^*, \quad i = 2, 3,$$
$$d\widetilde{U}_t(C) = a_t \, dt + b_t \, dW_t^*.$$

If, in addition, $dC_t = c_t dt$ then we obtain the following system of equations for $\phi = (\phi^1, \psi^2, \phi^3)$

$$\phi_t^1 \widetilde{Y}_t^1 = \widetilde{U}_t(C),$$
$$\phi_t^2 Y_t^2 + \phi_t^3 Y_t^3 = 0,$$
$$\phi_t^1 \mu_t^1 + \phi_t^2 \mu_t^2 + \phi_t^3 \mu_t^3 = a_t + c_t,$$
$$\phi_t^1 \sigma_t^1 + \phi_t^2 \sigma_t^2 + \phi_t^3 \sigma_t^3 = b_t.$$

4.4 Replication of a First-to-Default Claim

Until now, we have always postulated that a random time τ represents a common default time for all defaultable primary assets, as well as for a defaultable contingent claim under consideration. This simplifying assumptions manifestly fails to hold in the case of a credit derivative that explicitly depends on default times of several (possibly independent) reference entities. Consequently, the issue of replication of a so-called *first-to-default claim* is more challenging, and the approach presented in the preceding sections needs to be extended.

Let the random times $\tau_1, \ldots, \tau_m$ represent the default times of m reference entities that underlie a given first-to-default claim. We assume that $\mathbb{Q}^*\{\tau_i = \tau_j\} = 0$ for every $i \neq j$, and we denote by $\tau_{(1)}$ the random moment of the first default, that is, we set $\tau_{(1)} = \min\{\tau_1, \tau_2, \ldots, \tau_n\} = \tau_1 \wedge \tau_2 \wedge \cdots \wedge \tau_n$. A *first-to-default claim* $(X, C, Z^1, \ldots, Z^m, \tau_1, \ldots, \tau_m)$ with maturity date T can be described as follows. If $\tau_{(1)} = \tau_i \leq T$ for some $i = 1, \ldots, m$, then it pays at time $\tau_{(1)}$ the amount $Z^i_{\tau_{(1)}}$, where Z^i is an $\mathbb{F}$-predictable recovery process. Otherwise, that is, if $\tau_{(1)} > T$, the claim pays at time T an $\mathcal{F}_T$-measurable promised amount X. Finally, a claim pays promised dividends stream C prior to the default time $\tau_{(1)}$, more precisely, on the random interval $\mathbb{1}_{\{\tau_{(1)} \leq T\}}[\![0, \tau_{(1)}[\![\cup \mathbb{1}_{\{\tau_{(1)} > T\}}[\![0, T]\!]$. It is clear the dividend process of a generic first-to-default claim equals, for every $t \in [0, T]$,

$$D_t = X \mathbb{1}_{\{\tau_{(1)} > T\}} \mathbb{1}_{[T, \infty)}(t) + \int_{(0,t]} (1 - H_u^{(1)}) \, dC_u + \int_{(0,t]} Z_u^i \mathbb{1}_{\{\tau_{(1)} = \tau_i\}} \, dH_u^{(1)},$$

where $H_t^{(1)} = 1 - \prod_{i=1}^m (1 - H_t^i)$ or, equivalently, $H_t^{(1)} = \mathbb{1}_{\{\tau_{(1)} \leq t\}}$. Let $\mathbb{H}^i$ be the filtration generated by the process $H_t^i = \mathbb{1}_{\{\tau_i \leq t\}}$ for $i = 1, 2, \ldots, m$, and let the filtration $\mathbb{G}$ be given as $\mathbb{G} = \mathbb{F} \vee \mathbb{H}^1 \vee \mathbb{H}^2 \vee \cdots \vee \mathbb{H}^m$. Then, by definition, the (ex-dividend) price of $(X, C, Z^1, \ldots, Z^m, \tau_1, \ldots, \tau_m)$ equals, for every $t \in [0, T]$,

$$U_t = B_t \, \mathbb{E}_{\mathbb{Q}^*} \left(\int_{(t,T]} B_u^{-1} \, dD_u \, \Big| \, \mathcal{G}_t \right).$$

By a pre-default value of a claim we mean an $\mathbb{F}$-adapted process $\widetilde{U}$ such that $U_t = \widetilde{U}_t \mathbb{1}_{\{\tau_{(1)} > t\}}$ for every $t \in [0, T]$. The following definition is a direct extension of Definition 5 (thus, we maintain the assumption that $C = 0$). By a self-financing strategy we mean here a strategy which is self-financing prior to the first default (cf. Definition 4), and thus it is self-financing on $[0, T]$ as well.

Definition 6. *A self-financing strategy ϕ is a replicating strategy for a first-to-default contingent claim $(X, 0, Z^1, \ldots, Z^m, \tau_1, \ldots, \tau_m)$ if and only if the following hold:*
(i) $V_t(\phi) = \widetilde{U}_t$ on the random interval $[\![0, \tau_{(1)} \wedge T[\![$,
(ii) $V_\tau(\phi) = Z^i_\tau$ on the event $\{\tau_{(1)} = \tau_i \leq T\}$,
(iii) $V_T(\phi) = X$ on the event $\{\tau_{(1)} > T\}$.

In order to provide a replicating strategy for a first-to-default claim we postulate the existence of m defaultable primary assets $Y^1, \ldots, Y^m$ with the corresponding default times $\widetilde{\tau}_1, \ldots, \widetilde{\tau}_m$. It is natural to postulate that the default times $\widetilde{\tau}_1, \ldots, \widetilde{\tau}_m$ are also the default times of m reference entities that underlie a first-to-default claim under consideration, so that, $\widetilde{\tau}_i = \tau_i$ for $i = 1, 2, \ldots, m$. It should be stressed that, typically, the pre-default value $\widehat{Y}^j$ will follow a discontinuous process (for instance, it may have jumps at default times of other entities). Finally, let us recall that $\bar{Y}_t^i$ represents the recovery payoff of the ith defaultable asset if its default occurs at time t.

Case of zero promised dividends. We shall assume from now on that $C = 0$. For arbitrary $i \neq j$, let $\widehat{Y}_t^{ij}$ represent the pre-default value of the ith asset conditioned on the event $\{\tau_{(1)} = \tau_j = t\}$. More explicitly, $\widehat{Y}_t^{ij}$ is equal to $\widetilde{Y}_t^i$ on the random interval

$[\![\tau_{(1)} 1\!\!1_D, \tau_{(2)} 1\!\!1_D[\![$, where $D = \{\tau_{(1)} = \tau_j\}$ and $\tau_{(2)}$ is the time of the second default ($\widehat{Y}_t^{ij}$ is not defined outside the random interval introduced above). At the intuitive level, the process $\widehat{Y}_t^{ij}$ gives the value at time t of the ith defaultable asset, provided that the first default has occurred at time t, and the jth entity is the first defaulting entity. Hence, $\widehat{Y}_t^{ij}$ is not a new process, but rather an additional notation introduced in order to simplify the formulae that follow.

Remark. It is important to stress that the notion of a 'defaultable asset' should not be understood literally. For instance, if the case of the so-called *flight to quality* the price of a default-free bond is discontinuous, and it jumps at the moment τ associated with some 'default event' (see, e.g., Collin-Dufresne et al. (2003)). Thus, from the perspective of hedging a default-free bond may be formally classified as a 'defaultable asset'.

In order to find a replicating strategy ϕ for a first-to-default claim within the present setup, we need to impose the following m conditions on its components $\phi^1, \ldots, \phi^k$: for every $j = 1, \ldots, m$ and every $t \in [0, T]$

$$\sum_{i=1,\, i \neq j}^{m} \phi_t^i \widehat{Y}_t^{ij} + \phi_t^j \bar{Y}_t^j + \sum_{i=m+1}^{k} \phi_t^i Y_t^i = Z_t^j, \qquad (29)$$

where $Z^1, \ldots, Z^m$ is a given family of recovery processes. Recall that Z^j specifies the payoff received by the owner of a claim if the first default occurs prior to or at T, and the first defaulting entity is the jth entity.

For the sake of concreteness, assume that

$$Z_t^j = g_j(t, \widetilde{Y}_t^1, \ldots, \widetilde{Y}_t^m, \bar{Y}_t^1, \ldots, \bar{Y}_t^m, Y_t^{m+1}, \ldots, Y_t^k)$$

for some function $g : \mathbb{R}^{k+m+1} \to \mathbb{R}$. Under some additional assumptions, the system of equations (29) can be solved explicitly for $\phi^1, \ldots, \phi^m$. In the second step, we need to choose processes $\phi^{m+1}, \ldots, \phi^k$ in such a way that a strategy ϕ is self-financing prior to the first default, and thus also on the random interval $[\![0, \tau_{(1)} \wedge T]\!]$. Finally, the wealth of a strategy ϕ should match the promised payoff X at time T on the event $\{\tau_{(1)} > T\}$. Equivalently, the wealth of ϕ should coincide with the value of a considered claim prior to and at default, or up to time T if there is no default in $[0, T]$. It is apparent that the problem of existence of a replicating strategy is non-trivial, but it can be solved in some circumstances.

A detailed analysis of an explicit replication result for a particular example of a first-to-default claim is given in Section 5.2.

5 Vulnerable Claims and Credit Derivatives

In this section, we present a few examples of models and simple defaultable claims for which there exists explicit replicating strategy. We maintain our assumption that the default time τ admits a continuous hazard process Γ with respect to $\mathbb{F}$ under $\mathbb{Q}^*$, where $\mathbb{F} = \mathbb{F}^{W^*}$ is generated by a Brownian motion W^*. Recall that Γ is also assumed to be an increasing process.

5.1 Vulnerable Claims

Let us fix $T > 0$. We postulate that the T-maturity default-free bond and defaultable zero-coupon bond with zero recovery are also traded assets. As before, we assume that the risk-neutral dynamics of the discount default-free bond are

$$dB(t,T) = B(t,T)\left(r_t\,dt + b(t,T)\,dW_t^*\right)$$

for some $\mathbb{F}$-predictable volatility process $b(t,T)$.

Vulnerable Call Options

For a fixed $U > T$, we assume that the U-maturity default-free bond is also traded, and we consider a vulnerable European call option with the terminal payoff

$$\widehat{C}_T = \mathbb{1}_{\{\tau>T\}}(B(T,U) - K)^+ = \mathbb{1}_{\{\tau>T\}}X.$$

We thus deal with a defaultable claim $(X, 0, 0, \tau)$ with the promised payoff $X = (B(T,U) - K)^+$. The same method can be applied to an arbitrary $\mathcal{F}_T$-measurable promised payoff $X = g(B(T,U))$, where a function $g : \mathbb{R} \to \mathbb{R}$ satisfies usual technical assumptions.

We consider here the situation when one defaultable asset and two default-free assets are traded; we thus place ourselves within the framework of Corollary 3. Specifically, we take $Y_t^1 = D^0(t,T)$, $Y_t^2 = B(t,U)$ and $Y_t^3 = B(t,T)$. Consider a strategy $\phi = (\phi^1, \phi^2, \phi^3)$ such that $V_t(\phi) = \phi_t^1 \widetilde{D}^0(t,T)$ and $\phi_t^2 B(t,U) + \phi_t^3 B(t,T) = 0$ for every $t \in [0,T]$. Observe that in view of the definition of $\Gamma(t,T)$ (see Section 2.3) we have

$$\widetilde{Y}_t^{1,3} = \widetilde{D}^0(t,T)(B(t,T))^{-1} = \Gamma(t,T).$$

Moreover, $Y_t^{2,3} = F(t,U,T)$ and $\widehat{Y}_t^{2,3,1} = F(t,U,T)e^{-\widetilde{\alpha}_t^{2,3,1}}$, where we denote $F(t,U,T) = B(t,U)(B(t,T))^{-1}$ and, by virtue of formula (2),

$$\widetilde{\alpha}_t^{2,3,1} = \langle \ln F(\cdot,U,T), \ln \Gamma(\cdot,T)\rangle_t = \int_0^t \big(b(u,U) - b(u,T)\big)\beta(u,T)\,du.$$

Therefore, the dynamics of $\widehat{Y}^{2,3,1}$ under $\mathbb{Q}_T$ are

$$d\widehat{Y}_t^{2,3,1} = \widehat{Y}_t^{2,3,1}\Big(\big(b(t,T) - b(t,U)\big)\beta(t,T)\,dt + \big(b(t,U) - b(t,T)\big)\,dW_t^T\Big)$$

$$= \widehat{Y}_t^{2,3,1}\big(b(t,U) - b(t,T)\big)\big(dW_t^T - \beta(t,T)\,dt\big).$$

Let $\widetilde{\mathbb{Q}}$ be a probability measure such that $\widehat{Y}^{2,3,1}$ is a martingale under $\widetilde{\mathbb{Q}}$. By virtue of Girsanov's theorem, it is clear that the process $\widetilde{W}$, given by the formula

$$\widetilde{W}_t = W_t^T - \int_0^t \beta(u,T)\,du, \quad \forall\, t \in [0,T],$$

is a Brownian motion under $\widetilde{\mathbb{Q}}$. Thus, the process $F(t,U,T)$ satisfies under $\widetilde{\mathbb{Q}}$

$$dF(t,U,T) = F(t,U,T)\big(b(t,U) - b(t,T)\big)\big(d\widetilde{W}_t + \beta(t,T)\,dt\big). \tag{30}$$

Since $\widetilde{D}^0(T,T) = 1$, equation (19) becomes

$$\widetilde{C}_0 + \int_0^T \widetilde{\phi}_u^{2,3,1}\,d\widehat{Y}_u^{2,3,1} = X = (F(T,U,T) - K)^+. \tag{31}$$

By a simple extension of (21), for any $t \in [0,T]$ the pre-default value of the option equals

$$\widetilde{C}_t = \widetilde{D}^0(t,T)\,\mathbb{E}_{\widetilde{\mathbb{Q}}}\big((F(T,U,T) - K)^+ \,|\, \mathcal{F}_t\big), \tag{32}$$

provided that the integral in (31) is a $\widetilde{\mathbb{Q}}$-martingale, rather than a $\widetilde{\mathbb{Q}}$-local martingale. Let us denote

$$f(t) = \beta(t,T)(b(t,U) - b(t,T)), \quad \forall\, t \in [0,T], \tag{33}$$

and let us assume that f is a deterministic function. Then we have the following result, which extends the valuation formula for a call option written on a default-free zero-coupon bond within the framework of the Gaussian HJM model.

Proposition 7. *The pre-default price $\widetilde{C}_t$ of a vulnerable call option written on a default-free zero-coupon bond equals*

$$\widetilde{C}_t = \widetilde{D}^0(t,T)\left(F(t,U,T)e^{\int_t^T f(u)\,du}N\big(h_+(t,U,T)\big) - KN\big(h_-(t,U,T)\big)\right),$$

where

$$h_\pm(t,U,T) = \frac{\ln F(t,U,T) - \ln K + \int_t^T f(u)\,du \pm \frac{1}{2}v^2(t,T)}{v(t,T)}$$

and

$$v^2(t,T) = \int_t^T |b(u,U) - b(u,T)|^2\,du.$$

The replicating strategy $\phi = (\phi^1, \phi^2, \phi^3)$ for the option satisfies

$$\phi_t^1 = \widetilde{C}_t(\widetilde{D}^0(t,T))^{-1},$$
$$\phi_t^2 = e^{\widetilde{\alpha}_T^{2,3,1} - \widetilde{\alpha}_t^{2,3,1}}\,\Gamma(t,T)N\big(h_+(t,U,T)\big),$$
$$\phi_t^3 = -\phi_t^2 F(t,U,T).$$

Proof. Considering the Itô differential $d(\widetilde{C}_t/\widetilde{D}^0(t,T))$, and identifying terms in expression (31), we obtain that the process $\widehat{\phi}^{2,3,1}$ in the integral representation (31) is given by the formula

$$\widetilde{\phi}_t^{2,3,1} = e^{\int_0^T f(u)\,du} N\big(h_+(t,U,T)\big) = e^{\widetilde{\alpha}_T^{2,3,1}} N\big(h_+(t,U,T)\big).$$

Consequently the valuation formula presented in the proposition is a rather straightforward consequence of (30) and (32). □

Remark. Although we consider here the bond $B(t,U)$ as the underlying asset, it is apparent that the method (and thus also the result) can be applied to a much wider class of underlying assets. For instance, a zero-coupon bond can be substituted with a non-dividend paying stock with the price S (this case was examined in Jeanblanc and Rutkowski (2003)). A suitable modification of formulae established in Proposition 7 can also be used to the valuation and hedging of vulnerable caplets, swaptions, and other vulnerable derivatives in lognormal market models of (non-defaultable) LIBORs and swap rates.

Case of a deterministic hazard process. Assume now that the $\mathbb{F}$-hazard process Γ of τ is deterministic. Then $\beta(t,T) = 0$ for every $t \in [0,T]$, and thus $\widetilde{\alpha}_t^{2,3,1} = 0$ and $\widehat{Y}_t^{2,3,1} = F(t,U,T)$ for every $t \in [0,T]$. We thus obtain the following result.

Corollary 6. *Let the $\mathbb{F}$-hazard process Γ and the volatility $b(t,U) - b(t,T)$ of the forward price $F(t,U,T)$ be deterministic. Then the pre-default price $\widetilde{C}_t$ of a vulnerable option satisfies $\widetilde{C}_t = \Gamma(t,T)C_t$, where C_t is the price of an equivalent non-vulnerable option*

$$C_t = B(t,U)N\big(h_+(t,U,T)\big) - KB(t,T)N\big(h_-(t,U,T)\big),$$

where

$$h_\pm(t,U,T) = \frac{\ln F(t,U,T) - \ln K \pm \tfrac{1}{2}v^2(t,T)}{v(t,T)}$$

and

$$v^2(t,T) = \int_t^T |b(u,U) - b(u,T)|^2\,du.$$

The replicating strategy $\phi = (\phi^1, \phi^2, \phi^3)$ is given by

$$\phi_t^1 = \widetilde{C}_t(\Gamma(t,T)B(t,T))^{-1},$$
$$\phi_t^2 = \Gamma(t,T)N\big(h_+(t,U,T)\big),$$
$$\phi_t^3 = -\phi_t^2 F(t,U,T).$$

Vulnerable Bonds

Let us consider the payoff of the form $\mathbb{1}_{\{\tau>T\}}$ which occurs at some date $U > T$. This payoff is, of course, equivalent to the payoff $B(T,U)\mathbb{1}_{\{\tau>T\}}$ at time T. We interpret this claim as a *vulnerable bond*; Vaillant (2001) proposes to term such a *delayed defaultable bond*. Although vulnerable bonds are not traded, under suitable assumptions one can show that they can be replicated by other liquid assets. Indeed, to replicate this claim within the framework of this section, it suffices to assume that default-free bonds with maturities T and U, as well as the defaultable bond with maturity T are among primary traded assets.

Specifically, we postulate that $\phi_t^2 B(t, U) + \phi_t^3 B(t, T) = 0$ for every $t \in [0, T]$ and thus the total wealth is invested in defaultable bonds of maturity T, so that $\phi_t^1 \widetilde{D}^0(t, T) = \widetilde{U}_t(X)$ for every $t \in [0, T]$, where $X = B(T, U) = F(T, U, T)$. Let $\widetilde{D}^0(t, T, U)$ stand for the pre-default value of a vulnerable bond at time $t < T$. Then formulae (31) and (32) become

$$\widetilde{D}^0(0, T, U) + \int_0^T \widetilde{\phi}_u^{2,3,1} \, d\widehat{Y}_u^{2,3,1} = F(T, U, T)$$

and

$$\widetilde{D}^0(t, T, U) = \widetilde{D}^0(t, T) \, \mathbb{E}_{\widetilde{\mathbb{Q}}} \big(F(T, U, T) \,|\, \mathcal{F}_t \big),$$

respectively. Using dynamics (30), we obtain

$$
\begin{aligned}
\widetilde{D}^0(t, T, U) &= \widetilde{D}^0(t, T) F(t, T, U) \, e^{\int_t^T f(u) \, du} \\
&= \widetilde{D}^0(t, T) F(t, T, U) \, e^{\widetilde{\alpha}_T^{2,3,1} - \widetilde{\alpha}_t^{2,3,1}}
\end{aligned}
\tag{34}
$$

provided that $\widetilde{\alpha}^{2,3,1}$ is deterministic.

5.2 Credit Derivatives

The most widely traded credit derivatives are credit default swaps and swaptions, total rate of return swaps and credit linked notes. Furthermore, a large class of basket credit derivatives have a special feature of being linked to the default risk of several reference entities. We shall consider here only two examples: a credit default swap and a first-to-default contract. Before proceeding to the analysis of more complex contract, we shall first examine a standard (non-vulnerable) option written on a defaultable asset.

Options on a Defaultable Asset

We shall now consider a non-vulnerable call option written on a defaultable bond with maturity date U and zero recovery. Let T be the expiration date and let $K > 0$ stand for the strike. Formally, we deal with the terminal payoff $\bar{C}_T$ given by

$$\bar{C}_T = (D^0(T, U) - K)^+.$$

To replicate this option, we postulate that defaultable bonds of maturities U and T are primary assets. Notice also that

$$\bar{C}_T = \left(\mathbb{1}_{\{\tau>T\}}\tilde{D}^0(T,U) - K\right)^+ = \mathbb{1}_{\{\tau>T\}}\left(\tilde{D}^0(T,U) - K\right)^+ = \mathbb{1}_{\{\tau>T\}}X,$$

where $X = (\tilde{D}^0(T,U) - K)^+$, so that once again we deal with a defaultable claim of the form $(X, 0, 0, \tau)$. It should be stressed, however, that since the underlying asset is now defaultable, the valuation result will differ from Proposition 7.

We shall use two defaultable primary assets for replication. Specifically, we shall now apply Corollary 4, by choosing $Y_t^1 = D^0(t,T)$ and $Y_t^2 = D^0(t,U)$ as primary assets. As before, we denote by $\tilde{C}_t$ the pre-default value of the option under consideration. By virtue of Corollary 4, it suffices to show that there exists a process ϕ^2 such that

$$\tilde{C}_0 + \int_0^T \phi_u^2 \, d\tilde{Y}_u^{2,1} = X = (\tilde{D}^0(T,U) - K)^+ = (\tilde{Y}_T^{2,1} - K)^+, \qquad (35)$$

where $\tilde{Y}_t^{2,1} = \tilde{D}^0(t,U)(\tilde{D}^0(t,T))^{-1}$. Then the trading strategy $\phi = (\phi^1, \phi^2)$ where

$$\phi_t^1 = (\tilde{C}_t - \phi_t^2\tilde{D}^0(t,U))(\tilde{D}^0(t,T))^{-1}$$

is self-financing and it replicates the option. To derive the valuation formula, it suffices to find the probability measure $\tilde{\mathbb{Q}}$ such that the process $\tilde{Y}^{2,1}$ is a $\tilde{\mathbb{Q}}$-martingale, and to use the generic representation

$$\tilde{C}_t = \tilde{D}^0(t,T) \, \mathbb{E}_{\tilde{\mathbb{Q}}}\big((\tilde{Y}_T^{2,1} - K)^+ \,|\, \mathcal{F}_t\big).$$

Notice that, with the notation $\tilde{D}^0(t,U) = \Gamma(t,U)B(t,T)$, the price process $D^0(t,U)$ admits the representation $D^0(t,U) = \mathbb{1}_{\{\tau>t\}}\tilde{D}^0(t,U)$. Assume that τ has a stochastic intensity γ. Then we have (see (3))

$$d\tilde{D}^0(t,U) = \tilde{D}^0(t,U)\Big(\big(r_t + \gamma_t + \beta(t,U)b(t,U)\big) \, dt + \big(\beta(t,U) + b(t,U)\big) \, dW_t^*\Big),$$

and the dynamics of $\tilde{Y}_t^{2,1} = \tilde{D}^0(t,U)(\tilde{D}^0(t,T))^{-1}$ under $\mathbb{Q}^*$ are

$$d\tilde{Y}_t^{2,1} = \tilde{Y}_t^{2,1}\Big(\big(r_t + \gamma_t + \beta(t,U)b(t,U)\big) \, dt$$
$$+ \big(\beta(t,U) + b(t,U) - b(t,T)\big) \big(dW_t^* - b(t,T)dt\big)\Big).$$

As we said above, it suffices to find the probability measure $\tilde{\mathbb{Q}}$ such that the process $\tilde{Y}^{2,1}$ is a $\tilde{\mathbb{Q}}$-martingale. By applying standard Girsanov's transformation, we can construct a measure $\tilde{\mathbb{Q}}$ so that we have

$$d\tilde{Y}_t^{2,1} = \tilde{Y}_t^{2,1}\big(\beta(t,U) + b(t,U) - b(t,T)\big) \, d\tilde{W}_t,$$

where $\tilde{W}$ is a Brownian motion under $\tilde{\mathbb{Q}}$.

Proposition 8. *Assume that $\beta(t,U)+b(t,U)-b(t,T)$, $t\in[0,T]$, is a deterministic function. Then the pre-default price $\widetilde{C}_t$ of a call option written on a U-maturity defaultable bond equals*

$$\widetilde{C}_t = \widetilde{D}^0(t,U)N\big(k_+(t,U,T)\big) - K\widetilde{D}^0(t,T)N\big(k_-(t,U,T)\big),$$

where

$$k_\pm(t,U,T) = \frac{\ln\widetilde{D}^0(t,U) - \ln\widetilde{D}^0(t,T) - \ln K \pm \frac{1}{2}\widetilde{v}^2(t,T)}{\widetilde{v}(t,T)}$$

and

$$\widetilde{v}^2(t,T) = \int_t^T |\beta(u,U)+b(u,U)-b(u,T)|^2\,du.$$

The replicating strategy $\phi=(\phi^1,\phi^2)$ for the option is given by

$$\phi_t^1 = (\widetilde{C}_t - \phi_t^2\widetilde{D}^0(t,U))(\widetilde{D}^0(t,T))^{-1}, \quad \phi_t^2 = N\big(k_+(t,U,T)\big).$$

Case of a deterministic hazard process. Assume that the $\mathbb{F}$-hazard process Γ and the volatility $b(t,U)-b(t,T)$, $t\in[0,T]$, of the forward price $F(t,U,T)$ are deterministic.

Corollary 7. *The pre-default price $\widetilde{C}_t$ of a call option written on a U-maturity defaultable bond equals*

$$\widetilde{C}_t = e^{-\int_t^U \gamma(u)\,du}B(t,U)N\big(k_+(t,U,T)\big)$$
$$\qquad - Ke^{-\int_t^T \gamma(u)\,du}B(t,T)N\big(k_-(t,U,T)\big),$$

where

$$k_\pm(t,U,T) = \frac{\ln B(t,U) - \ln B(t,T) - \ln K - \int_T^U \gamma(u)\,du \pm \frac{1}{2}v^2(t,T)}{v(t,T)}$$

and

$$v^2(t,T) = \int_t^T |b(u,U)-b(u,T)|^2\,du.$$

The replicating strategy $\phi=(\phi^1,\phi^2)$ for the option is given by

$$\phi_t^1 = (\widetilde{C}_t - \phi_t^2\widetilde{D}^0(t,U))(\widetilde{D}^0(t,T))^{-1}, \quad \phi_t^2 = N\big(k_+(t,U,T)\big).$$

Notice that this is exactly the same result as in the case of a call option written on a zero-coupon bond in a default-free term structure model with the interest rate r_t substituted with the default-risk adjusted rate $r_t + \gamma(t)$.

Credit Default Swaps

A generic *credit default swap* (CDS, for short) is a derivative contract which allows to directly transfer the credit risk of the reference entity from one party (the risk seller) to another party (the risk buyer). The contingent payment is triggered by the pre-specified default event, provided that it happens before the maturity date T. The standard version of a credit default swap stipulates that the contract is settled at default time τ of the reference entity, and the recovery payoff equals $Z_\tau = 1 - \delta B(\tau, T)$ where δ represents the recovery rate at default of a reference entity. It is usually assumed that $0 \le \delta < 1$ is non-random, and known in advance. This convention corresponds to the fractional recovery of Treasury value scheme for a defaultable bond issued by the reference entity. Otherwise, that is, in case of no default prior to or at T, the contract expires at time T worthless. The following alternative market conventions are encountered in practice:

- The buyer of the insurance pays a lump sum at inception, and the contract is termed a *default option*,
- The buyer of the insurance pays annuities κ at the predetermined dates $0 < T_1 < \cdots < T_{n-1} < T_n = T$ prior to τ, so that the contract represents a plain-vanilla *default swap*.

In the former case, the (pre-default) value $\widetilde{U}_0(Z)$ at time 0 of the default option equals

$$\widetilde{U}_0(Z) = \mathbb{E}_{\mathbb{Q}^*}\left(B_\tau^{-1}\left(1 - \delta B(\tau, T)\right)\mathbb{1}_{\{\tau \le T\}}\right). \tag{36}$$

In the latter case, the level of the annuity κ should be chosen in such a way that the value of the contract at time 0 equals zero. The annuity κ can thus be specified by solving the following equation

$$\widetilde{U}_0(Z) = \kappa\,\mathbb{E}_{\mathbb{Q}^*}\left(\sum_{i=1}^{n} B_{T_i}^{-1}\,\mathbb{1}_{\{\tau > T_i\}}\right),$$

where the value $\widetilde{U}_0(Z)$ is given by (36).

Digital credit default swap. The fixed leg of a CDS can be represented as the sequence of payoffs $c_i = \kappa\mathbb{1}_{\{\tau > T_i\}}$ at the dates T_i for $i = 1, \ldots, n$. The fixed leg of a CDS can thus be seen as a portfolio of defaultable zero-coupon bonds with zero recovery, and thus the valuation of the fixed leg is rather straightforward. To simplify the valuation of the floating leg, we shall consider a digital CDS. Specifically, we postulate that the constant payoff δ is received at time T_{i+1} if default occurs between T_i and T_{i+1}. Therefore, the floating leg is represented by the following sequence of payoffs:

$$d_i = \delta\mathbb{1}_{\{T_i < \tau \le T_{i+1}\}} = \delta\mathbb{1}_{\{\tau \le T_{i+1}\}} - \delta\mathbb{1}_{\{\tau \le T_i\}}$$

at the dates T_{i+1} for $i = 1, \ldots, n - 1$. Clearly

$$d_i = \delta(1 - \mathbb{1}_{\{\tau > T_{i+1}\}}) - \delta(1 - \mathbb{1}_{\{\tau > T_i\}}).$$

We conclude that in order to analyze the floating leg of a digital CDS, it suffices to focus on the valuation and replication of the payoff $\mathbb{1}_{\{\tau > T_i\}}$ that occurs at time T_{i+1}, that is, a vulnerable bond. The latter problem was already examined in Section 5.1, however (see, in particular, the valuation formula (34)).

First-to-Default Claims

We shall now focus on the issue of modeling dependent ("correlated") defaults, which arises in the context of *basket credit derivatives*. In order to model dependent default times, we shall employ Kusuoka's (1999) setting with $n = 2$ default times (for related results, see Jarrow and Yu (2001), Gregory and Laurent (2002, 2003), Bielecki and Rutkowski (2003), or Collin-Dufresne et al. (2003)). Our main goal is to show that the jump risk of a first-to-default claim can be perfectly hedged using the underlying defaultable zero-coupon bonds. Recovery schemes and the associated values of (deterministic) recovery rates should be specified a priori.

Construction of dependent defaults. Following Kusuoka (1999), we postulate that under the original probability $\mathbb{Q}$ the random times τ_i, $i = 1, 2$, given on a probability space $(\Omega, \mathcal{G}, \mathbb{Q})$, are assumed to be mutually independent random variables with exponential laws with parameters λ_1 and λ_2, resp. Let $\mathbb{F}$ be some reference filtration (generated by a Wiener process W, say) such that τ_1 and τ_2 are independent of $\mathbb{F}$ under $\mathbb{Q}$. We write $\mathbb{H}^i$ to denote the filtration generated by the process $H_t^i = \mathbb{1}_{\{\tau_i \leq t\}}$ for $i = 1, 2$, and we set $\mathbb{G} = \mathbb{F} \vee \mathbb{H}^1 \vee \mathbb{H}^2$. Notice that the process $M_t^i = H_t^i - \int_0^{t \wedge \tau_i} \lambda_i \, du = H_t^i - \lambda(\tau_i \wedge t)$ is a $\mathbb{G}$-martingale for $i = 1, 2$.

For a fixed $T > 0$, we define a probability measure $\mathbb{Q}^*$ on $(\Omega, \mathcal{G}_T)$ by setting

$$\frac{d\mathbb{Q}^*}{d\mathbb{Q}} = \eta_T, \quad \mathbb{Q}\text{-a.s.,}$$

where the Radon-Nikodym density process η_t, $t \in [0, T]$, satisfies

$$\eta_t = 1 + \sum_{i=1}^{2} \int_{(0,t]} \eta_{u-} \kappa_u^i \, dM_u^i$$

with auxiliary processes κ^1, κ^2 given by

$$\kappa_t^1 = \mathbb{1}_{\{\tau_2 < t\}} \left(\frac{\alpha_1}{\lambda_1} - 1 \right), \quad \kappa_t^2 = \mathbb{1}_{\{\tau_1 < t\}} \left(\frac{\alpha_2}{\lambda_2} - 1 \right).$$

Let $B(t, T)$ be the price of zero-coupon bond, and let $\mathbb{Q}_T$ be the forward martingale measure for the date T. It appears that the 'martingale intensities' under $\mathbb{Q}^*$ and under $\mathbb{Q}_T$ are

$$\lambda_t^1 = \lambda_1 \mathbb{1}_{\{\tau_2 > t\}} + \alpha_1 \mathbb{1}_{\{\tau_2 \leq t\}}, \quad \lambda_t^2 = \lambda_2 \mathbb{1}_{\{\tau_1 > t\}} + \alpha_2 \mathbb{1}_{\{\tau_1 \leq t\}}.$$

Specifically, the process $\bar{M}_t^i = H_t^i - \int_0^{t \wedge \tau_i} \lambda_u^i \, du$ is a $\mathbb{G}$-martingale under $\mathbb{Q}^*$ and under $\mathbb{Q}_T$ for $i = 1, 2$. Moreover, it is easily seen that the random times τ_1 and τ_2 are independent of the filtration $\mathbb{F}$ under $\mathbb{Q}^*$ and $\mathbb{Q}_T$. The following result shows that intensities λ^1 and λ^2 can be interpreted as *local intensities* of default with respect to the information available at time t. Therefore, the model can be reformulated as a two-dimensional Markov chain.

Proposition 9. *For $i = 1, 2$ and every $t \in [0, T]$ we have*

$$\lambda_i = \lim_{h \downarrow 0} h^{-1} \mathbb{Q}_T \{ t < \tau_i \leq t + h \,|\, \mathcal{F}_t, \tau_1 > t, \tau_2 > t \}.$$

Moreover

$$\alpha_1 = \lim_{h \downarrow 0} h^{-1} \mathbb{Q}_T \{ t < \tau_1 \leq t + h \,|\, \mathcal{F}_t, \tau_1 > t, \tau_2 \leq t \}$$

and

$$\alpha_2 = \lim_{h \downarrow 0} h^{-1} \mathbb{Q}_T \{ t < \tau_2 \leq t + h \,|\, \mathcal{F}_t, \tau_2 > t, \tau_1 \leq t \}.$$

Assume that defaultable zero-coupon bonds are subject to zero recovery rule. Then the price of the bond issued by the ith entity is given by

$$D_i^0(t, T) = B(t, T) \, \mathbb{Q}_T \{ \tau_i > T \,|\, \mathcal{G}_t \} = \mathbb{1}_{\{\tau_i > t\}} \tilde{D}_i^0(t, T),$$

where, as usual, $\tilde{D}_i^0(t, T)$ stands for the pre-default value of the bond. Let us denote $\lambda = \lambda_1 + \lambda_2$ and let us assume that $\lambda - \alpha_1 \neq 0$. Then straightforward calculations lead to an explicit formula for $\tilde{D}_i^0(t, T)$ (for details, see Bielecki and Rutkowski (2003)). Of course, an analogous expression holds for the pre-default price $\tilde{D}_2^0(t, T)$ provided that $\lambda - \alpha_2 \neq 0$.

Proposition 10. *Assume that $\lambda - \alpha_1 \neq 0$. Then for every $t \in [0, T]$ the pre-default price $\tilde{D}_1^0(t, T)$ equals*

$$\tilde{D}_1^0(t, T) = \mathbb{1}_{\{\tau_2 > t\}} D_1^*(t, T) + \mathbb{1}_{\{\tau_2 \leq t\}} \hat{D}_1(t, T),$$

where

$$D_1^*(t, T) = \frac{B(t, T)}{\lambda - \alpha_1} \left(\lambda_2 e^{-\alpha_1(T-t)} + (\lambda_1 - \alpha_1) e^{-\lambda(T-t)} \right)$$

represents the value of the bond prior to the first default, that is, on the random interval $[\![0, \tau_{(1)} \wedge T[\![$, and $\hat{D}_1(t, T) = B(t, T) e^{-\alpha_1(T-t)}$ is the value of the bond after the default of the second entity, but prior to default of the issuer, that is, on $[\![\tau_2 \wedge T, \tau_1 \wedge T[\![$.

Let $\tau_{(1)} = \tau_1 \wedge \tau_2$ be the date of the first default. Consider a first-to-default claim with the terminal payoff $X \mathbb{1}_{\{\tau_{(1)} > T\}}$, where X is an $\mathcal{F}_T$-adapted random variable, and $\mathbb{F}$-predictable recovery processes Z^1 and Z^2. As primary traded assets, we take defaultable zero-coupon bonds $D_1^0(t, T)$ and $D_2^0(t, T)$ with respective default times τ_1 and τ_2, as well as the default-free zero-coupon bond $B(t, T)$.

In Section 4.4, we have examined the basic features of a replicating strategy for a first-to-default claim. Under the present assumptions, (29) yields

$$\phi_t^1 B(t, T) e^{-\alpha_1(T-t)} + \phi_t^3 B(t, T) = Z_t^2$$

and

$$\phi_t^2 B(t, T) e^{-\alpha_2(T-t)} + \phi_t^3 B(t, T) = Z_t^1.$$

A strategy ϕ should be self-financing prior to the first default (and thus also on the random interval $[\![0, \tau_{(1)} \wedge T]\!]$). In other words, we are looking for ϕ such that the pre-default wealth process $\widetilde{V}(\phi)$, given by the formula

$$\widetilde{V}_t(\phi) = \phi_t^1 D_1^*(t, T) + \phi_t^2 D_2^*(t, T) + \phi_t^3 B(t, T), \quad \forall t \in [0, T],$$

satisfies

$$d\widetilde{V}_t(\phi) = \phi_t^1 \, dD_1^*(t, T) + \phi_t^2 \, dD_2^*(t, T) + \phi_t^3 \, dB(t, T). \tag{37}$$

Finally, at time T the wealth of ϕ should coincide with the promised payoff X on the event $\{\tau_{(1)} > T\}$. This means that the pre-default wealth needs to satisfy $\widetilde{V}_T(\phi) - X$, so that (37) becomes

$$\widetilde{V}_0(\phi) + \int_0^T \phi_t^1 \, dD_1^*(t, T) + \int_0^T \phi_t^2 \, dD_2^*(t, T) + \int_0^T \phi_t^3 \, dB(t, T) = X.$$

Equivalently, the pre-default wealth should coincide with the pre-default value of a first-to-default claim on the random interval $[\![0, \tau_{(1)} \wedge T[\![$ and the jump of the wealth at default time $\tau_{(1)}$ should adequately reproduce the behavior at $\tau_{(1)}$ of a first-to-default claim.

First-to-default credit swap. For the sake of concreteness, let us consider a *first-to-default credit swap*. Specifically, we shall examine replication of a first-to-default claim with $X = 0$ and $Z_t^i = \delta B(t, T)$ for $i = 1, 2$, where $0 \leq \delta \leq 1$. Let U_t be the value of this claim at time $t \in [0, T]$. It can be shown that

$$\mathbb{Q}_T\{\tau_{(1)} > T \mid \mathcal{G}_t\} = \mathbb{1}_{\{\tau_{(1)} > t\}} e^{-\lambda(T-t)}.$$

Consequently, for every $t \in [0, T]$ we have

$$U_t = \mathbb{1}_{\{\tau_{(1)} > t\}} \delta\left(1 - e^{-\lambda(T-t)}\right) B(t, T) + \mathbb{1}_{\{\tau_{(1)} \leq t\}} \delta B(t, T),$$

and thus the pre-default value equals

$$\widetilde{U}_t = \delta(1 - e^{-\lambda(T-t)}) B(t, T).$$

To find the replicating strategy ϕ, we first observe that ϕ needs to satisfy, for every $t \in [0, T]$,

$$\phi_t^1 e^{-\alpha_1(T-t)} + \phi_t^3 = \delta, \quad \phi_t^2 e^{-\alpha_2(T-t)} + \phi_t^3 = \delta, \tag{38}$$

Moreover, the pre-default wealth process $\widetilde{V}(\phi)$, given by

$$\widetilde{V}_t(\phi) = \phi_t^1 D_1^*(t, T) + \phi_t^2 D_2^*(t, T) + \phi_t^3 B(t, T), \tag{39}$$

should satisfy $\widetilde{V}_t(\phi) = \widetilde{U}_t$ and

$$d\widetilde{V}_t(\phi) = \phi_t^1 \, dD_1^*(t, T) + \phi_t^2 \, dD_2^*(t, T) + \phi_t^3 \, dB(t, T). \tag{40}$$

It is convenient to work with relative prices, by taking $B(t, T)$ as a numeraire, so that (39)-(40) become

$$\widetilde{V}_t^B(\phi) = \phi_t^1 Y_t^1 + \phi_t^2 Y_t^2 + \phi_t^3 = \delta\left(1 - e^{-\lambda(T-t)}\right) \tag{41}$$

and

$$\widetilde{V}_t^B(\phi) = \widetilde{V}_0^B(\phi) + \int_0^t \phi_u^1 \, dY_u^1 + \int_0^t \phi_u^2 \, dY_u^2, \tag{42}$$

where $\widetilde{V}_t^B(\phi) = \widetilde{V}_t(\phi) B^{-1}(t, T)$ and

$$Y_t^1 = \frac{D_1^*(t, T)}{B(t, T)} = \frac{1}{\lambda - \alpha_1}\left(\lambda_2 e^{-\alpha_1(T-t)} + (\lambda_1 - \alpha_1)e^{-\lambda(T-t)}\right)$$

and

$$Y_t^2 = \frac{D_2^*(t, T)}{B(t, T)} = \frac{1}{\lambda - \alpha_2}\left(\lambda_1 e^{-\alpha_2(T-t)} + (\lambda_2 - \alpha_2)e^{-\lambda(T-t)}\right).$$

Working with relative values is here equivalent to setting $B(t, T) = 1$ for every $t \in [0, T]$ in equations (39)-(40), as well as in the pricing formulae of Proposition 10.

From (38) it follows that ϕ^3 equals

$$\phi_t^3 = \delta - \phi_t^1 e^{-\alpha_1(T-t)} = \delta - \phi_t^2 e^{-\alpha_2(T-t)}, \tag{43}$$

where ϕ^1 and ϕ^2 are related to each other through the formula

$$\phi_t^2 = \phi_t^1 e^{(\alpha_2 - \alpha_1)(T-t)}, \quad \forall t \in [0, T]. \tag{44}$$

By substituting the last equality in (41), we obtain the following expression for ϕ^1

$$\phi_t^1 = -\delta e^{-\lambda(T-t)}\left(Y_t^1 + Y_t^2 e^{(\alpha_2 - \alpha_1)(T-t)} - e^{-\alpha_1(T-t)}\right)^{-1}.$$

More explicitly,

$$\phi_t^1 = -\delta \xi_1 \xi_2 e^{-\xi_1(T-t)}(g(t))^{-1}, \tag{45}$$

where we denote $\xi_i = \lambda - \alpha_i$ for $i = 1, 2$ and where $g(t)$ equals

$$g(t) = \lambda_2\xi_2 + (\lambda_1 - \alpha_1)\xi_2 e^{-\xi_1(T-t)} + \lambda_1\xi_1 + (\lambda_2 - \alpha_2)\xi_1 e^{-\xi_2(T-t)} - \xi_1\xi_2.$$

To determine ϕ^2 we may either use (44) with (45), or to observe that by the symmetry of the problem

$$\phi_t^2 = -\delta e^{-\lambda(T-t)} \left(Y_t^2 + Y_t^1 e^{(\alpha_1-\alpha_2)(T-t)} - e^{-\alpha_2(T-t)} \right)^{-1}.$$

Of course, both methods yield, as expected, the same expression for ϕ^2, namely,

$$\phi_t^2 = -\delta\xi_1\xi_2 e^{-\xi_2(T-t)}(g(t))^{-1}.$$

Moreover, straightforward calculations show that for ϕ^1, ϕ^2 as above, we have

$$\phi_t^1 \, dY_t^1 + \phi_t^2 \, dY_t^2 = d\widetilde{V}_t^B(\phi) = -\delta\lambda e^{-\lambda(T-t)}.$$

Finally, the component ϕ^3 can be found from (43), and thus the calculation of a replicating strategy for the considered example of first-to-default credit swap is completed.

6 PDE Approach

Let us assume that two (defaultable, in general) assets are tradeable, with respective price processes

$$dY_t^1 = Y_{t-}^1 (\nu_1 \, dt + \sigma_1 \, dW_t + \varrho_1 \, dM_t), \quad Y_0^1 > 0, \tag{46}$$

$$dY_t^2 = Y_{t-}^2 (\nu_2 \, dt + \sigma_2 \, dW_t + \varrho_2 \, dM_t), \quad Y_0^2 > 0, \tag{47}$$

under the real-world probability $\mathbb{Q}$, where W is a one-dimensional standard Brownian motion and the $\mathbb{G}$-martingale M is given by

$$M_t = H_t - \int_0^t \mathbb{1}_{\{\tau>u\}}\varsigma_u \, du, \quad \forall t \in [0, T],$$

and the $\mathbb{F}$-adapted intensity ς of the default time τ is strictly positive. We postulate that the interest rate is equal to a constant r, so that the money market account equals $Y_t^3 = B_t = e^{rt}$. We assume that $\sigma_1 \neq 0$, $\sigma_2 \neq 0$ and the constants ϱ_1 and ϱ_2 are greater or equal to -1 so that the price process Y^i is non-negative for $i = 1, 2$.

Remark. It may happen that either ϱ_1 or ϱ_2 equals 0, and thus the corresponding asset is default-free. The case when $\varrho_1 = \varrho_2 = 0$ will be excluded, however (see condition (48) below).

We shall now examine the no-arbitrage property of this market. Specifically, we shall impose additional conditions on the model's coefficients that will ensure the existence of an equivalent martingale measure. From Kusuoka's (1999) representation theorem, any equivalent martingale measure $\mathbb{Q}^*$ on $(\Omega, \mathcal{G}_T)$ is of the form $d\mathbb{Q}^*|_{\mathcal{G}_t} = \widetilde{\eta}_t \, d\mathbb{Q}|_{\mathcal{G}_t}$ for $t \in [0, T]$, where

$$d\widetilde{\eta}_t = \widetilde{\eta}_{t-}(\psi_t \, dW_t + \kappa_t \, dM_t), \quad \widetilde{\eta}_0 = 1,$$

for some $\mathbb{G}$-predictable processes ψ and κ. By applying Itô's formula, we obtain for $i = 1, 2$,

$$Y_t^i \widetilde{\eta}_t e^{-rt} = Y_0^i + \int_0^t Y_u^i \widetilde{\eta}_u e^{-ru}\big(\nu_i - r + \psi_u \sigma_1 + \kappa_u \varrho_i \xi_u\big) \, du + \text{martingale},$$

where we denote $\xi_t = \varsigma_t \mathbb{1}_{\{\tau > t\}}$. Hence, the process $Y_t^i \widetilde{\eta}_t e^{-rt}$ is a (local) $\mathbb{G}$-martingale under $\mathbb{Q}$ for $i = 1, 2$ if and only if

$$\nu_i - r + \psi_t \sigma_i + \kappa_t \varrho_i \xi_t = 0$$

for $i = 1, 2$ and almost every $t \in [0, T]$. Hence, a density process $\widetilde{\eta}$ determines an equivalent martingale measure $\mathbb{Q}^*$ for the processes $Y_t^i e^{-rt}$, $i = 1, 2$ if and only if the processes ψ and κ are such that for every $t \in [0, T]$

$$\nu_1 - r + \psi_t \sigma_1 + \kappa_t \varrho_1 \xi_t = 0,$$
$$\nu_2 - r + \psi_t \sigma_2 + \kappa_t \varrho_2 \xi_t = 0.$$

Assume that $\varrho_1 \sigma_2 - \varrho_2 \sigma_1 \neq 0$. Then the unique solution is the pair of processes (ψ_t, κ_t), $t \in [0, T]$, such that

$$\psi_t = \frac{(\nu_2 - r)\varrho_1 - (\nu_1 - r)\varrho_2}{\varrho_1 \sigma_2 - \varrho_2 \sigma_1}$$

and

$$\kappa_t \xi_t = \frac{(\nu_2 - r)\sigma_1 - (\nu_1 - r)\sigma_2}{\varrho_1 \sigma_2 - \varrho_2 \sigma_1}.$$

Since $\widetilde{\eta}$ is a strictly positive process, we restrict our attention to parameters such that the process κ is greater than -1. Obviously, the value of the process κ after the default time τ is irrelevant. However, the pre-default value of κ is uniquely given as

$$\kappa_t = \frac{(\nu_2 - r)\sigma_1 - (\nu_1 - r)\sigma_2}{\varsigma_t(\varrho_1 \sigma_2 - \varrho_2 \sigma_1)},$$

and thus we postulate that the last formula holds for every $t \in [0, T]$. We thus have the following auxiliary result. Let us set $\gamma_t = \varsigma_t(1 + \kappa_t)$.

Lemma 8. *Assume that* $\varrho_1 \sigma_2 - \varrho_2 \sigma_1 \neq 0$ *and*

$$\frac{(\nu_2 - r)\sigma_1 - (\nu_1 - r)\sigma_2}{\varsigma_t(\varrho_1 \sigma_2 - \varrho_2 \sigma_1)} > -1, \quad \forall t \in [0, T]. \tag{48}$$

Then the market model defined by (46)-(47) and the money market account $Y_t^3 = e^{rt}$ is complete and arbitrage-free. Moreover, under the unique equivalent martingale measure $\mathbb{Q}^*$ we have

$$
\begin{aligned}
dY_t^1 &= Y_{t-}^1 \left(r\, dt + \sigma_1\, dW_t^* + \varrho_1\, dM_t^* \right), \\
dY_t^2 &= Y_{t-}^2 \left(r\, dt + \sigma_2\, dW_t^* + \varrho_2\, dM_t^* \right), \\
dY_t^3 &= rY_t^3\, dt,
\end{aligned}
\tag{49}
$$

where W^* is a Brownian motion under $\mathbb{Q}^*$, and where the process M^*, given by

$$
M_t^* = M_t - \int_0^t \xi_u \kappa_u\, du = H_t - \int_0^t \mathbb{1}_{\{\tau > u\}} \gamma_u\, du,
$$

follows a martingale under $\mathbb{Q}^*$.

From now on, we shall conduct the analysis of the model given by (49) under the martingale measure $\mathbb{Q}^*$.

6.1 Markovian Case

To proceed further it would be convenient to assume that ς, and thus also κ, are deterministic functions of the time parameter. In this case, the default intensity γ under $\mathbb{Q}^*$ would be a deterministic function as well. More generally, it suffices to postulate that the $\mathbb{F}$-intensity of default under $\mathbb{Q}^*$ is of the form $\gamma_t = \gamma(t, Y_t^1, Y_t^2)$ for some sufficiently smooth function γ. For instance, $\gamma(t, x, y)$ may be assumed to be piecewise continuous with respect to t and Lipschitz continuous with respect to x and y. Under this assumption, the process (Y^1, Y^2, H), where the two-dimensional process (Y^1, Y^2) is the unique solution to the SDE (49), is Markovian under $\mathbb{Q}^*$ (since Y^3 is deterministic, it is not essential here).

For the sake of concreteness, we shall frequently focus on a defaultable claim represented by the following payoff at the maturity date T

$$
Y = \mathbb{1}_{\{\tau > T\}} g(Y_T^1, Y_T^2) + \mathbb{1}_{\{\tau \le T\}} h(Y_T^1, Y_T^2)
\tag{50}
$$

for some functions $g, h : \mathbb{R}_+^2 \to \mathbb{R}$ satisfying suitable integrability conditions. Hence, the price of Y is given by the risk-neutral valuation formula

$$
\pi_t(Y) = B_t\, \mathbb{E}_{\mathbb{Q}^*}(B_T^{-1} Y \mid \mathcal{G}_t), \quad \forall t \in [0, T].
\tag{51}
$$

Notice that $\pi_t(Y)$ represents the standard (cum-dividend) price of a European contingent claim Y, which settles at time T. Our goal is to find a quasi-explicit representation for a self-financing trading strategy ψ such that $\pi_t(Y) = V_t(\psi)$ for every $t \in [0, T]$, where $V_t(\psi) = \sum_{i=1}^3 \psi_t^i Y_t^i$ (see Section 6.3).

We shall first prove an auxiliary result, which shows that the arbitrage price of the claim Y splits in a natural way into the pre-default price and the post-default price.

Lemma 9. *The price $\pi_t(Y)$ of the claim Y given by (50) satisfies*

$$\pi_t(Y) = (1 - H_t)\widetilde{v}(t, Y_t^1, Y_t^2) + H_t\bar{v}(t, Y_t^1, Y_t^2), \quad \forall t \in [0, T], \qquad (52)$$

for some functions $\widetilde{v}, \bar{v} : [0, T] \times \mathbb{R}_+^2 \to \mathbb{R}$ such that $\widetilde{v}(T, x, y) = g(x, y)$ and $\bar{v}(T, x, y) = h(x, y)$.

Proof. We have

$$\begin{aligned}
\pi_t(Y) &= B_t\, \mathbb{E}_{\mathbb{Q}^*}\big(B_T^{-1}Y \,\big|\, \mathcal{G}_t\big) \\
&= B_t\, \mathbb{E}_{\mathbb{Q}^*}\big(B_T^{-1}\mathbb{1}_{\{\tau>T\}}g(Y_T^1, Y_T^2) \,\big|\, \mathcal{G}_t\big) + B_t\, \mathbb{E}_{\mathbb{Q}^*}\big(B_T^{-1}\mathbb{1}_{\{\tau\leq T\}}h(Y_T^1, Y_T^2) \,\big|\, \mathcal{G}_t\big) \\
&= \mathbb{1}_{\{\tau>t\}}B_t\, \mathbb{E}_{\mathbb{Q}^*}\big(\mathbb{1}_{\{\tau>T\}}B_T^{-1}g(Y_T^1, Y_T^2) + \mathbb{1}_{\{t<\tau\leq T\}}B_T^{-1}h(Y_T^1, Y_T^2) \,\big|\, \mathcal{G}_t\big) \\
&\quad + \mathbb{1}_{\{\tau\leq t\}}B_t\, \mathbb{E}_{\mathbb{Q}^*}\big(\mathbb{1}_{\{\tau\leq t\}}B_T^{-1}h(Y_T^1, Y_T^2) \,\big|\, \mathcal{G}_t\big).
\end{aligned}$$

This shows that

$$\pi_t(Y) = \mathbb{1}_{\{\tau>t\}}\widetilde{u}(t, Y_t^1, Y_t^2, 0) + \mathbb{1}_{\{\tau\leq t\}}\bar{u}(t, Y_t^1, Y_t^2, 1),$$

where

$$\begin{aligned}
\widetilde{u}(t, Y_t^1, Y_t^2, H_t) &= B_t\, \mathbb{E}_{\mathbb{Q}^*}\big(\mathbb{1}_{\{\tau>T\}}B_T^{-1}g(Y_T^1, Y_T^2) \,\big|\, \mathcal{G}_t\big) \\
&\quad + B_t\, \mathbb{E}_{\mathbb{Q}^*}\big(\mathbb{1}_{\{t<\tau\leq T\}}B_T^{-1}h(Y_T^1, Y_T^2) \,\big|\, \mathcal{G}_t\big) \\
&= B_t\, \mathbb{E}_{\mathbb{Q}^*}\big((1 - H_T)B_T^{-1}g(Y_T^1, Y_T^2) \\
&\quad + (H_T - H_t)B_T^{-1}h(Y_T^1, Y_T^2) \,\big|\, Y_t^1, Y_t^2, H_t\big)
\end{aligned}$$

and

$$\begin{aligned}
\bar{u}(t, Y_t^1, Y_t^2, H_t) &= B_t\, \mathbb{E}_{\mathbb{Q}^*}\big(\mathbb{1}_{\{\tau\leq t\}}B_T^{-1}h(Y_T^1, Y_T^2) \,\big|\, \mathcal{G}_t\big) \\
&= B_t\, \mathbb{E}_{\mathbb{Q}^*}\big(H_t B_T^{-1}h(Y_T^1, Y_T^2) \,\big|\, Y_t^1, Y_t^2, H_t\big).
\end{aligned}$$

Let us set

$$\begin{aligned}
\widetilde{v}(t, Y_t^1, Y_t^2) &= \widetilde{u}(t, Y_t^1, Y_t^2, 0) \\
&= B_t\, \mathbb{E}_{\mathbb{Q}^*}\big(B_T^{-1}Y \,\big|\, Y_t^1, Y_t^2, H_t = 0\big)
\end{aligned} \qquad (53)$$

and

$$\begin{aligned}
\bar{v}(t, Y_t^1, Y_t^2) &= \bar{u}(t, Y_t^1, Y_t^2, 1) \\
&= B_t\, \mathbb{E}_{\mathbb{Q}^*}\big(B_T^{-1}h(Y_T^1, Y_T^2) \,\big|\, Y_t^1, Y_t^2, H_t = 1\big).
\end{aligned} \qquad (54)$$

It is clear that $\widetilde{v}(T, Y_T^1, Y_T^2) = g(Y_T^1, Y_T^2)$ and $\bar{v}(T, Y_T^1, Y_T^2) = h(Y_T^1, Y_T^2)$. We conclude that the price of the claim Y is of the form $v(t, Y_t^1, Y_t^2)$, where

$$v(t, Y_t^1, Y_t^2) = \mathbb{1}_{\{\tau>t\}}\widetilde{v}(t, Y_t^1, Y_t^2) + \mathbb{1}_{\{\tau\leq t\}}\bar{v}(t, Y_t^1, Y_t^2).$$

Notice that $\tilde{v}(t, Y_t^1, Y_t^2)$ and $\bar{v}(t, Y_t^1, Y_t^2)$ represent the pre-default and post-default values of Y, respectively. □

Post-default value. It should be stressed that the conditional expectation in (53) is to be evaluated using the dynamics of (Y^1, Y^2, Y^3) given by (49). To compute the conditional expectation in (54), however, it is manifestly sufficient to make use of the post-default dynamics of (Y^1, Y^2, Y^3), which is given by the following expressions, which are valid if $\varrho_1 > -1$ and $\varrho_2 > -1$,

$$
\begin{aligned}
dY_t^1 &= Y_{t-}^1 \left(r\, dt + \sigma_1\, dW_t^* \right), \\
dY_t^2 &= Y_{t-}^2 \left(r\, dt + \sigma_2\, dW_t^* \right), \\
dY_t^3 &= rY_t^3\, dt.
\end{aligned}
\tag{55}
$$

Using standard arguments, we conclude that if the function $\bar{v} = \bar{v}(t, x, y)$ is sufficiently regular then it satisfies the following PDE:

$$
-r\bar{v} + \partial_t \bar{v} + rx\partial_x \bar{v} + ry\partial_y \bar{v} + \tfrac{1}{2}\left(\sigma_1^2 x^2 \partial_{xx}^2 \bar{v} + \sigma_2^2 y^2 \partial_{yy}^2 \bar{v}\right)
$$
$$
+ \sigma_1\sigma_2 xy \partial_{xy}^2 \bar{v} = 0
\tag{56}
$$

with the terminal condition $\bar{v}(T, x, y) = h(x, y)$. Hence, the equation (56) can be referred to as the *post-default pricing* PDE for our claim. Of course, since after the default time our model becomes a default-free model, the use of a such a PDE to arbitrage valuation of path-independent European claims is fairly standard

If $\varrho_1 > -1$ and $\varrho_2 = -1$, then the process Y^2 jumps to zero at time of default, and thus the post-default pricing PDE becomes:

$$
-r\bar{v} + \partial_t \bar{v} + rx\partial_x \bar{v} + \tfrac{1}{2}\sigma_1^2 x^2 \partial_{xx}^2 \bar{v} = 0
\tag{57}
$$

with the terminal condition $\bar{v}(T, x) = \bar{h}(x)$ for some function $\bar{h} : \mathbb{R}_+ \to \mathbb{R}$ (formally, $\bar{h}(x) = h(x, 0)$).

Recovery process. Following Jamshidian (2002) (see Theorem 2.1), one may check that for any $t \in [0, T]$ we have

$$
B_t\, \mathbb{E}_{\mathbb{Q}^*} \left(B_T^{-1} \mathbb{1}_D h(Y_T^1, Y_T^2) \,\big|\, \mathcal{G}_t \right) = B_t\, \mathbb{E}_{\mathbb{Q}^*} \left(B_\tau^{-1} \mathbb{1}_D \bar{v}(\tau, Y_\tau^1, Y_\tau^2) \,\big|\, \mathcal{G}_t \right),
$$

where $D = \{t < \tau \leq T\}$. Hence, if we wish to compute the pre-default value of Y, it is tempting to consider the process $\bar{v}(t, Y_t^1, Y_t^2)$ as the recovery process Z. According to our convention, the recovery process Z should necessarily be an $\mathbb{F}$-predictable process, and the process $\bar{v}(t, Y_t^1, Y_t^2)$ is not $\mathbb{F}$-predictable, in general. Therefore, we formally define the recovery process Z associated with the claim Y by setting

$$
Z_t = z(t, \tilde{Y}_t^1(1 + \varrho_1), \tilde{Y}_t^2(1 + \varrho_2)) = \bar{v}(t, \tilde{Y}_t^1(1 + \varrho_1), \tilde{Y}_t^2(1 + \varrho_2))),
\tag{58}
$$

where $\tilde{Y}^i$ is the pre-default value of the ith asset (so that $\tilde{Y}^i$ is manifestly an $\mathbb{F}$-adapted, continuous process). It is clear that

$$Z_\tau = \bar{v}(\tau, \widetilde{Y}_\tau^1(1 + \varrho_1), \widetilde{Y}_\tau^2(1 + \varrho_2)) = \bar{v}(\tau, Y_\tau^1, Y_\tau^2), \quad \mathbb{Q}^*\text{-a.s.}$$

Notice that the pre-default value of the claim Y given by (50) coincides with the pre-default value of $(X, Z, 0, \tau)$, where the promised payoff $X = g(Y_T^1, Y_T^2)$ and the $\mathbb{F}$-predictable recovery process Z is given by (58).

6.2 Pricing PDE for the Pre-Default Value

Recall that the price process of the claim Y given by (50) admits the following representation, for every $t \in [0, T]$,

$$\pi_t(Y) = (1 - H_t)\widetilde{v}(t, Y_t^1, Y_t^2) + H_t\bar{v}(t, Y_t^1, Y_t^2) \tag{59}$$

for some functions $\widetilde{v}, \bar{v} : [0, T] \times \mathbb{R}_+^2 \rightarrow \mathbb{R}$ such that $\widetilde{v}(T, x, y) = g(x, y)$ and $\bar{v}(T, x, y) = h(x, y)$. We assume that processes $\widetilde{v}(t, Y_t^1, Y_t^2)$ and $\bar{v}(t, Y_t^1, Y_t^2)$ are semimartingales. We shall need the following simple version of the Itô integration by parts formula for (discontinuous) semimartingales.

Lemma 10. *Assume that Z is a semimartingale and A is a bounded process of finite variation. Then*

$$Z_t A_t = Z_0 A_0 + \int_0^t Z_{u-} \, dA_u + \int_0^t A_u \, dZ_u$$

$$= Z_0 A_0 + \int_0^t Z_u \, dA_u + \int_0^t A_{u-} \, dZ_u.$$

Proof. Both formulae are almost immediate consequences of the general Itô formula for semimartingales (see, for instance, Protter (2003)), and the fact that under the present assumptions we have $[Z, A]_t = \sum_{0 < s \leq t} \Delta Z_s \Delta A_s$. $\qquad\square$

Our next goal is to derive the partial differential equation satisfied by the pre-default pricing function $\widetilde{v}$. The post-default pricing function $\bar{v}$ (or, equivalently, the recovery function z) is taken here as an input. Hence, the only unknown function at this stage is the pre-default pricing function $\widetilde{v}$.

In view of the financial interpretation of the function $\widetilde{v}$, the PDE derived in Proposition 11 will be referred to as the *pre-default pricing PDE* for a defaultable claim Y. For a related result, see Proposition 3.4 in Lukas (2001).

Proposition 11. *We now assume that the function $\widetilde{v} = \widetilde{v}(t, x, y)$ belong to the class $C^{1,2,2}([0, T] \times \mathbb{R}_+ \times \mathbb{R}_+)$, and that in addition, it satisfies the PDE*

$$-r\widetilde{v} + \partial_t\widetilde{v} + rx\partial_x\widetilde{v} + ry\partial_y\widetilde{v} + \tfrac{1}{2}(\sigma_1^2 x^2 \partial_{xx}^2 \widetilde{v} + \sigma_2^2 y^2 \partial_{yy}^2 \widetilde{v}) + \sigma_1\sigma_2 xy \partial_{xy}^2 \widetilde{v}$$

$$+ \gamma(t, x, y)\big(\bar{v}(t, x(1 + \varrho_1), y(1 + \varrho_2)) - \widetilde{v} - \varrho_1 x \partial_x \widetilde{v} - \varrho_2 y \partial_y \widetilde{v}\big) = 0$$

with the terminal condition $\widetilde{v}(T, x, y) = g(x, y)$. Let the process $\pi(Y)$ be given by (59). Then the process $V_t^* = B_t^{-1}\pi_t(Y)$ stopped at τ is a $\mathbb{G}$-martingale under $\mathbb{Q}^*$.

Proof. By applying the Itô integration by parts formula to both terms in the right-hand side of (59), we obtain

$$
\begin{aligned}
d\pi_t(Y) &= (1 - H_t)\, d\tilde{v}(t, Y_t^1, Y_t^2) - \tilde{v}(t, Y_{t-}^1, Y_{t-}^2)\, dH_t \\
&\quad + H_{t-}\, d\bar{v}(t, Y_t^1, Y_t^2) + \bar{v}(t, Y_t^1, Y_t^2)\, dH_t \\
&= \mathbb{1}_{\{\tau > t\}}\, d\tilde{v}(t, Y_t^1, Y_t^2) + \big(\bar{v}(t, Y_t^1, Y_t^2) - \tilde{v}(t, Y_{t-}^1, Y_{t-}^2)\big)\, dH_t \\
&\quad + \mathbb{1}_{\{\tau < t\}}\, d\bar{v}(t, Y_t^1, Y_t^2).
\end{aligned}
$$

Hence, the process $V_t^* = e^{-rt}\pi_t(Y)$ satisfies for every $t \in [0, T]$

$$
\begin{aligned}
V_t^* &= V_0^* - \int_0^{\tau \wedge t} r e^{-ru} \tilde{v}(u, Y_u^1, Y_u^2)\, du + \int_{(0, \tau \wedge t)} e^{-ru}\, d\tilde{v}(u, Y_u^1, Y_u^2) \\
&\quad + \int_{(0, \tau \wedge t]} e^{-ru} \big(\bar{v}(u, Y_u^1, Y_u^2) - \tilde{v}(u, Y_{u-}^1, Y_{u-}^2) \big)\, dH_u \\
&\quad + \int_{(\tau \wedge t, t]} e^{-ru}\, d\bar{v}(u, Y_u^1, Y_u^2).
\end{aligned}
$$

It is clear that if $\pi(Y)$ is given by (51) then the process V^* is a $\mathbb{G}$-martingale under $\mathbb{Q}^*$ (see also Corollary 8 below). To derive the pre-default pricing PDE, it suffices to make use of the martingale property of the stopped process

$$
V_{\tau \wedge t}^* = \mathbb{1}_{\{\tau > t\}} e^{-rt} \tilde{v}(t, Y_t^1, Y_t^2) + \mathbb{1}_{\{\tau \le t\}} e^{-r\tau} \bar{n}(\tau, Y_\tau^1, Y_\tau^2).
$$

By applying Itô's formula to $\tilde{v}(t, Y_t^1, Y_t^2)$ on $\{\tau > t\}$, we obtain

$$
\begin{aligned}
V_{\tau \wedge t}^* &= V_0^* \\
&\quad + \int_0^{\tau \wedge t} e^{-ru} \Big(-r\tilde{v}_u + \partial_t \tilde{v}_u + rY_u^1 \partial_x \tilde{v}_u + rY_u^2 \partial_y \tilde{v}_u \Big)\, du \\
&\quad + \int_0^{\tau \wedge t} \frac{1}{2} e^{-ru} \Big(\sigma_1^2 (Y_u^1)^2 \partial_{xx} \tilde{v}_u + \sigma_2^2 (Y_u^2)^2 \partial_{yy} \tilde{v}_u + 2\sigma_1 \sigma_2 Y_u^1 Y_u^2 \partial_{xy} \tilde{v}_u \Big)\, du \\
&\quad - \int_0^{\tau \wedge t} e^{-ru} \Big(\varrho_1 Y_u^1 \partial_x \tilde{v}_u + \varrho_2 Y_u^2 \partial_y \tilde{v}_u \Big) \gamma_u\, du \\
&\quad + \int_{(0, \tau \wedge t]} e^{-ru} \Big(\bar{v}(u, Y_{u-}^1(1 + \varrho_1), Y_{u-}^2(1 + \varrho_2)) - \tilde{v}(u, Y_{u-}^1, Y_{u-}^2) \Big)\, dH_u \\
&\quad + \int_0^{\tau \wedge t} e^{-ru} \Big(\sigma_1 Y_u^1 \partial_x \tilde{v}_u + \sigma_2 Y_u^2 \partial_y \tilde{v}_u \Big)\, dW_u^*,
\end{aligned}
$$

where $\tilde{v}_u = \tilde{v}(u, Y_u^1, Y_u^2)$, $\partial_x \tilde{v}_u = \partial_x \tilde{v}(u, Y_u^1, Y_u^2)$, $\gamma_u = \gamma(u, Y_u^1, Y_u^2)$, etc. The last formula can be rewritten as follows:

$$V_{\tau \wedge t}^* = V_0^*$$

$$+ \int_0^{\tau \wedge t} e^{-ru} \Big(-r\widetilde{v}_u + \partial_t \widetilde{v}_u + rY_u^1 \partial_x \widetilde{v}_u + rY_u^2 \partial_y \widetilde{v}_u \Big) du$$

$$+ \int_0^{\tau \wedge t} \frac{1}{2} e^{-ru} \Big(\sigma_1^2 (Y_u^1)^2 \partial_{xx} \widetilde{v}_u + \sigma_2^2 (Y_u^2)^2 \partial_{yy} \widetilde{v}_u + 2\sigma_1 \sigma_2 Y_u^1 Y_u^2 \partial_{xy} \widetilde{v}_u \Big) du$$

$$- \int_0^{\tau \wedge t} e^{-ru} \Big(\varrho_1 Y_u^1 \partial_x \widetilde{v}_u + \varrho_2 Y_u^2 \partial_y \widetilde{v}_u \Big) \gamma_u \, du$$

$$+ \int_0^{\tau \wedge t} e^{-ru} \Big(\bar{v}(u, Y_{u-}^1(1+\varrho_1), Y_{u-}^2(1+\varrho_2)) - \widetilde{v}(u, Y_{u-}^1, Y_{u-}^2) \Big) \gamma_u \, du$$

$$+ \int_{(0,\tau \wedge t]} e^{-ru} \Big(\bar{v}(u, Y_{u-}^1(1+\varrho_1), Y_{u-}^2(1+\varrho_2)) - \widetilde{v}(u, Y_{u-}^1, Y_{u-}^2) \Big) dM_u^*$$

$$+ \int_0^{\tau \wedge t} e^{-ru} \Big(\sigma_1 Y_u^1 \partial_x \widetilde{v}_u + \sigma_2 Y_u^2 \partial_y \widetilde{v}_u \Big) dW_u^*.$$

Recall that the processes W^* and M^* are $\mathbb{G}$-martingales under $\mathbb{Q}^*$. Thus, the stopped process $V_{t \wedge \tau}^*$ is a $\mathbb{G}$-martingale if and only if for every $t \in [0, T]$

$$\int_0^{\tau \wedge t} e^{-ru} \Big(-r\widetilde{v}_u + \partial_t \widetilde{v}_u + rY_u^1 \partial_x \widetilde{v}_u + rY_u^2 \partial_y \widetilde{v}_u \Big) du$$

$$+ \int_0^{\tau \wedge t} \frac{1}{2} e^{-ru} \Big(\sigma_1^2 (Y_u^1)^2 \partial_{xx} \widetilde{v}_u + \sigma_2^2 (Y_u^2)^2 \partial_{yy} \widetilde{v}_u + 2\sigma_1 \sigma_2 Y_u^1 Y_u^2 \partial_{xy} \widetilde{v}_u \Big) du$$

$$- \int_0^{\tau \wedge t} e^{-ru} \Big(\varrho_1 Y_u^1 \partial_x \widetilde{v}_u + \varrho_2 Y_u^2 \partial_y \widetilde{v}_u \Big) \gamma_u \, du$$

$$+ \int_0^{\tau \wedge t} e^{-ru} \Big(\bar{v}(u, Y_u^1(1+\varrho_1), Y_u^2(1+\varrho_2)) - \widetilde{v}(u, Y_u^1, Y_u^2) \Big) \gamma_u \, du = 0.$$

The last equality is manifestly satisfied if the function $\widetilde{v}$ solves the PDE given in the statement of the proposition. Conversely, if the function $\widetilde{v}$ in representation (59) is sufficiently regular, then it necessarily satisfies the last equation.   □

Corollary 8. *Assume that the pricing functions $\bar{v}$ and $\widetilde{v}$ belong to the class $C^{1,2,2}$ ($[0, T] \times \mathbb{R}_+ \times \mathbb{R}_+$) and satisfy the post-default and pre-default pricing PDEs, respectively. Then the discounted price process V_t^*, $t \in [0, T]$, is a $\mathbb{G}$-martingale under $\mathbb{Q}^*$ and the dynamics of V^* under $\mathbb{Q}^*$ are*

$$dV_t^* = \mathbb{1}_{\{\tau > t\}} e^{-rt} (\sigma_1 Y_t^1 \partial_x \widetilde{v}_t + \sigma_2 Y_t^2 \partial_y \widetilde{v}_t) \, dW_t^*$$

$$+ \mathbb{1}_{\{\tau < t\}} e^{-rt} (\sigma_1 Y_t^1 \partial_x \bar{v}_t + \sigma_2 Y_t^2 \partial_y \bar{v}_t) \, dW_t^*$$

$$+ e^{-rt} \big(\bar{v}(t, Y_{t-}^1(1+\varrho_1), Y_{t-}^2(1+\varrho_2)) - \widetilde{v}(t, Y_{t-}^1, Y_{t-}^2) \big) dM_t^*.$$

Generic defaultable claim. Technique described above can be applied to the case of a general defaultable claim. Consider a generic defaultable claim $(X, Z, 0, \tau)$ with the promised payoff $X = g(Y_T^1, Y_T^2)$ and the recovery process $Z_t = z(t, Y_t^1, Y_t^2)$, where z is a continuous function. Then the discounted price process stopped at τ equals

$$V^*_{\tau \wedge t} = \mathbb{1}_{\{\tau > t\}} e^{-rt} \tilde{v}(t, Y^1_t, Y^2_t) + \mathbb{1}_{\{\tau \leq t\}} e^{-r\tau} z(\tau, Y^1_\tau, Y^2_\tau).$$

The latter formula can also be rewritten as follows (note that the pre-default prices $\tilde{Y}^1$ and $\tilde{Y}^2$ are continuous)

$$V^*_{\tau \wedge t} = \mathbb{1}_{\{\tau > t\}} e^{-rt} \tilde{v}(t, \tilde{Y}^1_t, \tilde{Y}^2_t) + \mathbb{1}_{\{\tau \leq t\}} e^{-r\tau} z(\tau, \tilde{Y}^1_\tau(1 + \varrho_1), \tilde{Y}^2_\tau(1 + \varrho_2)).$$

In this case, the pre-default pricing PDE reads

$$-r\tilde{v} + \partial_t \tilde{v} + rx\partial_x \tilde{v} + ry\partial_y \tilde{v} + \tfrac{1}{2}\left(\sigma_1^2 x^2 \partial^2_{xx}\tilde{v} + \sigma_2^2 y^2 \partial^2_{yy}\tilde{v}\right) + \sigma_1\sigma_2 xy \partial^2_{xy}\tilde{v}$$
$$+ \gamma(t, x, y)\left(z(t, x(1 + \varrho_1), y(1 + \varrho_2)) - \tilde{v} - \varrho_1 x\partial_x\tilde{v} - \varrho_2 y\partial_y\tilde{v}\right) = 0$$

with the terminal condition $\tilde{v}(T, x, y) = g(x, y)$. According to our interpretation of the pre-default value $\tilde{U} = \tilde{U}(X) + \tilde{U}(Z)$ of the claim $(X, Z, 0, \tau)$, the solution to the last equation is expected to satisfy $\tilde{v}(t, \tilde{Y}^1_t, \tilde{Y}^2_t) = \tilde{U}_t$ for every $t \in [0, T]$.

6.3 Replicating Strategy

Consider a claim Y of the form (50), and assume that any $t \in [0, T]$ we have

$$\pi_t(Y) = \tilde{v}(t, \tilde{Y}^1_t, \tilde{Y}^2_t)\mathbb{1}_{\{\tau > t\}} + \bar{v}(t, Y^1_t, Y^2_t)\mathbb{1}_{\{\tau \leq t\}},$$

where the functions $\bar{v}$ and $\tilde{v}$ satisfy the post-default and pre-default pricing PDEs, respectively. It view of Corollary 8, we have (recall that the process M^* is stopped at τ and processes $\tilde{Y}^1$ and $\tilde{Y}^2$ are continuous)

$$dV^*_t = \mathbb{1}_{\{\tau \geq t\}} e^{-rt} \bar{V}_t \, dW^*_t + \mathbb{1}_{\{\tau < t\}} e^{-rt} \tilde{V}_t \, dW^*_t$$
$$+ e^{-rt}\left[\bar{v}(t, \tilde{Y}^1_t(1 + \varrho_1), \tilde{Y}^2_t(1 + \varrho_2)) - \tilde{v}(t, \tilde{Y}^1_t, \tilde{Y}^2_t)\right] dM^*_t,$$

where the $\mathbb{F}$-adapted process $\tilde{V}$ is given by

$$\tilde{V}_t = \sigma_1 \tilde{Y}^1_t \partial_x \tilde{v}(t, \tilde{Y}^1_t, \tilde{Y}^2_t) + \sigma_2 \tilde{Y}^2_t \partial_y \tilde{v}(t, \tilde{Y}^1_t, \tilde{Y}^2_t) \tag{60}$$

and $\bar{V}$ is the $\mathbb{G}$-adapted process:

$$\bar{V}_t = \sigma_1 Y^1_t \partial_x \bar{v}(t, Y^1_t, Y^2_t) + \sigma_2 Y^2_t \partial_y \bar{v}(t, Y^1_t, Y^2_t). \tag{61}$$

As before, we denote the discounted prices by

$$Y^{1,3}_t = Y^1_t / Y^3_t = Y^1_t e^{-rt}, \quad Y^{2,3}_t = Y^2_t / Y^3_t = Y^2_t e^{-rt}.$$

Some algebra leads to

$$dW_t^* = \frac{1}{\varrho_2\sigma_1 - \varrho_1\sigma_2}\left(\frac{\varrho_2}{Y_{t-}^{1,3}}\,dY_t^{1,3} - \frac{\varrho_1}{Y_{t-}^{2,3}}\,dY_t^{2,3}\right),$$

$$dM_t^* = \frac{1}{\varrho_1\sigma_2 - \varrho_2\sigma_1}\left(\frac{\sigma_2}{Y_{t-}^{1,3}}\,dY_t^{1,3} - \frac{\sigma_1}{Y_{t-}^{2,3}}\,dY_t^{2,3}\right).$$

It should be stressed that the above representation for W^* and M^* is always valid, under the present assumptions, on the stochastic interval $[\![0, \tau \wedge T]\!]$. It also holds after default, provided that neither Y^1 nor Y^2 jumps to zero at time τ. Hence, the case when Y^1 (or Y^2) becomes worthless at time τ (and thus also after τ) should be considered separately. It is worthwhile to emphasize that the strategy ϕ derived below is always the replicating strategy for the claim Y up to default time τ. Recall that we work under the standing assumption that $c = \varrho_2\sigma_1 - \varrho_1\sigma_2 \neq 0$. Hence, under the assumption that $\varrho_1 > -1$ and $\varrho_2 > -1$, we obtain

$$
\begin{aligned}
V_t^* = V_0^* &+ \frac{1}{c}\int_{(0,t]}\left[\varrho_2\left(\mathbb{1}_{\{\tau \geq u\}}\widetilde{V}_u + \mathbb{1}_{\{\tau < u\}}\bar{V}_u\right)\right.\\
&\left. - \sigma_2\left(\bar{v}(u, \widetilde{Y}_u^1(1+\varrho_1), \widetilde{Y}_u^2(1+\varrho_2)) - \widetilde{v}(u, \widetilde{Y}_u^1, \widetilde{Y}_u^2)\right)\right]\frac{dY_u^{1,3}}{Y_{u-}^1}\\
&- \frac{1}{c}\int_{(0,t]}\left[\varrho_1\left(\mathbb{1}_{\{\tau \geq u\}}\widetilde{V}_u + \mathbb{1}_{\{\tau < u\}}\bar{V}_u\right)\right.\\
&\left. - \sigma_1\left(\bar{v}(u, \widetilde{Y}_u^1(1+\varrho_1), \widetilde{Y}_u^2(1+\varrho_2)) - \widetilde{v}(u, \widetilde{Y}_u^1, \widetilde{Y}_u^2)\right)\right]\frac{dY_u^{2,3}}{Y_{u-}^2}\\
&= V_0^* + \int_{(0,t]}\psi_u^1\,dY_u^{1,3} + \int_{(0,t]}\psi_u^2\,dY_u^{2,3},
\end{aligned}
$$

where the processes ψ^1 and ψ^2 are $\mathbb{G}$-predictable. If we do not postulate that $\varrho_1 > -1$ and $\varrho_2 > -1$, then we obtain

$$
V_{t\wedge\tau}^* = V_0^* + \int_0^{t\wedge\tau}\widetilde{V}_u\,dW_u^*
$$
$$
+ \int_{(0,t\wedge\tau]}e^{-ru}\left[\bar{v}(u, \widetilde{Y}_u^1(1+\varrho_1), \widetilde{Y}_u^2(1+\varrho_2)) - \widetilde{v}(u, \widetilde{Y}_u^1, \widetilde{Y}_u^2)\right]dM_u^*
$$

or, equivalently,

$$
\begin{aligned}
V_{t\wedge\tau}^* = V_0^* &+ \frac{1}{c}\int_{(0,t\wedge\tau]}\varrho_2\widetilde{V}_u\frac{dY_u^{1,3}}{\widetilde{Y}_u^1} - \frac{1}{c}\int_{(0,t\wedge\tau]}\varrho_1\widetilde{V}_u\frac{dY_u^{2,3}}{\widetilde{Y}_u^2}\\
&- \frac{\sigma_2}{c}\int_{(0,t\wedge\tau]}\left(\bar{v}(u, \widetilde{Y}_u^1(1+\varrho_1), \widetilde{Y}_u^2(1+\varrho_2)) - \widetilde{v}(u, \widetilde{Y}_u^1, \widetilde{Y}_u^2)\right)\frac{dY_u^{1,3}}{\widetilde{Y}_u^1}\\
&+ \frac{\sigma_1}{c}\int_{(0,t\wedge\tau]}\left(\bar{v}(u, \widetilde{Y}_u^1(1+\varrho_1), \widetilde{Y}_u^2(1+\varrho_2)) - \widetilde{v}(u, \widetilde{Y}_u^1, \widetilde{Y}_u^2)\right)\frac{dY_u^{2,3}}{\widetilde{Y}_u^2}\\
&= V_0^* + \int_{(0,t\wedge\tau]}\phi_u^1\,dY_u^{1,3} + \int_{(0,t\wedge\tau]}\phi_u^2\,dY_u^{2,3},
\end{aligned}
$$

where the processes ϕ^1 and ϕ^2 are $\mathbb{F}$-predictable.

We are in a position to state the following result, which establishes the formula for the replicating strategy for Y.

Proposition 12. *Assume that $\varrho_1 > -1$ and $\varrho_2 > -1$. Then the replicating strategy for the defaultable claim Y defined by (50) is given as $\psi = (\psi^1, \psi^2, \pi(Y) - \psi^1 Y^1 - \psi^2 Y^2)$, where the $\mathbb{G}$-predictable processes ψ^1 and ψ^2 are given by the expressions*

$$\psi_t^1 = (cY_{t-}^1)^{-1}\Big(\varrho_2\big(\mathbb{1}_{\{\tau \geq t\}}\widetilde{V}_t + \mathbb{1}_{\{\tau < t\}}\overline{V}_t\big)$$
$$- \sigma_2\big(\overline{v}(t, \widetilde{Y}_t^1(1+\varrho_1), \widetilde{Y}_t^2(1+\varrho_2)) - \widetilde{v}(t, \widetilde{Y}_t^1, \widetilde{Y}_t^2))\big)\Big)$$

and

$$\psi_t^2 = -(cY_{t-}^2)^{-1}\Big(\varrho_1\big(\mathbb{1}_{\{\tau \geq t\}}\widetilde{V}_t + \mathbb{1}_{\{\tau < t\}}\overline{V}_t\big)$$
$$- \sigma_1\big(\overline{v}(t, \widetilde{Y}_t^1(1+\varrho_1), \widetilde{Y}_t^2(1+\varrho_2)) - \widetilde{v}(t, \widetilde{Y}_t^1, \widetilde{Y}_t^2))\big)\Big)$$

with the processes $\widetilde{V}$ and $\overline{V}$ given by (60) and (61), respectively. The wealth process of ψ satisfies $V_t(\psi) = \pi_t(Y)$ for every $t \in [0, T]$.

It is worthwhile to stress that the replicating strategy ψ is understood in the standard sense, that is, it duplicates the payoff Y at the maturity date T. If we wish instead to use the convention adopted in Section 4, then we should focus on the defaultable claim $(X, Z, 0, \tau)$ associated with Y through equality (58) (in this case, $z(t, x, y) = \overline{v}(t, x, y)$), and thus it is a replicating strategy for the associated defaultable claim (X, Z, τ). The latter convention is particularly convenient if the assumption that both ϱ_1 and ϱ_2 are strictly greater than -1 is relaxed. Let us focus on the replication of the claim $(X, Z, 0, \tau)$ with the pre-default value

$$\widetilde{U}_t = \widetilde{U}_t(X) + \widetilde{U}_t(Z) = \widetilde{v}(t, \widetilde{Y}_t^1, \widetilde{Y}_t^2).$$

Proposition 13. *Assume that either $\varrho_1 > -1$ or $\varrho_2 > -1$, and let the process $\widetilde{V}$ be given by (60). Then the replicating strategy for the defaultable claim $(X, Z, 0, \tau)$ is $\phi = (\phi^1, \phi^2, \widetilde{U} - \phi^1 \widetilde{Y}^1 - \phi^2 \widetilde{Y}^2)$, where the $\mathbb{F}$-predictable processes ϕ^1 and ϕ^2 are given by the formulae*

$$\phi_t^1 = (c\widetilde{Y}_t^1)^{-1}\Big(\varrho_2 \widetilde{V}_t - \sigma_2\big(z(t, \widetilde{Y}_t^1(1+\varrho_1), \widetilde{Y}_t^2(1+\varrho_2)) - \widetilde{U}_t\big)\Big)$$
$$\phi_t^2 = -(c\widetilde{Y}_t^2)^{-1}\Big(\varrho_1 \widetilde{V}_t - \sigma_1\big(z(t, \widetilde{Y}_t^1(1+\varrho_1), \widetilde{Y}_t^2(1+\varrho_2)) - \widetilde{U}_t\big)\Big).$$

Survival claim. Assume that the first tradeable asset is a default-free asset (that is, $\varrho_1 = 0$), and the second asset is a defaultable asset with zero recovery (hence, $\varrho_2 = -1$). Then we have

$$dY_t^1 = Y_t^1(r\,dt + \sigma_1\,dW_t^*), \quad Y_0^1 > 0,$$
$$dY_t^2 = Y_{t-}^2(r\,dt + \sigma_2\,dW_t^* - dM_t^*), \quad Y_0^2 > 0,$$
$$dY_t^3 = rY_t^3\,dt, \quad Y_0^3 = 1.$$

Notice that $c = \varrho_2\sigma_1 - \varrho_1\sigma_2 = -\sigma_1 \neq 0$. Consider a survival claim Y of the form $Y = \mathbb{1}_{\{\tau>T\}}g(Y_T^1)$, that is, a vulnerable claim with zero recovery written on the default-free asset Y^1. It is obvious that we may formally identify Y with the defaultable claim $(X, 0, 0, \tau)$ with the promised payoff $X = g(Y_T^1)$ and $Z = 0$. For the replicating strategy ϕ we obtain that

$$\phi_t^2 Y_t^2 = \tilde{v}(t, \tilde{Y}_t^1, \tilde{Y}_t^2) - \bar{v}(t, \tilde{Y}_t^1(1+\varrho_1), \tilde{Y}_t^2(1+\varrho_2)y) = \tilde{U}_t,$$

since $\bar{v}(t, x, y) = 0$. We conclude that the net investment in default-free assets equals 0 at any time $t \in [0, T]$. One can check, by inspection, that the strategy ϕ replicates the claim Y also after default (formally, we set $\psi_t^i = 0$ for $i = 1, 2, 3$ on the event $\{\tau > t\}$).

Suppose that the risk-neutral intensity of default is of the form $\gamma_t = \gamma(t, Y_t^1)$. In this case, it is rather obvious that the pre-default pricing function $\tilde{v}$ does not depend on the variable y. In particular, the volatility coefficient σ_2 of the second asset plays no role in the risk-neutral valuation of Y; only the properties of the default time τ really matter. This feature of the function $\tilde{v}$ can be formally deduced from the representation (53) and the observation that if $\gamma_t = \gamma(t, Y_t^1)$ then the two-dimensional process (Y^1, H) is Markovian with respect to the filtration $\mathbb{G}$. We conclude that the function $\tilde{v} = \tilde{v}(t, x)$ satisfies the following simple version of the pre-default pricing PDE

$$-r\tilde{v} + \partial_t\tilde{v} + rx\partial_x\tilde{v} + \tfrac{1}{2}\sigma_1^2 x^2 \partial_{xx}^2 \tilde{v} - \gamma(t, x)\tilde{v} = 0$$

with the terminal condition $\tilde{v}(T, x) = g(x)$.

6.4 Generalizations

For the sake of simplicity, we have postulated that the prices Y^1, Y^2 and Y^3 are given by the SDE (49) with constant coefficients. In order to cover a large class of defaultable assets, we should relax these restrictive assumptions by postulating, for instance, that the processes Y^1 and Y^2 are governed under $\mathbb{Q}^*$ by

$$dY_t^i = Y_{t-}^i(r_t\,dt + \sigma_t^i\,dW_t^* + \varrho_t^i\,dM_t^*), \quad Y_0^i > 0,$$

where

$$\sigma_t^i = \tilde{\sigma}_i(t, T)\mathbb{1}_{\{\tau<t\}} + \bar{\sigma}_2(t, T)\mathbb{1}_{\{\tau\geq t\}}$$

for some pre-default and post-default volatilities $\tilde{\sigma}_i(t, T)$ and $\bar{\sigma}_i(t, T)$, and where $\varrho_t^i = \varrho_i(t, Y_{t-}^1, Y_{t-}^2, Y_{t-}^3)$ for some functions $\varrho_i : [0, T] \times \mathbb{R}_+^3 \to [-1, \infty)$. The proposed dynamics for Y^1 and Y^2 has the following practical consequences. First, the choice of $\tilde{\sigma}_i$ and $\bar{\sigma}_i$ allows us to model the real-life fact that the character of a defaultable security may change essentially after default. Second, through a judicious specification of the function ϱ_i, we are able to examine various alternative recovery schemes at time of default. As the process Y^3, we may take the price of a zero-coupon default-free bond. Hence, $Y^3 = B(t, T)$ satisfies under $\mathbb{Q}^*$

$$dY_t^3 = Y_t^3 \left(r_t \, dt + b(t, T) \, dW_t^*\right).$$

Example 3. Suppose that the process Y^1 represents the price of a generic defaultable zero-coupon bond with maturity date T. Then the bond is subject to the fractional recovery of market value scheme with recovery rate $\delta_1 \in [0, 1]$ if the process ϱ^1 is constant, specifically,

$$\varrho_t^1 = \varrho_1(t, Y_{t-}^1, Y_{t-}^2, Y_{t-}^3) = \delta_1 - 1.$$

To model a defaultable bond with the fractional recovery of par value at default, we set

$$\varrho_t^1 = \varrho_1(t, Y_{t-}^1, Y_{t-}^2, Y_{t-}^3) = \delta_1 (Y_{t-}^1)^{-1} - 1.$$

Finally, the fractional recovery of Treasury value scheme corresponds to the following choice of the process ϱ_t^1 (recall that $Y_t^3 = B(t, T)$, and thus it is a continuous process)

$$\varrho_t^1 = \varrho_1(t, Y_{t-}^1, Y_{t-}^2, Y_{t-}^3) = \delta_1 Y_t^3 (Y_{t-}^1)^{-1} - 1.$$

In all cases, the post-default volatility $\bar{\sigma}_1(t, T)$ should coincide with the volatility of the default-free zero-coupon bond of maturity T. This corresponds to the natural interpretation that after default the recovery payoff is invested in default free bonds.

Part II. Mean-Variance Approach

In this part, we formulate a new paradigm for pricing and hedging financial risks in incomplete markets, rooted in the classical Markowitz mean-variance portfolio selection principle. We consider an underlying market of liquid financial instruments that are available to an investor (also called an agent) for investment. We assume that the underlying market is arbitrage-free and complete. We also consider an investor who is interested in dynamic selection of her portfolio, so that the expected value of her wealth at the end of the pre-selected planning horizon is no less then some floor value, and so that the associated risk, as measured by the variance of the wealth at the end of the planning horizon, is minimized.

When a new investment opportunity becomes available for the agent, in a form of some contingent claim, she needs to decide how much she is willing to pay for acquiring the opportunity. More specifically, she has to decide what portion of her current endowment she is willing to invest in a new opportunity. It is assumed that the new claim, if acquired, is held until the horizon date, and the remaining part of the endowment is dynamically invested in primary (liquid) assets. If the cash-flows generated by the new opportunity can be perfectly replicated by the existing liquid market instruments already available for trading, then the price of the opportunity will be uniquely determined by the wealth of the replicating strategy. However, if perfect replication is not possible, then the determination of a purchase (or bid) price that the investor is willing to pay for the opportunity, will become subject to the investor's overall attitude towards trading. In case of our investor, the bid price and the corresponding hedging strategy will be determined in accordance with the mean-variance paradigm. Analogous remarks apply to an investor who engages in creation of an investment opportunity and needs to decide about its selling (or ask) price.

As explained above, it suffices to focus on a situation when the newly available investment opportunity can not be perfectly replicated by the instruments existing in the underlying market. Thus, the emerging investment opportunity is not *attainable*, and consequently the market model (that is the underlying market and new investment opportunities) is *incomplete*.

It is well known (see, e.g., El Karoui and Quenez (1995) or Kramkov (1996)) that when a market is incomplete, then for any non-attainable contingent claim X there exists a non-empty interval of arbitrage prices, referred to as the *no-arbitrage interval*, determined by the maximum bid price $\pi^u(X)$ (the upper price) and the minimum ask price $\pi^l(X)$ (the lower price) The maximum bid price represents the cost of the most expensive dynamic portfolio that can be used to perfectly hedge the long position in the contingent claim. The minimum ask price represents the initial cost of the cheapest dynamic portfolio that can be used to perfectly hedge the short position in the contingent claim.

Put another way, the maximum bid price is the maximum amount that the agent purchasing the contingent claim can afford to pay for the claim, and still be sure to find an admissible portfolio that would fully manage her debt and repay it with cash flows generated by the strategy and the contingent claim, and end up with a non-negative wealth at the maturity date of the claim. Likewise, the minimum ask price is the minimum amount that the agent selling the claim can afford to accept to charge for the claim, and still be sure to find an admissible portfolio that would generate enough cash flow to make good on her commitment to buyer of the claim, and end up with a non-negative wealth at the maturity date of the claim.

As is well known, the arbitrage opportunities are precluded if and only if the actual price of the contingent claim belongs to the no-arbitrage interval. But this means, of course, that perfect hedging will not be accomplished by neither the short party, nor by the long party. Thus, any price that precludes arbitrage, enforces possibility of a financial loss for either party at the maturity date. This observation gave rise to quite abundant literature regarding the judicious choice of a specific price within the no-arbitrage interval by means of minimizing some functional that assesses the risk associated with potential losses.

We shall not be discussing this extensive literature here. Let us only observe that much work within this line of research has been done with regard to the so-called *mean-variance hedging*; we refer to the recent paper by Schweizer (2001) for an exhaustive survey of relevant results. It is worth stressing that the interpretation of the term "mean-variance hedging", as defined in these works, is entirely different from what is meant here by mean-variance hedging.

The optimization techniques used in this part are based on mean-variance portfolio selection in continuous time. Probably the first work in this area was the paper by Zhou and Li (2000) who used the embedding technique and linear-quadratic (LQ) optimal control theory to solve the continuous-time, mean-variance problem with assets having deterministic diffusion coefficients. They essentially ended up with a problem that was inherently an *indefinite stochastic LQ control problem*, the theory of which has been developed only very recently (see, e.g., Yong and Zhou (1999), Chapter 6). In subsequent works, the techniques of stochastic LQ optimal control were heavily exploited in order to solve more sophisticated variants of the mean-variance portfolio selection in continuous time. For instance, Li et al. (2001) introduced a constraint on short-selling, Lim and Zhou (2002) allowed for stocks which are modeled by processes having random drift and diffusion coefficients, Zhou and Yin (2004) featured assets in a regime switching market, and Bielecki et al. (2004b) solved the problem with positivity constraint imposed on the wealth process. An excellent survey of most of these results is presented in Zhou (2003), who also provided a number of examples that illustrate the similarities as well as differences between the continuous-time and single-period settings.

7 Mean-Variance Pricing and Hedging

We consider an economy in continuous time, $t \in [0, T^*]$, and the underlying probability space $(\Omega, \mathcal{G}, \mathbb{P})$ endowed with a one-dimensional standard Brownian motion W (with respect to its natural filtration). The probability $\mathbb{P}$ plays the role of the statistical probability. We denote by $\mathbb{F}$ the $\mathbb{P}$-augmentation of the filtration generated by W. Consider an agent who initially has two liquid assets available to invest in:

- a risky asset whose price dynamics are

$$dZ_t^1 = Z_t^1 \left(\nu \, dt + \sigma \, dW_t \right), \quad Z_0^1 > 0,$$

for some constants ν and $\sigma > 0$,

- a money market account whose price dynamics under $\mathbb{P}$ are

$$dZ_t^2 = r Z_t^2 \, dt, \quad Z_0^2 = 1,$$

where r is a constant interest rate.

Suppose for the moment that $\mathcal{G} = \mathcal{F}_{T^*}$. It is well known that in this case the *underlying market*, consisting of the two above assets, is complete. Thus the fair value of any claim contingent X which settles at time $T \leq T^*$, and thus is formally defined as an $\mathcal{F}_T$-measurable random variable, is the (unique) arbitrage price of X, denoted as $\pi_0(X)$ in what follows.

Now let $\mathbb{H}$ be another filtration in $(\Omega, \mathcal{G}, \mathbb{P})$, which satisfies the usual conditions. We consider the enlarged filtration $\mathbb{G} = \mathbb{F} \vee \mathbb{H}$ and we postulate that $\mathcal{G} = \mathcal{G}_{T^*}$. We shall refer to $\mathbb{G}$ as to the *full filtration*; the Brownian filtration $\mathbb{F}$ will be called the *reference filtration*. We make an important assumption that W is a standard Brownian motion with respect to the full filtration $\mathbb{G}$ under the probability $\mathbb{P}$.

Let ϕ_t^i represent the number of shares of asset i held in the agent's portfolio at time t. We consider trading strategies $\phi = (\phi^1, \phi^2)$, where ϕ^1 and ϕ^2 are $\mathbb{G}$-predictable processes. A strategy ϕ is *self-financing* if

$$V_t(\phi) = V_0(\phi) + \int_0^t \phi_u^1 \, dZ_u^1 + \int_0^t \phi_u^2 \, dZ_u^2, \quad \forall t \in [0, T^*],$$

where $V_t(\phi) = \phi_t^1 Z_t^1 + \phi_t^2 Z_t^2$ is the *wealth* of ϕ at time t. Thus, we postulate the absence of outside endowments and/or consumption.

Definition 7. *We say that a self-financing strategy ϕ is admissible on the interval $[0, T]$ if and only if for any $t \in [0, T]$ the wealth $V_t(\phi)$ is a $\mathbb{P}$-square-integrable random variable.*

The condition

$$\mathbb{E}_{\mathbb{P}} \left(\int_0^T (\phi_u^i Z_u^i)^2 du \right) < \infty, \quad i = 1, 2,$$

is manifestly sufficient for the admissibility of ϕ on $[0, T]$. Let us fix T and let us denote by $\Phi(\mathbb{G})$ the linear space of all admissible trading strategies on the finite interval $[0, T]$.

Suppose that the agent has at time $t = 0$ a positive amount $v > 0$ available for investment (we shall refer to v as the *initial endowment*). It is easily seen that for any $\phi \in \Phi(\mathbb{G})$ the wealth process satisfies the following SDE

$$dV_t^v(\phi) = rV_t^v(\phi) \, dt + \phi_t^1 (dZ_t^1 - rZ_t^1 dt), \quad V_0^v(\phi) = v.$$

This shows that the wealth at time t depends exclusively on the initial endowment v and the component ϕ^1 of a self-financing strategy ϕ.

Now, imagine that a new investment opportunity becomes available for the agent. Namely, the agent may purchase at time $t = 0$ a contingent claim X, whose corresponding cash-flow of X units of cash occurs at time T. We assume that X is not an $\mathcal{F}_T$-measurable random variable. Notice that this requirement alone may not suffice for the non-attainability of X. Indeed, in the present setup, we have the following definition of attainability.

Definition 8. *A contingent claim X is attainable if there exists a strategy $\phi \in \Phi(\mathbb{G})$ such that $X = V_T(\phi)$ or, equivalently,*

$$X = V_0(\phi) + \int_0^T \phi_u^1 \, dZ_u^1 + \int_0^T \phi_u^2 \, dZ_u^2.$$

If a claim X can be replicated by means of a trading strategy $\phi \in \Phi(\mathbb{F})$, we shall say that X is $\mathbb{F}$-*attainable*. According to the definition of admissibility, the square-integrability of X under $\mathbb{P}$ is a necessary condition for attainability. Notice, however, that it may happen that X is not an $\mathcal{F}_T$-measurable random variable, but it represents an attainable contingent claim according to the definition above.

Suppose now that a considered claim X is not attainable. The main question that we want to study is: how much would the agent be willing to pay at time $t = 0$ for X, and how the agent should hedge her investment? A symmetric study can be conducted for an agent creating such an investment opportunity by selling the claim. In what follows, we shall first present our results in a general framework of a generic $\mathcal{G}_T$-measurable claim; then we shall examine a particular case of defaultable claims.

7.1 Mean-Variance Portfolio Selection

We postulate that the agent's objective for investment is based on the classical *mean-variance portfolio selection*. Let $\mathbb{V}_\mathbb{P}(Z)$ be the variance under $\mathbb{P}$ of a random variable Z. For any fixed date T, any initial endowment $v > 0$, and any given $d \in \mathbb{R}$, the agent is interested in solving the following problem:

Problem MV(d, v): Minimize $\mathbb{V}_\mathbb{P}(V_T^v(\phi))$ over all strategies $\phi \in \Phi(\mathbb{G})$, subject to $\mathbb{E}_\mathbb{P} V_T^v(\phi) \geq d$.

We shall show that, given the parameters d and v satisfy certain additional conditions, the above problem admits a solution, so that there exists an optimal trading strategy, say $\phi^*(d, v)$. Let $V^*(d, v) = V(\phi^*(d, v))$ stand for the optimal wealth process, and let us denote by $\mathbf{v}^*(d, v)$ the value of the variance $\mathbb{V}_\mathbb{P}(V_T^*(d, v))$.

For simplicity of presentation, we did not postulate above that agent's wealth should be non-negative at any time. Problem MV(d, v) with this additional restriction has been recently studied in Bielecki et al. (2004b).

Remark. It is apparent that the problem MV(d, v) is non-trivial only if $d > ve^{rT}$. Otherwise, investing in the money market alone generates the wealth process $V_t^v(\phi) = ve^{rt}$, that obviously satisfies the terminal condition $\mathbb{E}_\mathbb{P} V_T^v(\phi) = ve^{rT} \geq d$, and for which the variance of the terminal wealth $V_T^v(\phi)$ is zero. Thus, when considering the problem MV(d, v) we shall always assume that $d > ve^{rT}$. Put another way, we shall only consider trading strategies ϕ for which the expected return satisfies $\mathbb{E}_\mathbb{P}(V_T^v(\phi)/v) \geq e^{rT}$, that is, it is strictly higher than the return on the money market account.

Assume that a claim X is available for purchase at time $t = 0$. We postulate that the random variable X is $\mathcal{G}_T$-measurable and square-integrable under $\mathbb{P}$. The agent shall decide whether to purchase X, and what is the maximal price she could offer for X. According to the mean-variance paradigm, her decision will be based on the following reasoning. First, for any $p \in [0, v]$ the agent needs to solve the related mean-variance problem.

Problem MV(d, v, p, X): Minimize $\mathbb{V}_{\mathbb{P}}(V_T^{v-p}(\phi) + X)$ over all trading strategies $\phi \in \Phi(\mathbb{G})$, subject to $\mathbb{E}_{\mathbb{P}}(V_T^{v-p}(\phi) + X) \geq d$.

We shall show that if d, v, p and X satisfy certain sufficient conditions, then there exists an optimal strategy for this problem. We shall denote by $\phi^*(d, v, p, X)$ this optimal strategy, and by $V_T^*(d, v, p, X)$ the corresponding value of $V_T^{v-p}(\phi^*(d, v, p, X))$ and we set $\mathbf{v}^*(d, v, p, X) = \mathbb{V}_{\mathbb{P}}(V_T^*(d, v, p, X) + X)$.

It is reasonable to expect that the agent will be willing to pay for the claim X the price that is no more than (by convention, $\sup \emptyset = -\infty$)

$$p^{d,v}(X) := \sup \{ p \in [0, v] : \text{MV}(d, v, p, X) \text{ admits a solution}$$
$$\text{and } \mathbf{v}^*(d, v, p, X) \leq \mathbf{v}^*(d, v) \}.$$

This leads to the following definition of mean-variance price and hedging strategy.

Definition 9. *The number $p^{d,v}(X)$ is called the buying agent's mean-variance price of X. The optimal trading strategy $\phi^*(d, v, p^{d,v}(X), X)$ is called the agent's mean-variance hedging strategy for X.*

Of course, in order to make the last definition operational, we need to be able to solve explicitly problems MV(d, v) and MV(d, v, p, X), at least in some special cases of a common interest. These issues will be examined in some detail in the remaining part of this note, first for the special case of $\mathbb{F}$-adapted trading strategies (see Section 8), and subsequently, in the general case of $\mathbb{G}$-adapted strategies (see Section 9).

Remark. Let us denote $\mu_X = \mathbb{E}_{\mathbb{P}} X$. Inequality $\mathbb{E}_{\mathbb{P}}(V_T^{v-p}(\phi) + X) \geq d$ is equivalent to $\mathbb{E}_{\mathbb{P}} V_T^{v-p}(\phi) \geq d - \mu_X$. Observe that, unlike as in the case of the problem MV(d, v), the problem MV(d, v, p, X) may be non-trivial even if $d - \mu_X \leq e^{rT}(v - p)$. Although investing in a money market alone will produce in this case a wealth process for which the condition $\mathbb{E}_{\mathbb{P}} V_T^{v-p}(\phi) \geq d - \mu_X$ is manifestly satisfied, the corresponding variance $\mathbb{V}_{\mathbb{P}}(V_T^{v-p}(\phi) + X) = \mathbb{V}_{\mathbb{P}}(X)$ is not necessarily minimal.

Financial Interpretation

Let us denote by $\mathcal{N}(X)$ the no-arbitrage interval for the claim X, that is, $\mathcal{N}(X) = [\pi^l(X), \pi^u(X)]$. It may well happen that the mean-variance price $p^{d,v}(X)$ is outside this interval. Since this possibility may appear as an unwanted feature of the approach to pricing and hedging presented in this note, we shall comment briefly on this issue. When we consider the valuation of a claim X from the perspective of the entire market, then we naturally apply the no-arbitrage paradigm.

According to the no-arbitrage paradigm, the financial market as a whole will accept only those prices of a financial asset, which fall into the no-arbitrage interval. Prices from outside this interval can't be sustained in a longer term due to market forces, which will tend to eliminate any arbitrage opportunity.

Now, let us consider the same issue from the perspective of an individual. Suppose that an individual investor is interested in putting some of her initial endowment $v > 0$ into an investment opportunity provided by some claim X. Thus, the investor needs to decide whether to acquire the investment opportunity, and if so then how much to pay for it, based on her overall attitude towards risk and reward.

The number $p^{d,v}(X)$ is the price that investor is willing to pay for the investment opportunity X, given her initial capital v, given her attitude towards risk and reward, and given the primary market. The investor "submits" her price to the market. Now, suppose that the market recognized no-arbitrage interval for X is $\mathcal{N}(X)$. If it happens that $p \in \mathcal{N}(X)$ then the investor's bid price for X can be accepted by the market. In the opposite case, the investor's bid price may not be accepted by the market, and the investor may not enter into the investment opportunity.

8 Strategies Adapted to the Reference Filtration

In this section, we shall solve the problem $\mathrm{MV}(d, v)$ under the restriction that trading strategies are based on the reference filtration $\mathbb{F}$. In other words, we postulate that ϕ belongs to the class $\Phi(\mathbb{F})$ of all admissible and $\mathbb{F}$-predictable strategies ϕ. In this case, we shall say that a strategy ϕ is $\mathbb{F}$-*admissible*. The assumption that ϕ is $\mathbb{F}$-admissible implies, of course, that the terminal wealth $V_T^v(\phi)$ is an $\mathcal{F}_T$-measurable random variable.

8.1 Solution to MV(d, v) in the Class $\Phi(\mathbb{F})$

A general version of the problem $\mathrm{MV}(d, v)$ has been studied in Bielecki et al. (2004b). Because our problem is a very special version of the general one, we give below a complete solution tailored to present set-up.

Reduction to Zero Interest Rate Case

Recall our standing assumption that $d > ve^{rT}$. Problem $\mathrm{MV}(d, v)$ is clearly equivalent to: minimize the variance $\mathbb{V}_\mathbb{P}(e^{-rT} V_T(\phi))$ under the constraint

$$\mathbb{E}_{\mathbb{P}}(e^{-rT}V_T(\phi)) \geq e^{-rT}d.$$

For the sake of notational simplicity, we shall write V_t instead of $V_t^v(\phi)$. We set $\widetilde{V}_t = V_t(Z_t^2)^{-1} = e^{-rt}V_t$, so that

$$d\widetilde{V}_t = \phi_t^1 \, d\widetilde{Z}_t^1 = \phi_t^1 \widetilde{Z}_t^1 (\widehat{\nu} \, dt + \sigma \, dW_t), \tag{62}$$

where we denote $\widehat{\nu} = \nu - r$. So we can and do restrict our attention to the case $r = 0$. Thus, in what follows, we shall have $Z_t^2 = 1$ for every $t \in \mathbb{R}_+$. In the rest of this note, unless explicitly stated otherwise, we assume that $d > v$.

Decomposition of Problem MV(d, v)

Let $\mathbb{Q}$ be a (unique) equivalent martingale measure on $(\Omega, \mathcal{F}_{T^*})$ for the underlying market. It is easily seen that

$$\frac{d\mathbb{Q}}{d\mathbb{P}}\Big|_{\mathcal{F}_t} = \eta_t, \quad \forall t \in [0, T^*],$$

where we denote by η the Radon-Nikodym density process. Specifically, we have

$$d\eta_t = -\theta \eta_t \, dW_t, \quad \eta_0 = 1, \tag{63}$$

or, equivalently,

$$\eta_t = \exp\left(-\theta W_t - \tfrac{1}{2}\theta^2 t\right),$$

where $\theta = \nu/\sigma$ (recall that we have formally reduced the problem to the case $r = 0$). The process η is a $\mathbb{F}$-martingale under $\mathbb{P}$. Moreover,

$$\mathbb{E}_{\mathbb{P}}(\eta_T^2 \mid \mathcal{F}_t) = \eta_t^2 \exp(\theta^2(T - t)), \tag{64}$$

and thus $\mathbb{E}_{\mathbb{P}}(\eta_t^2) = \exp(\theta^2 t)$ for $t \in [0, T^*]$. It is easily seen that the price Z^1 is an $\mathbb{F}$-martingale under $\mathbb{Q}$, since

$$dZ_t^1 = \sigma Z_t^1 \, d(W_t + \theta t) = \sigma Z_t^1 \, d\widetilde{W}_t \tag{65}$$

for the $\mathbb{Q}$-Brownian motion $\widetilde{W}_t = W_t + \theta t$. The measure $\mathbb{Q}$ is thus the equivalent martingale measure for our primary market.

From (62), we have that

$$V_t = v + \int_0^t \phi_u^1 \, dZ_u^1 = v + \int_0^t \phi_u^1 \sigma Z_u^1 \, d\widetilde{W}_u. \tag{66}$$

Recall that if ϕ is an $\mathbb{F}$-admissible strategy, that is, $\phi \in \Phi(\mathbb{F})$, then V_T is an $\mathcal{F}_T$-measurable random variable, which is $\mathbb{P}$-square-integrable.

Let X be a $\mathbb{P}$-square-integrable and $\mathcal{F}_T$-measurable random variable. It is easily seen that X is integrable with respect to $\mathbb{Q}$ (since η_T is square-integrable with respect to $\mathbb{P}$). The existence of a self-financing trading strategy that replicates X can be justified by the predictable representation theorem combined with the Bayes formula. We thus have the following result.

Lemma 11. *Let X be a $\mathbb{P}$-square-integrable and $\mathcal{F}_T$-measurable random variable. Then X is an $\mathbb{F}$-attainable contingent claim, i.e., there exists a strategy ϕ^X in $\Phi(\mathbb{F})$ such that $V_T(\phi) = X$.*

We shall argue that problem $MV(d, v)$ can be split into two problems (see also Pliska (2001) and Bielecki et al. (2004b) in this regard). We first focus on the optimal terminal wealth $V_T^*(d, v)$. Let $L^2(\Omega, \mathcal{F}_T, \mathbb{P})$ denote the collection of $\mathbb{P}$-square-integrable random variables that are $\mathcal{F}_T$-measurable. Thus the first problem we need to solve is:

Problem MV1: Minimize $\mathbb{V}_\mathbb{P}(\xi)$ over all $\xi \in L^2(\Omega, \mathcal{F}_T, \mathbb{P})$, subject to $\mathbb{E}_\mathbb{P}\xi \geq d$ and $\mathbb{E}_\mathbb{Q}\xi = v$.

Lemma 12. *Suppose that $\phi^* = \phi^*(d, v)$ solves the problem $MV(d, v)$, and let $V^*(d, v) = V(\phi^*)$. Then the random variable $\xi^* = V_T^*(d, v)$ solves the problem MV1.*

Proof. We argue by contradiction. Suppose that there exists a random variable $\widehat{\xi} \in L^2(\Omega, \mathcal{F}_T, \mathbb{P})$ such that $\mathbb{E}_\mathbb{P}\widehat{\xi} \geq d$, $\mathbb{E}_\mathbb{Q}\widehat{\xi} = v$ and $\mathbb{V}_\mathbb{P}(\widehat{\xi}) < \mathbb{V}_\mathbb{P}(\xi^*)$. Since $\widehat{\xi}$ is $\mathbb{P}$-square-integrable and $\mathcal{F}_T$-measurable, it represents an attainable contingent claim, so that there exists an $\mathbb{F}$-admissible strategy $\widehat{\phi}$ such that $\widehat{\xi} = V_T(\widehat{\phi})$. Of course, this contradicts the assumption that ϕ^* solves $MV(d, v)$. $\qquad\qquad\square$

Denoting by ξ^* the optimal solution to problem MV1, the second problem is:

Problem MV2: Find an $\mathbb{F}$ admissible strategy ϕ^* such that $V_T(\phi^*) = \xi^*$.

Since the next result is analogous to Theorem 2.1 in Bielecki et al. (2004b), its proof is omitted. It demonstrates that solving problem $MV(d, v)$ is indeed equivalent to successful solving problems MV1 and MV2. In the formulation of the result below we make use of a backward stochastic differential equation (BSDE). The reader can refer to El Karoui and Mazliak (1997), El Karoui and Quenez (1997), El Karoui et al. (1997), Ma and Yong (1999) or to the survey by Buckdahn (2000) for an introduction to the theory of backward stochastic differential equations and its applications in finance.

Proposition 14. *Suppose that the problem MV1 has a solution ξ^*. The following BSDE*

$$dv_t = -\theta z_t \, dt + z_t \, dW_t, \quad v_T = \xi^*, \quad t \in [0, T], \tag{67}$$

has a unique, $\mathbb{P}$-square-integrable solution, denoted as (v^*, z^*), which is adapted to $\mathbb{F}$. Moreover, if we define a process ϕ^{1*} by

$$\phi_t^{1*} = z_t^*(\sigma Z_t^1)^{-1}, \quad \forall t \in [0, T],$$

then the $\mathbb{F}$-admissible trading strategy $\phi^* = (\phi^{1*}, \phi^{2*})$ with the wealth process $V_t(\phi^*) = v_t^*$ solves the problem $MV(d, v)$.

For the last statement, recall that if the first component of a self-financing strategy ϕ and its wealth process $V(\phi)$ is known, then the component ϕ^2 is uniquely determined through the equality $V_t(\phi) = \phi_t^1 Z_t^1 + \phi_t^2 Z_t^2$.

Remark. In what follows, we shall derive closed-form expressions for ϕ^* and $V(\phi^*)$. It will be easily seen that the process $V(\phi^*)$ is not only $\mathbb{P}$-square-integrable, but also $\mathbb{Q}$-square-integrable. It should be stressed that Proposition 14 will not be used in the derivation of a solution to problem $\mathrm{MV}(d, v)$. In fact, we shall find a solution to $\mathrm{MV}(d, v)$ through explicit calculations.

Solution of Problem MV1

In order to make the problem MV1 non-trivial, we need to make an additional assumption that $\theta \neq 0$. Indeed, if $\theta = 0$ then we have $\mathbb{P} = \mathbb{Q}$, and thus the problem MV1 becomes:

Problem MV1: Minimize $\mathbb{V}_{\mathbb{P}}(\xi)$ over all $\xi \in L^2(\Omega, \mathcal{F}_T, \mathbb{P})$, subject to $\mathbb{E}_{\mathbb{P}}\xi \geq d$ and $\mathbb{E}_{\mathbb{Q}}\xi = v$.

It is easily seen that this problem admits a solution for $d = v$ only, and the optimal solution is trivial, in the sense that the optimal variance is null. Consequently, for $\theta = 0$, the solution to $\mathrm{MV}(d, v)$ exists if and only if $d = v$, and it is trivial: $\phi^* = (0, 1)$. Let us reiterate that we postulate that $d > v$ in order to avoid trivial solutions to $\mathrm{MV}(d, v)$.

From now on, we assume that $\theta \neq 0$. We begin with the following auxiliary problem:

Problem MV1A: Minimize $\mathbb{V}_{\mathbb{P}}(\xi)$ over all $\xi \in L^2(\Omega, \mathcal{F}_T, \mathbb{P})$, subject to $\mathbb{E}_{\mathbb{P}}\xi = d$ and $\mathbb{E}_{\mathbb{P}}\xi = v$.

The previous problem is manifestly equivalent to:

Problem MV1B: Minimize $\mathbb{E}_{\mathbb{P}}\xi^2$ over all $\xi \in L^2(\Omega, \mathcal{F}_T, \mathbb{P})$, subject to $\mathbb{E}_{\mathbb{P}}\xi = d$ and $\mathbb{E}_{\mathbb{Q}}\xi = v$.

Since $\mathbb{E}_{\mathbb{Q}}\xi = \mathbb{E}_{\mathbb{P}}(\eta_T \xi)$, the corresponding Lagrangian is

$$\mathbb{E}_{\mathbb{P}}(\xi^2 - \lambda_1 \xi - \lambda_2 \eta_T \xi) - d^2 + \lambda_1 d + \lambda_2 v.$$

The optimal random variable is given by $2\xi^* = \lambda_1 + \lambda_2 \eta_T$, where the Lagrange multipliers satisfy

$$2d = \lambda_1 + \lambda_2, \quad 2v = \lambda_1 + \lambda_2 \exp(\theta^2 T).$$

Hence, we have

$$\xi^* = \left(de^{\theta^2 T} - v + (v - d)\eta_T\right)\left(e^{\theta^2 T} - 1\right)^{-1}, \tag{68}$$

and the corresponding minimal variance is

$$\mathbb{V}_\mathbb{P}(\xi^*) = \mathbb{E}_\mathbb{P}(\xi^*)^2 - d^2 = (d-v)^2 \big(e^{\theta^2 T} - 1\big)^{-1}. \qquad (69)$$

Since we assumed that $d > v$, the minimal variance is an increasing function of the parameter d for any fixed value of the initial endowment v, we conclude that we have solved not only the problem MV1A, but the problem MV1 as well. We thus have the following result.

Proposition 15. *The solution ξ^* to problem MV1 is given by* (68) *and the minimal variance* $\mathbb{V}_\mathbb{P}(\xi^*)$ *is given by* (69).

For an alternative approach to Problem MV1, in a fairly general setup, see Jankunas (2001).

Solution of Problem MV2

We maintain the assumption that $\theta \neq 0$. Thus, the optimal wealth for the terminal time T is given by (68), that is, $V_T(\phi^*) = \xi^*$. Our goal is to determine an $\mathbb{F}$-admissible strategy ϕ^* for which the last equality is indeed satisfied. In view if (66), it suffices to find ϕ^{1*} such that the process V_t^* given by

$$V_t^* - v + \int_0^t \phi_u^{1*} \, dZ_u^1 \qquad (70)$$

satisfies $V_T = \xi^*$, and the strategy $\phi^* = (\phi^{1*}, \phi^{2*})$, where ϕ^{2*} is derived from $V_t = \phi_t^{1*} Z_t^1 + \phi_t^{2*} Z_t^2$, is $\mathbb{F}$-admissible.

To this end, let us introduce an $\mathbb{F}$-martingale V under $\mathbb{Q}$ by setting $V_t = \mathbb{E}_\mathbb{Q}(\xi^* \mid \mathcal{F}_t)$ (the integrability of ξ^* under $\mathbb{Q}$ is rather obvious).

It is easy to see that $V_T^* = \xi^*$ and $V_0^* = v$. It thus remains to find the process ϕ^{1*}. Using (64), we obtain

$$V_t^* = \big(de^{\theta^2 T} - v + (v-d)\eta_t e^{\theta^2(T-t)}\big)\big(e^{\theta^2 T} - 1\big)^{-1}.$$

Consequently, in view of (63) and (65), we have

$$\begin{aligned}
dV_t^* &= \frac{v-d}{e^{\theta^2 T} - 1}\big(e^{\theta^2(T-t)} d\eta_t - \eta_t e^{\theta^2(T-t)}\theta^2 \, dt\big) \\
&= e^{\theta^2(T-t)}\frac{\theta\eta_t(v-d)}{e^{\theta^2 T} - 1}(dW_t - \theta dt) \\
&= e^{\theta^2(T-t)}\frac{d-v}{e^{\theta^2 T} - 1}\frac{\nu\eta_t}{\sigma^2}\frac{dZ_t^1}{Z_t^1}.
\end{aligned}$$

This shows that we may choose

$$\phi_t^{1*} = e^{\theta^2(T-t)} \frac{d-v}{e^{\theta^2 T}-1} \frac{\nu}{\sigma^2} \frac{\eta_t}{Z_t^1}. \tag{71}$$

It is clear that ϕ^* is $\mathbb{F}$-admissible, since it is $\mathbb{F}$-adapted, self-financing, and $V_t(\phi^*)$ is $\mathbb{P}$-square-integrable for every $t \in [0, T]$.

Solution of Problem MV(d, v)

By virtue of Lemma 12, we conclude that ϕ^* solves $MV(d, v)$. In view of (69), the variance under $\mathbb{P}$ of the terminal wealth of the optimal strategy is

$$\mathbf{v}^*(d, v) = \mathbb{E}_{\mathbb{P}}(V_T^*)^2 - d^2 = \frac{(d-v)^2}{e^{\theta^2 T}-1}.$$

Let us stress that since we did not impose any no-bankruptcy condition, that is we do no require that the agent's wealth is non-negative, we see that d can be any number greater then v.

We are in a position to state the following result, which summarizes the analysis above. For a fixed $T > 0$, we denote $\rho(\theta) = e^{\theta^2 T}(e^{\theta^2 T}-1)^{-1}$ and $\eta_t(\theta) = \eta_t e^{-\theta^2 t}$, so that $\eta_0(\theta) = 1$.

Proposition 16. *Assume that $\theta \neq 0$ and let $d > v$. Then a solution $\phi^*(d, v) = (\phi^{*1}(d, v), \phi^{*2}(d, v))$ to $MV(d, v)$ is given by*

$$\phi_t^{1*}(d, v) = (d-v)\rho(\theta)\frac{\nu \eta_t(\theta)}{\sigma^2 Z_t^1} \tag{72}$$

and $V_t^*(d, v) = V_t(\phi^*(d, v)) = \phi_t^{*1}(d, v)Z_t^1 + \phi_t^{*2}(d, v)$, where the optimal wealth process equals

$$V_t^*(d, v) = v + (d-v)\rho(\theta)\big(1 - \eta_t(\theta)\big). \tag{73}$$

The minimal variance $\mathbf{v}^*(d, v)$ is given by

$$\mathbf{v}^*(d, v) = \mathbb{E}_{\mathbb{P}}(V_T^*(d, v))^2 - d^2 = \frac{(d-v)^2}{e^{\theta^2 T}-1}. \tag{74}$$

Notice that the optimal trading strategy $\phi^*(d, v)$, the minimal variance $\mathbf{v}^*(d, v)$ and the optimal gains process $G_t^*(d, v) = V_t^*(d, v) - v$ depend exclusively on the difference $d - v > 0$, rather than on parameters d and v themselves.

Efficient Portfolio

As it was observed above the function $f(d) := \mathbf{v}^*(d, v)$ is (strictly) increasing for $d \geq v$. Consider the following problem (as usual, for $d \geq v$):

Problem ME(d, v): Maximize $\mathbb{E}_{\mathbb{P}} V_T^v(\phi)$ over all strategies $\phi \in \Phi(\mathbb{G})$, subject to $\mathbb{V}_{\mathbb{P}}(V_T^v(\phi)) = \mathbf{v}^*(d, v)$.

Denote the maximal expectation in the above problem by $\mu^*(d, v)$. In view of the strict monotonicity of the function $f(d)$ for $d \geq v$, it is clear that $\mu^*(d, v) = d$. Consequently, the minimum variance portfolio ϕ^* is in fact an *efficient portfolio*.

8.2 Solution to MV(d, v, p, X) in the Class $\Phi(\mathbb{F})$

Consider first the special case of an attainable claim, which is $\mathcal{F}_T$-measurable. Subsequently, we shall show that in general it suffices to decompose a general claim X into an attainable component $\widetilde{X} = \mathbb{E}_{\mathbb{P}}(X \mid \mathcal{F}_T) \in L^2(\Omega, \mathcal{F}_T, \mathbb{P})$, and a component $X - \widetilde{X}$ which is orthogonal in $L^2(\Omega, \mathcal{G}_T, \mathbb{P})$ to the subspace $L^2(\Omega, \mathcal{F}_T, \mathbb{P})$ of admissible terminal wealths.

Case of an Attainable Claim

We shall verify that the mean-variance price coincides with the (unique) arbitrage price for any contingent claim that is attainable. Of course, this feature is a standard requirement for any reasonable valuation mechanism for contingent claims. Since in this section we consider only $\mathbb{F}$-adapted strategies, we postulate here that a claim X is $\mathcal{F}_T$-measurable; the general case of a $\mathcal{G}_T$-measurable claim is considered in Section 9.1. Let $\phi^X \in \Phi(\mathbb{F})$ be a replicating strategy for X, so that X is $\mathbb{F}$-attainable, and let $\pi_0(X) = \mathbb{E}_{\mathbb{Q}} X$ be the arbitrage price of X. Since $\Phi(\mathbb{F})$ is a linear space, it is easily seen that $\Phi(\mathbb{F}) = \Phi(\mathbb{F}) + \phi^X = \Phi(\mathbb{F}) - \phi^X$. The following lemma is thus easy to prove.

Lemma 13. *Let X be an $\mathbb{F}$-attainable contingent claim. In this situation, problem MV(d, v, p, X) is equivalent to problem MV$(d, \widehat{v})$ with $\widehat{v} = v - p + \pi_0(X)$.*

Equivalence of problems MV(d, v, p, X) and MV$(d, \widehat{v})$ is understood in the following way: first, the minimal variance for both problems is identical. Second, if a strategy ψ^* is a solution to MV$(d, v - p + \pi_0(X))$, then a strategy $\phi^* = \psi^* - \phi^X$ is a solution to the original problem MV(d, v, p, X).

Corollary 9. *Suppose that an $\mathcal{F}_T$-measurable random variable X represents an $\mathbb{F}$-attainable claim. (i) If the arbitrage price $\pi_0(X)$ satisfies $\pi_0(X) \in [0, v]$ then $p^{d,v}(X) = \pi_0(X)$.*
(ii) If the arbitrage price $\pi_0(X)$ is strictly greater than v then $p^{d,v}(X) = v$.

Proof. By definition, the mean-variance price of X is the maximal value of $p \in [0, v]$ for which $\mathbf{v}^*(d, v, p, X) = \mathbf{v}^*(d, \widehat{v}) \leq \mathbf{v}^*(d, v)$. Recall that we assume that $d > v$ so that, in view of (78),

$$\mathbf{v}^*(d, v) = \frac{(d - v)^2}{e^{\theta^2 T} - 1}.$$

By applying this result to $MV(d, \widehat{v})$ we obtain

$$\mathbf{v}^*(d, v, p, X) = \frac{(d - v + p - \pi_0(X))^2}{e^{\theta^2 T} - 1}$$

provided that $d > v - p + \pi_0(X)$. Assume that $p > \pi_0(X)$. Then $d > v - p + \pi_0(X)$ and thus $\mathbf{v}^*(d, v, p, X) > \mathbf{v}^*(d, v)$ since manifestly $(d - v)^2 > (d - v + p - \pi_0(X))^2$ in this case. This shows that $p^{d,v}(X) \leq \pi_0(X)$. Of course, for $p = \pi_0(X)$ we have the equality of minimal variances. We conclude that $p^{d,v}(X) = \pi_0(X)$ provided that $\pi_0(X) \in [0, v]$. This completes the proof of part (i).

To prove part (ii), let us assume that $\pi_0(X) > v$. In this case, it suffices to take $p = v$ and to check that $\mathbf{v}^*(d, v, v, X) = \mathbf{v}^*(d, \pi_0(X)) \leq \mathbf{v}^*(d, v)$. This is again rather obvious since for $v < \pi_0(X) < d$ we have $(d - \pi_0(X))^2 < (d - v)^2$, and for $\pi_0(X) \geq d$ we have $\mathbf{v}^*(d, \pi_0(X)) = 0$.   □

Case of a Generic Claim

Consider an arbitrary $\mathcal{G}_T$-measurable claim X, which is $\mathbb{P}$-square-integrable. Recall that our goal is to solve the following problem for $0 \leq p \leq v$.

Problem $MV(d, v, p, X)$: Minimize $\mathbb{V}_{\mathbb{P}}(V_T^{v-p}(\phi) + X)$ over all trading strategies $\phi \in \Phi(\mathbb{F})$, subject to $\mathbb{E}_{\mathbb{P}}(V_T^{v-p}(\phi) + X) \geq d$.

Let us denote by $\widetilde{X}$ the conditional expectation $\mathbb{E}_{\mathbb{P}}(X \mid \mathcal{F}_T)$. Then, of course, $\mathbb{E}_{\mathbb{P}}\widetilde{X} = \mathbb{E}_{\mathbb{P}}X$. Moreover, $\widetilde{X}$ is an attainable claim and its arbitrage price at time 0 equals

$$\pi_0(\widetilde{X}) = \mathbb{E}_{\mathbb{Q}}\widetilde{X} = \mathbb{E}_{\mathbb{P}}(\eta_T \mathbb{E}_{\mathbb{P}}(X \mid \mathcal{F}_T)) = \mathbb{E}_{\mathbb{P}}(\eta_T X) = \mathbb{E}_{\mathbb{Q}}X,$$

where $\mathbb{Q}$ is the martingale measure introduced in Section 8.1. Let $\phi^{\widetilde{X}}$ stand for the replicating strategy for $\widetilde{X}$ in the class $\Phi(\mathbb{F})$. Arguing as in the previous case, we conclude that the problem $MV(d, v, p, X)$ is equivalent to the following problem. We set here $\widetilde{p} = p - \pi_0(\widetilde{X})$.

Problem $MV(d, v, \widetilde{p}, X - \widetilde{X})$: Minimize $\mathbb{V}_{\mathbb{P}}(V_T^{v-\widetilde{p}}(\phi) + X - \widetilde{X})$ over all trading strategies $\phi \in \Phi(\mathbb{F})$, subject to $\mathbb{E}_{\mathbb{P}}(V_T^{v-\widetilde{p}}(\phi) + X - \widetilde{X}) \geq d$.

Recall that $\mathbb{E}_{\mathbb{P}}\widetilde{X} = \mathbb{E}_{\mathbb{P}}X$ and denote $\gamma_X = \mathbb{V}_{\mathbb{P}}(X - \widetilde{X})$. Observe that for any $\phi \in \Phi(\mathbb{F})$ we have

$$\mathbb{V}_{\mathbb{P}}(V_T^{v-\widetilde{p}}(\phi) + X - \widetilde{X}) = \mathbb{V}_{\mathbb{P}}(V_T^{v-\widetilde{p}}(\phi)) + \mathbb{V}_{\mathbb{P}}(X - \widetilde{X}) = \mathbb{V}_{\mathbb{P}}(V_T^{v-\widetilde{p}}(\phi)) + \gamma_X.$$

The problem $MV(d, v, \widetilde{p}, X - \widetilde{X})$ can thus be represented as follows. We denote $\widetilde{v} = v - \widetilde{p}$.

Problem MV$(d, \widetilde{v}; \gamma_X)$: Minimize $\mathbb{V}_\mathbb{P}(V_T^{\widetilde{v}}(\phi)) + \gamma_X$ over all trading strategies $\phi \in \Phi(\mathbb{F})$, subject to $\mathbb{E}_\mathbb{P}(V_T^{\widetilde{v}}(\phi)) \geq d$.

Observe that the problem $\mathrm{MV}(d, \widetilde{v}; \gamma_X)$ is formally equivalent to the original problem $\mathrm{MV}(d, v, p, X)$ in the following sense: first, the minimal variances for both problems are identical, more precisely, we have

$$\mathbf{v}^*(d, v, p, X) = \mathbf{v}^*(d, \widetilde{v}) + \gamma_X,$$

where $\mathbf{v}^*(d, \widetilde{v})$ is the minimal variance for $\mathrm{MV}(d, \widetilde{v})$. Second, if a strategy ψ^* is a solution to problem $\mathrm{MV}(d, \widetilde{v})$, then $\phi^* = \psi^* - \phi^{\widetilde{X}}$ is a solution to $\mathrm{MV}(d, v, p, X)$.

Remark. It is interesting to notice that a solution $\mathrm{MV}(d, \widetilde{v}; \gamma_X)$ does not depend explicitly on the expected value of X under $\mathbb{P}$. Hence, the minimal variance for the problem $\mathrm{MV}(d, v, p, X)$ is independent of μ_X as well, but, of course, it depends on the price $\pi_0(\widetilde{X}) = \mathbb{E}_\mathbb{Q}X$, which may in fact coincide with μ_X under some circumstances.

In view of the arguments above, it suffices to consider the problem $\mathrm{MV}(d, \widetilde{v})$, where $\widetilde{v} = v - p + \mathbb{E}_\mathbb{Q}X$. Since the problem of this form has been already solved in Section 8.1, we are in a position to state the following result, which is an immediate consequence of Proposition 16. Recall that $\rho(\theta) = e^{\theta^2 T}(e^{\theta^2 T} - 1)^{-1}$ and $\eta_t(\theta) = \eta_t e^{-\theta^2 t}$, so that $\eta_0(\theta) = 1$. Finally, $\widetilde{v} = v - p + \mathbb{E}_\mathbb{Q}X = v - p + \mathbb{E}_\mathbb{Q}\widetilde{X}$.

Proposition 17. *Assume that* $\theta \neq 0$. (i) *Suppose that* $d > \widetilde{v}$. *Then a solution* $\phi^*(d, v, p, X)$ *to MV*(d, v, p, X) *is given as* $\phi^*(d, v, p, X) = \psi^*(d, \widetilde{v}) - \phi^{\widetilde{X}}$, *where* $\psi^*(d, \widetilde{v}) = (\psi^{1*}(d, \widetilde{v}), \psi^{2*}(d, \widetilde{v}))$ *is such that* $\psi^{1*}(d, \widetilde{v})$ *equals*

$$\psi_t^{1*}(d, \widetilde{v}) = (d - \widetilde{v})\rho(\theta)\frac{\nu \eta_t(\theta)}{\sigma^2 Z_t^1} \tag{75}$$

and $\psi^{2*}(d, \widetilde{v})$ *satisfies* $\psi_t^{*1}(d, \widetilde{v})Z_t^1 + \psi_t^{*2}(d, \widetilde{v}) = V_t^*(d, \widetilde{v})$ *for* $t \in [0, T]$, *where in turn*

$$V_t^*(d, \widetilde{v}) = \widetilde{v} + (d - \widetilde{v})\rho(\theta)(1 - \eta_t(\theta)). \tag{76}$$

Thus the optimal wealth for the problem MV(d, v, p, X) *equals*

$$V_t^*(d, v, p, X) = v - p + (d - \widetilde{v})\rho(\theta)(1 - \eta_t(\theta)) + \mathbb{F}_\mathbb{Q}\widetilde{X} - \mathbb{E}_\mathbb{Q}(\widetilde{X} \mid \mathcal{J}_t) \tag{77}$$

and the minimal variance $\mathbf{v}^*(d, v, p, X)$ *is given by*

$$\mathbf{v}^*(d, v, p, X) = \frac{(d - \widetilde{v})^2}{e^{\theta^2 T} - 1} + \gamma_X. \tag{78}$$

(ii) *If* $d \leq \widetilde{v}$ *then the optimal wealth process equals*

$$V_t^*(d, v, p, X) = v - p + \mathbb{E}_\mathbb{Q}\widetilde{X} - \mathbb{E}_\mathbb{Q}(\widetilde{X} \mid \mathcal{F}_t)$$

and the minimal variance equals γ_X.

Remark. Let us comment briefly on the assumption $\theta \neq 0$. Recall that if it fails to hold, the problem $MV(d, \tilde{v})$ has no solution, unless $d = \tilde{v}$. Hence, for $\theta = 0$ we need to postulate that $d = v - p + \mathbb{E}_{\mathbb{P}} X$ (recall that $\theta = 0$ if and only if $\mathbb{Q} = \mathbb{P}$). The optimal strategy $\phi^* = (0, 1)$ and thus the solution to $MV(d, v, p, X)$ is exactly the same as in part (ii) of Proposition 17.

Mean-Variance Pricing and Hedging of a Generic Claim

Our next goal is to provide explicit representations for the mean-variance price of X. We maintain the assumption that the problem $MV(d, v, p, X)$ is examined in the class $\Phi(\mathbb{F})$. Thus, the mean-variance price considered in this section, denoted as $p_{\mathbb{F}}^{d,v}(X)$ in what follows, is relative to the reference filtration $\mathbb{F}$.

Assume that $d > \tilde{v} = v - p + \mathbb{E}_{\mathbb{Q}} X$ (recall that $\mathbb{E}_{\mathbb{Q}} X = \mathbb{E}_{\mathbb{Q}} \tilde{X} = \pi_0(\tilde{X})$). Then, by virtue of Proposition 17, we see that the minimal variance for the problem $MV(d, v, p, X)$ equals

$$\mathbf{v}^*(d, v, p, X) = \frac{(d - v + p - \mathbb{E}_{\mathbb{Q}} X)^2}{e^{\theta^2 T} - 1} + \gamma_X,$$

where

$$\gamma_X = \mathbb{V}_{\mathbb{P}}(X - \tilde{X}).$$

Of course, if $d \leq \tilde{v} = v - p + \mathbb{E}_{\mathbb{Q}} X$ then we have $\mathbf{v}^*(d, v, p, X) = \gamma_X$. Recall that we postulate that $d > v$, and thus the minimal variance for the problem $MV(d, v)$ equals

$$\mathbf{v}^*(d, v) = \frac{(d - v)^2}{e^{\theta^2 T} - 1}.$$

Let us denote

$$\kappa = d - v - \mathbb{E}_{\mathbb{Q}} X, \quad \rho = (d - v)^2 - \gamma_X(e^{\theta^2 T} - 1).$$

Proposition 18. (i) *Suppose that* $\pi_0(\tilde{X}) \geq d$ *so that* $\kappa \leq -v$. *If* $\gamma_X \leq \mathbf{v}^*(d, v)$ *then the mean variance price equals* $p_{\mathbb{F}}^{d,v}(X) = v$. *Otherwise,* $p_{\mathbb{F}}^{d,v}(X) = -\infty$.
(ii) *Suppose that* $d - v \leq \pi_0(\tilde{X}) < d$ *so that* $-v < \kappa \leq 0$. *If, in addition,* $\rho \geq 0$ *then we have*

$$p_{\mathbb{F}}^{d,v}(X) = \min\{-\kappa + \sqrt{\rho}, v\} \vee 0. \tag{79}$$

Otherwise, i.e., when $\rho < 0$, *we have* $p_{\mathbb{F}}^{d,v}(X) = -\kappa$ *if* $\gamma_X \leq \mathbf{v}^*(d, v)$, *and* $p^{d,v}(X) = -\infty$ *if* $\gamma_X > \mathbf{v}^*(d, v)$.
(iii) *Suppose that* $\pi_0(\tilde{X}) < d - v$ *so that* $\kappa > 0$. *If* $\rho \geq 0$ *then* $p_{\mathbb{F}}^{d,v}(X)$ *is given by* (79). *Otherwise, we have* $p_{\mathbb{F}}^{d,v}(X) = -\infty$.

Proof. In case (i), we have $d - v - \mathbb{E}_{\mathbb{Q}} X \leq -p$ for every $p \in [0, v]$. Thus $d \leq v - p + \mathbb{E}_{\mathbb{Q}} X$, so that $\mathbf{v}^*(d, v, p, X) = \gamma_X$. Therefore, if $\gamma_X \leq \mathbf{v}^*(d, v)$ it is clear that $p_{\mathbb{F}}^{d,v}(X) = v$. Otherwise, for every $p \in [0,v]$ we have $\mathbf{v}^*(d,v,p,X) = \gamma_X > \mathbf{v}^*(d,v)$ and thus $p_{\mathbb{F}}^{d,v}(X) = -\infty$.

In case (ii), it suffices to notice that $d \le v - p + \mathbb{E}_{\mathbb{Q}} X$ for any $p \in [0,-\kappa]$, and $d > v - p + \mathbb{E}_{\mathbb{Q}} X$ for any $p \in (-\kappa, v]$. Thus the maximal $p \in [0,v]$ for which $\mathbf{v}^*(d,v,p,X) \le \mathbf{v}^*(d,v)$ can be found from the equation

$$(\kappa + p)^2 + \gamma_X (e^{\theta^2 T} - 1) = (d - v)^2,$$

which admits the solution $p = -\kappa + \sqrt{\rho}$ provided that $\rho \ge 0$. If $\rho < 0$, then we need to examine the case $p \in [0,-\kappa]$, and we see that $p_{\mathbb{F}}^{d,v}(X)$ equals either $-\kappa$ or $-\infty$, depending on whether $\gamma_X \le \mathbf{v}^*(d,v)$ or $\gamma_X > \mathbf{v}^*(d,v)$.

In case (iii), we have $d - v - \mathbb{E}_{\mathbb{Q}} X > 0$, which yields $d > v - p + \mathbb{E}_{\mathbb{Q}} X$ for any $p \in [0,v]$. Inequality $\mathbf{v}^*(d,v,p,X) \le \mathbf{v}^*(d,v)$ becomes

$$(d - v + p - \mathbb{E}_{\mathbb{Q}} X)^2 + \gamma_X (e^{\theta^2 T} - 1) \le (d - v)^2$$

If $\rho \ge 0$ then $p_{\mathbb{F}}^{d,v}(X)$ is given by (79). Otherwise, we have $p_{\mathbb{F}}^{d,v}(X) = -\infty$.   □

The mean variance hedging strategy for a claim X is now obtained as $\phi^{\mathrm{MV}} = \phi^*(d,v,p_{\mathbb{F}}^{d,v}(X),X)$ for all cases above when $p_{\mathbb{F}}^{d,v}(X) \ne -\infty$.

8.3 Defaultable Claims

In order to provide a better intuition, we shall now examine in some detail two special cases. First, we shall assume that X is independent of the σ-field $\mathcal{F}_T$. Since X is $\mathcal{G}_T$-measurable, but obviously it is not $\mathcal{G}_T$-measurable, we shall refer to X as a *defaultable claim* (a more general interpretation of X is possible, however).

Although this case may look rather trivial at the first glance, we shall see that some interesting conclusions can be obtained. Second, we shall analyze the case of a defaultable zero-coupon bond with fractional recovery of Treasury value. Of course, both examples are merely simple illustrations of Proposition 17, and thus they should not be considered as real-life applications.

Claim Independent of the Reference Filtration

Consider a $\mathcal{G}_T$-measurable contingent claim X, such that X is independent of the σ-field $\mathcal{F}_T$. Then for any strategy $\phi \in \Phi(\mathbb{F})$, the terminal wealth $V_T(\phi)$ and the payoff X are independent random variables, so that

$$\mathbb{V}_{\mathbb{P}}(V_T(\phi) + X) = \mathbb{V}_{\mathbb{P}}(V_T(\phi)) + \mathbb{V}_{\mathbb{P}}(X).$$

It is clear that if the variance $\mathbb{V}_{\mathbb{P}}(X)$ satisfies $\mathbb{V}_{\mathbb{P}}(X) > \mathbf{v}^*(d,v)$, then $p_{\mathbb{F}}^{d,v}(X) = -\infty$ for every $v > 0$. Moreover, if $\mathbb{V}_{\mathbb{P}}(X) \le \mathbf{v}^*(d,v)$ and $\mathbb{E}_{\mathbb{P}} X \ge d$, then $p^{d,v}(X) = v$ for every $v > 0$.

It thus remains to examine the case when $\mathbb{V}_{\mathbb{P}}(X) \leq \mathbf{v}^*(d, v)$ and $\mathbb{E}_{\mathbb{P}}X < d$. Notice that $\widetilde{X} = \mathbb{E}_{\mathbb{P}}X$ and thus $\pi_0(\widetilde{X}) = \mathbb{E}_{\mathbb{P}}X$. In particular, since $\widetilde{X}$ is constant, its replicating strategy is trivial, i.e. $\phi^{\widetilde{X}} = 0$.

In view of Proposition 17, if $d > v - p + \mathbb{E}_{\mathbb{P}}X$ then the minimal variance for the problem $MV(d, v, p, X)$ equals

$$\mathbf{v}^*(d, v, p, X) = \frac{(d - v + p - \mu_X)^2}{e^{\theta^2 T} - 1} + \sigma_X^2,$$

where $\mu_X = \mathbb{E}_{\mathbb{P}}X$ and $\sigma_X^2 = \mathbb{V}_{\mathbb{P}}(X) = \gamma_X$. Let us denote

$$\widetilde{p}^{d,v}(X) = -d + v + \mu_X + \sqrt{(d - v)^2 - \sigma_X^2(e^{\theta^2 T} - 1)}.$$

Proposition 19. *The mean variance price of the claim X equals*

$$p_{\mathbb{F}}^{d,v}(X) = \min\{\widetilde{p}^{d,v}(X), v\} \vee 0$$

if $(d - v)^2 - \sigma_X^2(e^{\theta^2 T} - 1) \geq 0$, and $-\infty$ otherwise. The mean-variance hedging strategy $\phi^{MV} = \psi^*$, where ψ^* is such that

$$\psi_t^{1*} = e^{\theta^2(T-t)} \frac{d - v + p^{d,v}(X) - \mu_X}{e^{\theta^2 T} - 1} \frac{\nu}{\sigma^2} \frac{\eta_t}{Z_t^1}, \quad \forall t \in [0, T].$$

The mean-variance price depends, of course, on the initial value v of the investor's capital. This dependence has very intuitive and natural properties, though. Let us denote

$$k = d - \sqrt{(d - \mu_X)^2 + \sigma_X^2(e^{\theta^2 T} - 1)}, \quad l = d - \sigma_X\sqrt{e^{\theta^2 T} - 1}.$$

We fix all parameters, except for v. Notice that the function $p(v) = p_{\mathbb{F}}^{d,v}(X)$ is non-negative and finite for $v \in [0, l \vee 0]$. Moreover, the function $p(v)$ is increasing for $v \in [0, k \vee 0)$, and it is decreasing on the interval $[k \vee 0, l \vee 0]$. Specifically,

$$p(v) = \begin{cases} v & \text{if } 0 \leq v < k \vee 0, \\ \mu_X - d + v + \sqrt{(d - v)^2 - \sigma_X^2(e^{\theta^2 T} - 1)}, & \text{if } k \vee 0 \leq v \leq l \vee 0. \end{cases}$$

This conclusion is quite intuitive: once the initial level of investor's capital is big enough (that is, $v \geq l$) the investor is less and less interested in purchasing the claim X. This is because when the initial endowment is sufficiently close to the expected terminal wealth level, the investor has enough leverage to meet this terminal objective at minimum risk; therefore, the investor is increasingly reluctant to purchase the claim X as this would introduce unwanted additional risk (unless of course $\sigma_X = 0$). For example, if $v = d$ then the investor is not at all interested in purchasing the claim ($p_{\mathbb{F}}^{v,v}(X) = -\infty$ if $\sigma_X > 0$ and $\theta \neq 0$). For further properties of the mean-variance price of a claim X independent of $\mathcal{F}_T$, we refer to Bielecki and Jeanblanc (2003).

Defaultable Bond

Let τ be a random time on the underlying probability space $(\Omega, \mathcal{G}, \mathbb{P})$. We define the indicator process H associated with τ by setting $H_t = \mathbb{1}_{\{\tau \leq t\}}$ for $t \in \mathbb{R}_+$, and we denote by $\mathbb{H}$ the natural filtration of H ($\mathbb{P}$-completed). We take $\mathbb{H}$ to serve as the auxiliary filtration, so that $\mathbb{G} = \mathbb{F} \vee \mathbb{H}$. We assume that the default time τ is defined as follows:

$$\tau = \inf \{\, t \in \mathbb{R}_+ \, : \, \Gamma_t > \zeta \,\}, \tag{80}$$

where Γ is an increasing, $\mathbb{F}$-adapted process, with $\Gamma_0 = 0$, and ζ is an exponentially distributed random variable with parameter 1, independent of $\mathbb{F}$. It is well known that any Brownian motion W with respect to $\mathbb{F}$ is also a Brownian motion with respect to $\mathbb{G}$ within the present setup (the latter property is closely related to the so-called hypothesis (H) frequently used in the modeling of default event, see Jeanblanc and Rutkowski (2000) or Bielecki et al. (2004)).

Now, suppose that a new investment opportunity becomes available for the agent. Namely, the agent may purchase a defaultable bond that matures at time $T \in (0, T^*]$. We postulate that the terminal payoff at time T of the bond is $X = L\mathbb{1}_{\{\tau > T\}} + \delta L\mathbb{1}_{\{\tau \leq T\}}$, where $L > 0$ is the bond's notional amount and $\delta \in [0, 1)$ is the (constant) recovery rate. In other words, we deal with a defaultable zero-coupon bond that is subject to the fractional recovery of Treasury value.

Notice that the payoff X can be represented as follows $X = \delta L + Y$, where $Y = L(1 - \delta)\mathbb{1}_{\{\tau > T\}}$. According to our general definition, we associate to X an $\mathcal{F}_T$-measurable random variable $\widetilde{X}$ by setting

$$\widetilde{X} = \mathbb{E}_\mathbb{P}(X \,|\, \mathcal{F}_T) = \delta L + \mathbb{E}_\mathbb{P}(Y \,|\, \mathcal{F}_T).$$

In view of (80), we have

$$\mathbb{E}_\mathbb{P}(Y \,|\, F_T) = \mathbb{P}\{\tau > T \,|\, \mathcal{F}_T\} = e^{-\Gamma_T},$$

and thus the arbitrage price at time 0 of the attainable claim $\widetilde{X}$ equals (recall that we have reduced our problem to the case $r = 0$)

$$\pi_0(\widetilde{X}) = \mathbb{E}_\mathbb{Q}\widetilde{X} = \delta L + \mathbb{E}_\mathbb{P}(\eta_T e^{-\Gamma_T}).$$

Since clearly

$$X - \widetilde{X} = L(1 - \delta)\big(\mathbb{1}_{\{\tau > T\}} - \mathbb{P}\{\tau > T \,|\, \mathcal{F}_T\}\big),$$

we obtain

$$\gamma_X = \mathbb{V}_\mathbb{P}(X - \widetilde{X}) = L^2(1 - \delta)^2 \, \mathbb{E}_\mathbb{P}\big(\mathbb{1}_{\{\tau > T\}} - e^{-\Gamma_T}\big)^2.$$

In order to find the mean-variance price $p_\mathbb{F}^{d,v}(X)$ at time 0 of a defaultable bond with respect to the reference filtration $\mathbb{F}$, it suffices to make use of Proposition 17 (or Proposition 18). If we wish to describe the mean-variance hedging strategy with respect to $\mathbb{F}$, we need also to know an explicit representation for the replicating strategy $\phi^{\widetilde{X}}$ for the claim $\widetilde{X}$. To this end, it suffices to find the integral representation of the random variable $\mathbb{E}_{\mathbb{P}}(Y \mid \mathcal{F}_T)$ with respect to the price process Z^1 or, equivalently, to find a process $\phi^{\widetilde{X}}$ for which

$$\widetilde{X} = \pi_0(\widetilde{X}) + \int_0^T \phi_t^{\widetilde{X}} \, dZ_t^1.$$

Example 4. In practical applications of the reduced-form approach, it is fairly common to postulate that the $\mathbb{F}$-hazard process Γ is given as $\Gamma_t = \int_0^t \gamma_t \, dt$, where γ is a non-negative process, progressively measurable with respect to $\mathbb{F}$, referred to as the $\mathbb{F}$-intensity of default. Suppose, for the sake of simplicity, that the intensity of default γ is deterministic, and let us set

$$p_\gamma = \mathbb{P}\{\tau > T\} = \widetilde{\mathbb{Q}}\{\tau > T\} = \exp\left(-\int_0^T \gamma(t) \, dt\right).$$

Then we get

$$\pi_0(\widetilde{X}) = \mathbb{E}_{\mathbb{Q}}\widetilde{X} = \delta L + p_\gamma$$

and

$$\gamma_X = L^2(1-\delta)^2 p_\gamma (1 - p_\gamma).$$

Of course, in the case of a deterministic default intensity γ, in order to replicate the claim $\widetilde{X}$, it suffices to invest the amount $\pi_0(\widetilde{X})$ in the savings account. For a more detailed analysis of the mean-variance price of a defaultable bond, the reader may consult Bielecki and Jeanblanc (2003).

9 Strategies Adapted to the Full Filtration

In this section, the mean-variance hedging and pricing is examined in the case of trading strategies adapted to the full filtration. Recall that W is assumed to be a one-dimensional Brownian motion with respect to $\mathbb{F}$ under $\mathbb{P}$. We postulated, in addition, that W is also a Brownian motion with respect to the filtration $\mathbb{G}$ under the probability $\mathbb{P}$. We define a new probability $\widetilde{\mathbb{Q}}$ on $(\Omega, \mathcal{G}_{T^*})$ by setting

$$\left.\frac{d\widetilde{\mathbb{Q}}}{d\mathbb{P}}\right|_{\mathcal{G}_t} = \eta_t, \quad \forall t \in [0, T^*],$$

where the process η is given by (63). Clearly, $\widetilde{\mathbb{Q}}$ is an equivalent martingale probability for our primary market and the process η is a $\mathbb{G}$-martingale under $\mathbb{P}$. Moreover, we have (cf. (64))

$$\mathbb{E}_{\mathbb{P}}(\eta_T^2 \mid \mathcal{G}_t) = \eta_t^2 e^{\theta^2(T-t)},$$

and thus $\mathbb{E}_{\mathbb{P}}(\eta_t^2) = \exp(\theta^2 t)$ for every $t \in [0, T^*]$. It is easy to check that the process $\widetilde{W}_t = W_t - \theta t$ is a martingale, and thus a Brownian motion, with respect to $\mathbb{G}$ under $\widetilde{\mathbb{Q}}$.

From the $\mathbb{P}$-square-integrability of η_T, it follows that for any strategy $\phi \in \Phi(\mathbb{G})$ the terminal wealth $V_T(\phi)$ is $\widetilde{\mathbb{Q}}$-integrable. In fact, we have the following useful result. Recall that a $\mathbb{G}$-predictable process ϕ^1 uniquely determines a self-financing strategy $\phi = (\phi^1, \phi^2)$, and thus we may formally identify ϕ^1 with the associated strategy ϕ (and vice versa). The following lemma will prove useful.

Lemma 14. *Let $\mathcal{A}(\widetilde{\mathbb{Q}})$ be the linear space of all $\mathbb{G}$-predictable processes ψ such that the process $\int_0^t \psi_u \, dZ_u^1$ is a $\widetilde{\mathbb{Q}}$-martingale and the integral $\int_0^T \psi_u \, dZ_u^1$ is in $L^2(\Omega, \mathcal{G}_T, \mathbb{P})$. Then $\mathcal{A}(\widetilde{\mathbb{Q}}) = \Phi(\mathbb{G})$.*

Proof. It is clear that $\mathcal{A}(\widetilde{\mathbb{Q}}) \subseteq \Phi(\mathbb{G})$. For the proof of the inclusion $\Phi(\mathbb{G}) \subseteq \mathcal{A}(\widetilde{\mathbb{Q}})$, see Lemma 9 in Rheinländer and Schweizer (1997). □

It is worthwhile to note that the class $\mathcal{A}(\widetilde{\mathbb{Q}})$ corresponds to the set Θ_{GLP} ($\widetilde{\Theta}$, respectively) considered in Schweizer (2001) (in Rheinländer and Schweizer (1997), respectively). The class $\Phi(\mathbb{G})$ corresponds with the class Θ_S (Θ, respectively) considered in Schweizer (2001) (in Rheinländer and Schweizer (1997), respectively).

Let us denote by $\mathbb{G}^1$ the filtration generated by all wealth processes:

$$V_t^v(\phi) = v + \int_0^t \phi_u^1 \, dZ_u^1,$$

where $v \in \mathbb{R}$ and $\phi = (\phi^1, \phi^2)$ belongs to $\Phi(\mathbb{G})$. Equivalently, $\mathbb{G}^1$ is generated by the processes

$$x + \int_0^t \psi_u \, dZ_u^1$$

with $x \in \mathbb{R}$ and $\psi \in \mathcal{A}(\widetilde{\mathbb{Q}})$. Also, we denote by $\mathcal{P}^0$ the following set of random variables:

$$\mathcal{P}^0 = \Big\{ \xi \in L^2(\Omega, \mathcal{G}_T^1, \mathbb{P}) \, \Big| \, \xi = \int_0^T \psi_u \, dZ_u^1, \, \psi \in \mathcal{A}(\widetilde{\mathbb{Q}}) \Big\}.$$

We write $\Pi_{\mathbb{P}}^0$ to denote the orthogonal projection (in the norm of the space $L^2(\Omega, \mathcal{G}_T, \mathbb{P})$) from $L^2(\Omega, \mathcal{G}_T, \mathbb{P})$ on the space $\mathcal{P}^0$. A similar notation will be also used for orthogonal projections on $\mathcal{P}^0$ under $\widetilde{\mathbb{Q}}$. Let us mention that, in general, we shall have $\Pi_{\mathbb{P}}^0(Y) \neq \mathbb{E}_{\mathbb{P}}(Y \,|\, \mathcal{G}_T^1)$ for $Y \in L^2(\Omega, \mathcal{G}_T, \mathbb{P})$ and $\Pi_{\widetilde{\mathbb{Q}}}^0(Y) \neq \mathbb{E}_{\widetilde{\mathbb{Q}}}(Y \,|\, \mathcal{G}_T^1)$ for $Y \in L^2(\Omega, \mathcal{G}_T, \widetilde{\mathbb{Q}})$ (see Section 9.3 for more details).

9.1 Solution to MV(d, v) in the Class $\Phi(\mathbb{G})$

Recall that our basic mean-variance problem has the following form:

Problem MV(d, v)**:** Minimize $\mathbb{V}_{\mathbb{P}}(V_T^v(\phi))$ over all strategies $\phi \in \Phi(\mathbb{G})$, subject to $\mathbb{E}_{\mathbb{P}}V_T^v(\phi) \geq d$.

As in Section 8.1, we postulate that $d > v$, since otherwise the problem is trivial. We shall argue that it suffices to solve a simpler problem:

Problem MVA(d, v)**:** Minimize $\mathbb{E}_{\mathbb{P}}(V_T^v(\phi))^2$ over all strategies $\phi \in \Phi(\mathbb{G})$, subject to $\mathbb{E}_{\mathbb{P}}V_T^v(\phi) = d$.

In view of the definition of class $\mathcal{A}(\widetilde{\mathbb{Q}})$, Lemma 14, and the fact that $\mathbb{E}_{\mathbb{Q}}\xi = 0$ for any $\xi \in \mathcal{P}^0$, we see that it suffices to solve the problem

Problem MVB(d, v)**:** Minimize $\mathbb{E}_{\mathbb{P}}(v + \xi)^2$ over all random variables $\xi \in \mathcal{P}^0$, subject to $\mathbb{E}_{\mathbb{P}}\xi = d - v$.

Solution to the last problem is exactly the same as in the case of strategies from $\Phi(\mathbb{F})$. Indeed, by solving the last problem in the class $L^2(\Omega, \mathcal{G}_T, \mathbb{P})$ (rather than in $\mathcal{P}^0$), and with additional constraint $\mathbb{E}_{\mathbb{Q}}\xi = 0$, we see that the optimal solution, given by (68), is in fact $\mathcal{F}_T$-measurable, and thus it belongs to the class $\mathcal{P}^0$ as well. In view of (69), the same random variable is a solution to $\mathrm{MV}(d, v)$, that is, it represents the optimal terminal wealth. We conclude that a solution to $\mathrm{MV}(d, v)$ in the class $\Phi(\mathbb{G})$ is given by the formulae (72)-(74) of Proposition 16, i.e., it coincides with a solution in the class $\Phi(\mathbb{F})$.

Assume that X is an attainable contingent claim, in the sense that there exists a trading strategy $\phi \in \Phi(\mathbb{G})$ which replicates X. Then, arguing along the same lines as in Section 8.2, we get the following result.

Corollary 10. *Let a $\mathcal{G}_T$-measurable random variable X represent an attainable contingent claim. Then*
(i) If the arbitrage price $\pi_0(X)$ satisfies $\pi_0(X) \in [0, v]$ then $p^{d,v}(X) = \pi_0(X)$.
(ii) If the arbitrage price $\pi_0(X)$ is strictly greater than v then $p^{d,v}(X) = v$.

9.2 Solution to MV(d, v, p, X) in the Class $\Phi(\mathbb{G})$

We shall study the problem $\mathrm{MV}(d, v, p, X)$ for an arbitrary $\mathcal{G}_T$-measurable claim X, which is $\mathbb{P}$-square-integrable. Recall that we deal with the following problem:

Problem MV(d, v, p, X)**:** Minimize $\mathbb{V}_{\mathbb{P}}(V_T^{v-p}(\phi) + X)$ over all trading strategies $\phi \in \Phi(\mathbb{G})$, subject to $\mathbb{E}_{\mathbb{P}}(V_T^{v-p}(\phi) + X) \geq d$.

Basic idea of solving the problem $\mathrm{MV}(d, v, p, X)$ with respect to $\mathbb{G}$-predictable strategies is similar to that used in the case of $\mathbb{F}$-predictable strategies. The main difference is that the auxiliary random variable $\widetilde{X}$ will now be defined as the orthogonal projection $\Pi_{\mathbb{P}}(X)$ of X on $\mathcal{P}^0$, rather than the conditional expectation $\mathbb{E}_{\mathbb{P}}(X \mid \mathcal{G}_T)$.

Let us denote $\widehat{d} = d - v + p$. The problem $\mathrm{MV}(d, v, p, X)$ can be reformulated as follows:

Problem $\mathrm{MV}(\widehat{d}, 0, 0, X)$: Minimize $\mathbb{V}_{\mathbb{P}}(V_T^0(\phi) + X)$ over all trading strategies $\phi \in \Phi(\mathbb{G})$, subject to $\mathbb{E}_{\mathbb{P}}(V_T^0(\phi) + X) \geq \widehat{d}$.

That is, if $V_T^{0,*}$ is the optimal wealth in problem $\mathrm{MV}(\widehat{d}, 0, 0, X)$ then $V_T^{v-p,*} = V_T^{0,*} + v - p$ is the optimal wealth in problem $\mathrm{MV}(d, v, p, X)$, and the optimal strategies as well as the optimal variances are the same in both problems.

Let $\widetilde{X}^0 = \Pi_{\mathbb{P}}^0(X)$ stand for the orthogonal projection of X on $\mathcal{P}^0$, so that $\psi^{\widetilde{X}^0}$ is a process from $\mathcal{A}(\widetilde{\mathbb{Q}}) = \Phi(\mathbb{G})$, for which

$$\widetilde{X}^0 = \int_0^T \psi_t^{1, \widetilde{X}^0} \, dZ_t^1 \tag{81}$$

and $X - \widetilde{X}^0 = X - \Pi_{\mathbb{P}}^0(X)$ is orthogonal to $\mathcal{P}^0$. The price of $\widetilde{X}^0$ equals

$$\pi_t(\widetilde{X}^0) = \int_0^t \psi_u^{1, \widetilde{X}^0} \, dZ_u^1 = \mathbb{E}_{\widetilde{\mathbb{Q}}}(\widetilde{X}^0 \mid \mathcal{G}_t), \quad \forall t \in [0, T]. \tag{82}$$

Let $\psi^{\widetilde{X}^0} \in \Phi(\mathbb{G})$ be a replicating strategy for the claim $\widetilde{X}^0$. Explicitly, $\psi^{\widetilde{X}^0} = (\psi^{1, \widetilde{X}^0}, \psi^{2, \widetilde{X}^0})$, where $\psi^{2, \widetilde{X}^0}$ satisfies $\psi_t^{1, \widetilde{X}^0} Z_t^1 + \psi_t^{2, \widetilde{X}^0} = \pi_t(\widetilde{X}^0)$. Notice that $\pi_0(\widetilde{X}^0) = \mathbb{E}_{\widetilde{\mathbb{Q}}} \widetilde{X}^0 = 0$ and, of course, $\pi_T(\widetilde{X}^0) = \widetilde{X}^0$. It thus suffices to consider the following problem:

Problem $\mathrm{MV}(\widehat{d}, 0, 0, X - \widetilde{X}^0)$: Minimize $\mathbb{V}_{\mathbb{P}}(V_T^0(\phi) + X - \widetilde{X}^0)$ over all trading strategies $\phi \in \Phi(\mathbb{G})$, subject to $\mathbb{E}_{\mathbb{P}}(V_T^0(\phi) + X - \widetilde{X}^0) \geq \widehat{d}$.

Since $X - \widetilde{X}^0$ is orthogonal to $\mathcal{P}^0$, for any strategy $\phi \in \Phi(\mathbb{G})$ we have

$$\mathbb{V}_{\mathbb{P}}(V_T^0(\phi) + X - \widetilde{X}^0) = \mathbb{V}_{\mathbb{P}}(V_T^0(\phi)) + \mathbb{V}_{\mathbb{P}}(X - \widetilde{X}^0) = \mathbb{V}_{\mathbb{P}}(V_T^0(\phi)) + \gamma_X^0,$$

where $\gamma_X^0 = \mathbb{V}_{\mathbb{P}}(X - \widetilde{X}^0)$. Let us denote $\widetilde{d} = d - v + p - \mathbb{E}_{\mathbb{P}} X + \mathbb{E}_{\mathbb{P}} \widetilde{X}^0$. Then the problem $\mathrm{MV}(\widehat{d}, 0, 0, X - \widetilde{X}^0)$ can thus be simplified as follows:

Problem $\mathrm{MV}(\widetilde{d}, 0, \gamma_X^0)$: Minimize $\mathbb{V}_{\mathbb{P}}(V_T^0(\phi)) + \gamma_X^0$ over all trading strategies $\phi \in \Phi(\mathbb{G})$, subject to $\mathbb{E}_{\mathbb{P}}(V_T^0(\phi)) \geq \widetilde{d} = d - v + p - \mathbb{E}_{\mathbb{P}} X + \mathbb{E}_{\mathbb{P}} \widetilde{X}^0$.

Let us write $\widetilde{v} = v - p - \mathbb{E}_{\mathbb{P}} X + \mathbb{E}_{\mathbb{P}} \widetilde{X}^0$, so that $\widetilde{d} = d - \widetilde{v}$. Then the minimal variance for the problem $\mathrm{MV}(d, v, p, X)$ equals

$$\mathbf{v}^*(d, v, p, X) = \mathbf{v}^*(\widetilde{d}, 0) + \gamma_X^0 = \mathbf{v}^*(d, \widetilde{v}) + \gamma_X^0.$$

Moreover, if ψ^* is an optimal strategy to $\mathrm{MV}(\widetilde{d}, 0)$, then $\phi^{1*} = \psi^{1*} - \psi^{\widetilde{X}^0}$ is a solution to $\mathrm{MV}(d, v, p, X)$. The proof of the next proposition is based on the considerations above, combined with Proposition 16. We use the standard notation $\rho(\theta) = e^{\theta^2 T}(e^{\theta^2 T} - 1)^{-1}$ and $\eta_t(\theta) = \eta_t e^{-\theta^2 t}$, so that $\eta_0(\theta) = 1$. Recall that $\mathbb{E}_{\widetilde{\mathbb{Q}}}\widetilde{X}^0 = 0$.

Proposition 20. *Assume that $\theta \neq 0$ and let $\psi^{\widetilde{X}^0} \in \Phi(\mathbb{G})$ be a replicating strategy for $\widetilde{X}^0 = \Pi_{\mathbb{P}}^0(X)$.*
(i) Suppose that $d > \widetilde{v}$. Then an optimal strategy $\phi^*(d, v, p, X)$ for the problem $MV(d, v, p, X)$ is given as $\phi^{1*}(d, v, p, X) = \psi^{1*}(\widetilde{d}, 0) - \psi^{1,\widetilde{X}^0}$ with $\psi^*(\widetilde{d}, 0) = (\psi^{1*}(\widetilde{d}, 0), \psi^{2*}(\widetilde{d}, 0))$ such that $\psi^{1*}(\widetilde{d}, 0)$ equals

$$\psi_t^{1*}(\widetilde{d}, 0) = (d - \widetilde{v})\rho(\theta)\frac{\nu\eta_t(\theta)}{\sigma^2 Z_t^1} \tag{83}$$

and $\psi^{2*}(\widetilde{d}, 0)$ satisfies $\psi_t^{*1}(\widetilde{d}, 0)Z_t^1 + \psi^{*2}(\widetilde{d}, 0) = V_t^*(\widetilde{d}, 0)$, where in turn

$$V_t^*(\widetilde{d}, 0) = (d - \widetilde{v})\rho(\theta)\big(1 - \eta_t(\theta)\big). \tag{84}$$

Thus the optimal wealth for the problem $MV(d, v, p, X)$ equals

$$V_t^*(d, v, p, X) = v - p + (d - \widetilde{v})\rho(\theta)\big(1 - \eta_t(\theta)\big) - \mathbb{E}_{\widetilde{\mathbb{Q}}}(\widetilde{X}^0 \mid \mathcal{G}_t). \tag{85}$$

The minimal variance $\mathbf{v}^*(d, v, p, X)$ is given by

$$\mathbf{v}^*(d, v, p, X) = \frac{(d - \widetilde{v})^2}{e^{\theta^2 T} - 1} + \gamma_X^0. \tag{86}$$

(ii) If $d \leq \widetilde{v}$ then the optimal wealth process equals

$$V_t^*(d, v, p, X) = v - p - \mathbb{E}_{\widetilde{\mathbb{Q}}}(\widetilde{X}^0 \mid \mathcal{G}_t)$$

and the minimal variance equals γ_X^0.

Remark. It is natural to expect that the optimal variance given in (86) is not greater than the optimal variance given in (78). In fact, this is the case (see Proposition 5.4 in Bielecki and Jeanblanc (2003)).

Of course, the practical relevance of the last result hinges on the availability of explicit representation for the orthogonal projection $\widetilde{X}^0 = \Pi_{\mathbb{P}}^0(X)$ of X on the space $\mathcal{P}^0$. This important issue will be examined in the next section in a general setup. We shall continue the study of this question in the framework of defaultable claims in Section 9.5.

9.3 Projection of a Generic Claim

Let us first recall two well-known result concerning the decomposition of a $\mathcal{G}_T$-measurable random variable, which represents a generic contingent claim in our financial model.

Galtchouk-Kunita-Watanabe decomposition under $\widetilde{\mathbb{Q}}$. Suppose first that we work under $\mathbb{Q}$, so that the process Z^1 is a continuous martingale. Recall that by assumption W is a Brownian motion with respect to $\mathbb{G}$ under $\mathbb{P}$; hence, the process $\widetilde{W}$ is a Brownian motion with respect to $\mathbb{G}$ under $\widetilde{\mathbb{Q}}$.

It is well known that any random variable $Y \in L^2(\Omega, \mathcal{G}_T, \widetilde{\mathbb{Q}})$ can be represented by means of the *Galtchouk-Kunita-Watanabe decomposition* with respect to the martingale Z^1 under $\widetilde{\mathbb{Q}}$. To be more specific, for any random variable $Y \in L^2(\Omega, \mathcal{G}_T, \widetilde{\mathbb{Q}})$ there exists a $\mathbb{G}$-martingale $N^{Y,\widetilde{\mathbb{Q}}}$, which is strongly orthogonal in the martingale sense to Z^1 under $\widetilde{\mathbb{Q}}$, and a $\mathbb{G}$-adapted process $\psi^{Y,\widetilde{\mathbb{Q}}}$, such that Y can be represented as follows:

$$Y = \mathbb{E}_{\widetilde{\mathbb{Q}}} Y + \int_0^T \psi_t^{Y,\widetilde{\mathbb{Q}}} \, dZ_t^1 + N_T^{Y,\widetilde{\mathbb{Q}}}. \tag{87}$$

Furthermore, the process $\psi^{Y,\widetilde{\mathbb{Q}}}$ can be represented as follows:

$$\psi_t^{Y,\widetilde{\mathbb{Q}}} = \frac{d\langle \mathcal{Y}, Z^1 \rangle_t}{d\langle Z^1 \rangle_t}, \tag{88}$$

where the $\mathbb{G}$-martingale $\mathcal{Y}$ is defined as $\mathcal{Y}_t = \mathbb{E}_{\widetilde{\mathbb{Q}}}(Y \mid \mathcal{G}_t)$.

Föllmer-Schweizer decomposition under $\mathbb{P}$. Let us now consider the same issue, but under the original probability $\mathbb{P}$. The process Z^1 is a (continuous) semimartingale with respect to $\mathbb{G}$ under $\mathbb{P}$, and thus it admits a unique continuous martingale part under $\mathbb{P}$.

Any random variable $Y \in L^2(\Omega, \mathcal{G}_T, \mathbb{P})$ can be represented by means of the *Föllmer-Schweizer decomposition*. Specifically, there exists a $\mathbb{G}$-adapted process $\psi^{Y,\mathbb{P}}$, a $(\mathbb{G}, \mathbb{P})$-martingale $N^{Y,\mathbb{P}}$, strongly orthogonal in the martingale sense to the continuous martingale part of Z^1, and a constant $y^{Y,\mathbb{P}}$, so that

$$Y = y^{Y,\mathbb{P}} + \int_0^T \psi_t^{Y,\mathbb{P}} \, dZ_t^1 + N_T^{Y,\mathbb{P}}. \tag{89}$$

We shall see that it will be not necessary to compute the process $\psi^{Y,\mathbb{P}}$ for the purpose of finding a hedging strategy for the problem considered in this section.

Projection on $\mathcal{P}^0$. As already mentioned, $\Pi_{\widetilde{\mathbb{Q}}}^0(Y) \neq \mathbb{E}_{\widetilde{\mathbb{Q}}}(Y \mid \mathcal{G}_T^1)$ for random variables Y in $L^2(\Omega, \mathcal{G}_T, \widetilde{\mathbb{Q}})$, as well as $\Pi_{\mathbb{P}}^0(Y) \neq \mathbb{E}_{\mathbb{P}}(Y \mid \mathcal{G}_T^1)$ for $Y \in L^2(\Omega, \mathcal{G}_T, \mathbb{P})$, in general. For instance, for any random variable Y as in (87) we get $\Pi_{\widetilde{\mathbb{Q}}}^0(Y) = \int_0^T \psi_t^{Y,\widetilde{\mathbb{Q}}} \, dZ_t^1$, whereas

$$\mathbb{E}_{\widetilde{\mathbb{Q}}}(Y \mid \mathcal{G}_T^1) = Y = \Pi_{\widetilde{\mathbb{Q}}}^0(Y) - \mathbb{E}_{\widetilde{\mathbb{Q}}} Y.$$

The projection $\Pi_{\widetilde{\mathbb{Q}}}^0(Y)$ differs here from the conditional expectation just by the expected value $\mathbb{E}_{\widetilde{\mathbb{Q}}} Y$. Consequently, we have $\Pi_{\widetilde{\mathbb{Q}}}^0(Y) = \mathbb{E}_{\widetilde{\mathbb{Q}}}(Y \mid \mathcal{G}_T^1)$ for any $Y \in$

$L^2(\Omega, \mathcal{G}_T, \widetilde{\mathbb{Q}})$ with $\mathbb{E}_{\widetilde{\mathbb{Q}}} Y = 0$. More importantly, observe that for Y as in (89) we shall have, in general,

$$\Pi_{\mathbb{P}}^0(Y) \neq \int_0^T \psi_t^{Y,\mathbb{P}} \, dZ_t^1,$$

so that, in particular, $\Pi_{\mathbb{P}}^0(Y) \neq \mathbb{E}_{\mathbb{P}}(Y \mid \mathcal{G}_T^1)$ even if $\mathbb{E}_{\mathbb{P}} Y = 0$.

Our next goal is to compute the projection $\Pi_{\mathbb{P}}^0(Y)$ for any random variable $Y \in L^2(\Omega, \mathcal{G}_T, \mathbb{Q})$. We know that any such Y can be represented as in (87). Due to linearity of the projection, it is enough to compute the projection of each component in the right-hand side of (87). Let us set $\widetilde{\eta}_t = \mathbb{E}_{\widetilde{\mathbb{Q}}}(\eta_T \mid \mathcal{G}_t)$ for every $t \in [0, T]$, so that, in particular, $\widetilde{\eta}_T = \eta_T$. Since $\widetilde{\eta}$ is a square-integrable $\mathbb{G}$-martingale under $\widetilde{\mathbb{Q}}$, there exists a process $\widetilde{\psi}$ in $\mathcal{A}(\widetilde{\mathbb{Q}})$ such that

$$\widetilde{\eta}_t = \mathbb{E}_{\widetilde{\mathbb{Q}}} \widetilde{\eta}_T + \int_0^t \widetilde{\psi}_u \, dZ_u^1 = \mathbb{E}_{\widetilde{\mathbb{Q}}} \widetilde{\eta}_T + Z_t^\eta, \quad \forall t \in [0, T], \tag{90}$$

where we denote

$$Z_t^\eta = \int_0^t \widetilde{\psi}_u \, dZ_u^1.$$

Lemma 15. *We have*

$$\widetilde{\psi}_t = -\frac{\theta \widetilde{\eta}_t}{\sigma Z_t^1} = -\frac{\theta e^{\theta^2 T}}{\sigma Z_t^1} \exp\left(-\theta \widetilde{W}_t - \tfrac{1}{2}\theta^2(t - 2T)\right) \tag{91}$$

and the process $\widetilde{W}_t = W_t + \theta t$ is a Brownian motion under $\widetilde{\mathbb{Q}}$.

Proof. Direct calculations show that for every $t \in [0, T]$

$$\widetilde{\eta}_t = \exp\left(-\frac{\theta}{\sigma} \int_0^t \frac{dZ_u^1}{Z_u^1} - \frac{1}{2}\theta^2(t - 2T)\right) = e^{\theta^2 T} \exp\left(-\theta \widetilde{W}_t - \tfrac{1}{2}\theta^2 t\right). \tag{92}$$

Hence, $\widetilde{\eta}$ solves the SDE

$$d\widetilde{\eta}_t = -\theta \widetilde{\eta}_t \, d\widetilde{W}_t = -\frac{\theta}{\sigma} \frac{\widetilde{\eta}_t}{Z_t^1} \, dZ_t^1$$

with the initial condition $\widetilde{\eta}_0 = \mathbb{E}_{\widetilde{\mathbb{Q}}} \widetilde{\eta}_T = \mathbb{E}_{\widetilde{\mathbb{Q}}} \eta_T = e^{\theta^2 T}$. $\qquad\square$

In the next result, we provide a general representation for the projection $\Pi_{\mathbb{P}}^0(Y)$ for a $\mathcal{G}_T$-measurable random variable Y, which is $\mathbb{P}$-square-integrable.

Proposition 21. *Let $Y \in L^2(\Omega, \mathcal{G}_T, \mathbb{P})$. Then we have*

$$\Pi_{\mathbb{P}}^0(Y) = \int_0^T \widetilde{\psi}_t^{Y,\mathbb{P}} \, dZ_t^1,$$

where

$$\widetilde{\psi}_t^{Y,\mathbb{P}} = \psi_t^{Y,\widetilde{\mathbb{Q}}} - \widetilde{\psi}_t \left(\widetilde{\eta}_0^{-1} \mathbb{E}_{\widetilde{\mathbb{Q}}} Y + \int_0^t \widetilde{\eta}_u^{-1} \, dN_u^{Y,\widetilde{\mathbb{Q}}} \right) \tag{93}$$

and where processes $\psi^{Y,\widetilde{\mathbb{Q}}}$ and $\underline{N}^{Y,\widetilde{\mathbb{Q}}}$ are given by the Galtchouk-Kunita-Watanabe decomposition (87) of Y under $\widetilde{\mathbb{Q}}$.

Proof. First, we compute projection of the constant $c = \mathbb{E}_{\widetilde{\mathbb{Q}}} Y$. To this end, recall that $\widetilde{\eta}_T = \eta_T$ and by virtue of (90) we have $\widetilde{\eta}_T = \widetilde{\eta}_0 + Z_T^\eta$. Hence, for any $\psi \in \mathcal{A}(\widetilde{\mathbb{Q}})$ we obtain

$$\mathbb{E}_{\mathbb{P}} \left((1 + \widetilde{\eta}_0^{-1} Z_T^\eta) \int_0^T \psi_t \, dZ_t^1 \right) = \widetilde{\eta}_0^{-1} \mathbb{E}_{\mathbb{P}} \left(\eta_T \int_0^T \psi_t \, dZ_t^1 \right)$$

$$= \widetilde{\eta}_0^{-1} \mathbb{E}_{\widetilde{\mathbb{Q}}} \left(\int_0^T \psi_t \, dZ_t^1 \right) = 0,$$

and thus $\varPi_{\mathbb{P}}^0(1) = -\widetilde{\eta}_0^{-1} Z_T^\eta$. We conclude that for any $c \in \mathbb{R}$

$$\varPi_{\mathbb{P}}^0(c) = c\,\varPi_{\mathbb{P}}^0(1) = -c\widetilde{\eta}_0^{-1} Z_T^\eta = -c\widetilde{\eta}_0^{-1} \int_0^T \widetilde{\psi}_t \, dZ_t^1. \tag{94}$$

Next, it is obvious that the projection of the second term, that is, the projection of $\int_0^T \psi_t^{Y,\widetilde{\mathbb{Q}}} \, dZ_t^1$, on $\mathcal{P}^0$ is equal to itself, so that

$$\varPi_{\mathbb{P}}^0 \left(\int_0^T \psi_t^{Y,\widetilde{\mathbb{Q}}} \, dZ_t^1 \right) = \int_0^T \psi_t^{Y,\widetilde{\mathbb{Q}}} \, dZ_t^1. \tag{95}$$

Finally, we shall compute the projection $\varPi_{\mathbb{P}}^0(N_T^{Y,\widetilde{\mathbb{Q}}})$. Recall that the process $N^{Y,\widetilde{\mathbb{Q}}}$ is a $\widetilde{\mathbb{Q}}$-martingale strongly orthogonal to Z^1 under $\widetilde{\mathbb{Q}}$. Hence, for any $N^{Y,\widetilde{\mathbb{Q}}}$-integrable process ν and any process $\psi \in \mathcal{A}(\widetilde{\mathbb{Q}})$ we have

$$\mathbb{E}_{\mathbb{P}} \left(\eta_T \int_0^T \nu_t \, dN_t^{Y,\widetilde{\mathbb{Q}}} \int_0^T \psi_t \, dZ_t^1 \right) = 0.$$

Thus, it remains to find processes $\widehat{\nu}$ and $\widehat{\psi} \in \mathcal{A}(\widetilde{\mathbb{Q}})$ for which

$$\eta_T \int_0^T \widehat{\nu}_t \, dN_t^{Y,\widetilde{\mathbb{Q}}} = N_T^{Y,\widetilde{\mathbb{Q}}} - \int_0^T \widehat{\psi}_t \, dZ_t^1, \tag{96}$$

in which case we shall have that $\varPi_{\mathbb{P}}^0(N_T^{Y,\widetilde{\mathbb{Q}}}) = \int_0^T \widehat{\psi}_t \, dZ_t^1$.

Let us set $U_t = \widetilde{\eta}_t \int_0^t \nu_u \, dN_u^{Y,\widetilde{\mathbb{Q}}}$ for every $t \in [0,T]$. Recall that (see (90)) there exists a process $\widetilde{\psi}$ in $\varPhi(\mathbb{G}) = \mathcal{A}(\widetilde{\mathbb{Q}})$ such that $d\widetilde{\eta}_t = \widetilde{\psi}_t \, dZ_t^1$. Using the product rule, and taking into account the orthogonality of $\widetilde{\eta}$ and $N^{Y,\widetilde{\mathbb{Q}}}$ under $\widetilde{\mathbb{Q}}$, we find that U is a local martingale under $\widetilde{\mathbb{Q}}$, and it satisfies

$$U_t = \int_0^t \widetilde{\eta}_{u-}\nu_u \, dN_u^{Y,\widetilde{\mathbb{Q}}} + \int_0^t \left(\int_0^u \nu_s \, dN_s^{Y,\widetilde{\mathbb{Q}}} \right) \widetilde{\psi}_u \, dZ_u^1. \tag{97}$$

Consequently, upon letting

$$\widehat{\nu}_t = (\widetilde{\eta}_{t-})^{-1}, \quad \forall\, t \in [0, T], \tag{98}$$

we obtain from (97)

$$U_t = N_t^{Y,\widetilde{\mathbb{Q}}} + \int_0^t \widetilde{\psi}_u \left(\int_0^u \widehat{\nu}_s \, dN_s^{Y,\widetilde{\mathbb{Q}}} \right) dZ_u^1. \tag{99}$$

Note that the left-hand side of (96) is equal to U_T. Thus, comparing (96) and (99), we see that we may take

$$\widehat{\psi}_t = -\widetilde{\psi}_t \int_0^t \widehat{\nu}_u \, dN_u^{Y,\widetilde{\mathbb{Q}}} = -\widetilde{\psi}_t \int_0^t (\widetilde{\eta}_{u-})^{-1} \, dN_u^{Y,\widetilde{\mathbb{Q}}}. \tag{100}$$

It is clear that with $\widehat{\nu}$ defined in (98) the integral $\int_0^t \widehat{\nu}_u \, dN_u^{Y,\widetilde{\mathbb{Q}}}$ is a $\widetilde{\mathbb{Q}}$-martingale. Thus, the process U is a martingale, rather than a local martingale, under $\widetilde{\mathbb{Q}}$. Together with (99) this implies that the process

$$\int_0^t \widetilde{\psi}_u \left(\int_0^u \widehat{\nu}_s \, dN_s^{Y,\widetilde{\mathbb{Q}}} \right) dZ_u^1$$

is a $\widetilde{\mathbb{Q}}$-martingale. Consequently, the process $\widehat{\psi}$ defined in (100) belongs to the class $\mathcal{A}(\widetilde{\mathbb{Q}})$. To complete the proof, it suffices to combine (94), (95) and (100).   □

It should be acknowledged that the last result is not new. In fact, it is merely a special case of Theorem 6 in Rheinländer and Schweizer (1997). We believe, however, that our derivation of the result sheds a new light on the structure of the orthogonal projection computed above.

Remark. Although the above proposition provides us with the structure of the projection $\Pi_{\mathbb{P}}^0(Y)$, it is not easy in general to obtain closed-form expressions for the components on the right-hand side of (93) in terms of the initial data for the problem. Thus, one may need to resort to numerical approximations, which in principle can be obtained by solving the following problem

$$\min_{\xi \in \mathcal{P}^0} \mathbb{E}_{\mathbb{P}}(Y - \xi)^2. \tag{101}$$

An approximate solution to the last problem yields a process, say $\psi^{Y,\mathbb{P}}$, so that $\Pi_{\mathbb{P}}^0(Y) \approx \int_0^T \psi_t^{Y,\mathbb{P}} \, dZ_t^1$.

9.4 Mean-Variance Pricing and Hedging of a Generic Claim

Let us define

$$\widetilde{\kappa} = \widetilde{d} - v = d - v - \mathbb{E}_\mathbb{P} X + \mathbb{E}_\mathbb{P} \widetilde{X}^0.$$

For simplicity, we shall only consider the case when $\widetilde{\kappa} > 0$. This is equivalent to assuming that $\widetilde{d} > v - p$ for all $p \in [0, v]$. Thus, the results of Proposition 20 (i) apply. Consequently, denoting

$$\widetilde{\rho} = (d - v)^2 - \gamma_X^0 (e^{\theta^2 T} - 1),$$

we obtain the following result.

Proposition 22. *Suppose that* $\gamma_X^0 \leq (d - v)^2 (e^{\theta^2 T} - 1)^{-1}$. *Then the buyer's mean variance price is*

$$p^{d,v}(X) = \min\{-\widetilde{\kappa}_1 + \sqrt{\widetilde{\rho}}, v\} \vee 0. \tag{102}$$

Otherwise, $p^{d,v}(X) = -\infty$.

In case when $\gamma_X^0 \leq (d - v)^2 (e^{\theta^2 T} - 1)^{-1}$, the mean-variance hedging strategy for a generic claim X is given by $\phi^*(d, v, p^{d,v}(X), X)$, where the process ϕ^* is defined in Proposition 20. The projection part of the strategy $\phi^*(d, v, p^{d,v}(X), X)$, that is, the process $\psi^{1,\widetilde{X}^0}$, can be computed according to (93).

9.5 Projections of Defaultable Claims

In this section, we adopt the framework of Section 8.3. In particular, the default time τ is a random time on $(\Omega, \mathcal{G}, \mathbb{P})$ given by formula (80), and the process H is given as $H_t = \mathbb{1}_{\{\tau \leq t\}}$ for every $t \in [0, T]$. The natural filtration $\mathbb{H}$ of H is an auxiliary filtration, so that $\mathbb{G} = \mathbb{F} \vee \mathbb{H}$. Recall that we have assumed that τ admits the $\mathbb{F}$-hazard process Γ under $\mathbb{P}$ and thus also, in view of the construction (80), under $\widetilde{\mathbb{Q}}$. Suppose, in addition, that the hazard process Γ is an increasing continuous process. Then the process $M_t = H_t - \Gamma_{t \wedge \tau}$ is known to be a $\mathbb{G}$-martingale under $\widetilde{\mathbb{Q}}$. Any $\mathcal{G}_T$-measurable random variable X is referred to as a *defaultable claim*.

Recall that the process $\widetilde{W}_t = W_t + \theta t$ is a Brownian motion with respect to $\mathbb{F}$ under $\widetilde{\mathbb{Q}}$, and thus the process Z^1 is a square-integrable $\mathbb{G}$-martingale under $\widetilde{\mathbb{Q}}$, since

$$dZ_t^1 = Z_t^1 \sigma \, d\widetilde{W}_t, \quad Z_0^1 > 0.$$

The following proposition is an important technical result.

Proposition 23. *The filtration* $\mathbb{G}^1$ *is equal to the filtration* $\mathbb{G}$, *that is,* $\mathcal{G}_t^1 = \mathcal{G}_t$ *for every* $t \in \mathbb{R}_+$.

Proof. It is clear that $\mathbb{G}^1 \subseteq \mathbb{G}$. For a fixed $T > 0$, let $y_1, y_2 \in \mathbb{R}$ and let the processes ψ^1, ψ^2 belong to $\mathcal{A}(\widetilde{\mathbb{Q}})$. Thus the processes

$$Y_t^1 = y_1 + \int_0^t \psi_u^1 \, dZ_u^1, \quad Y_t^2 = y_2 + \int_0^t \psi_u^2 \, dZ_u^1$$

be $\mathbb{G}^1$-adapted processes. Then the process

$$Y_t^1 Y_t^2 = y_1 y_2 + \int_0^t Y_u^1 \psi_u^2 \, dZ_u^1 + \int_0^t Y_u^2 \psi_u^1 \, dZ_u^1 + \int_0^t \psi_u^1 \psi_u^2 \, d\langle Z^1 \rangle_u$$

is also $\mathbb{G}^1$-adapted. It is easy to check that the processes

$$\int_0^t Y_u^1 \psi_u^2 \, dZ_u^1, \quad \int_0^t Y_u^2 \psi_u^1 \, dZ_u^1$$

are $\mathbb{G}^1$-adapted. We thus conclude that for any processes ϕ and ψ from $\mathcal{A}(\widetilde{\mathbb{Q}})$, the process

$$\int_0^t \psi_u^1 \psi_u^2 \, d\langle Z^1 \rangle_u = \int_0^t \psi_u^1 \psi_u^2 (Z_u^1)^2 \sigma^2 \, du$$

is $\mathbb{G}^1$-adapted as well. In particular, it follows that for any bounded $\mathbb{G}$-adapted process ζ the integral $\int_0^t \zeta_u \, du$ defines a $\mathbb{G}^1$-adapted process. Let us take $\zeta_u = H_u$. Then we obtain that the process $\tau \wedge t$ is $\mathbb{G}^1$-adapted. Hence, it is easily seen that $\mathcal{G}_t \subseteq \mathcal{G}_t^1$ for $t \in [0, T]$. Since T was an arbitrary positive number, we have shown that $\mathbb{G} = \mathbb{G}^1$.   □

Projection of a Survival Claim

We shall now compute the process $\psi^{Y,\mathbb{P}}$, which occurs in the projection $\Pi_{\mathbb{P}}^0(Y)$ for a random variable $Y = Z \mathbb{1}_{\{\tau > T\}}$, where $Z \in L^2(\Omega, \mathcal{F}_T, \widetilde{\mathbb{Q}})$. It is known that any random variable Y from $L^2(\Omega, \mathcal{G}_T, \widetilde{\mathbb{Q}}) = L^2(\Omega, \mathcal{G}_T^1, \widetilde{\mathbb{Q}})$, which vanishes on the set $\{\tau > T\}$, can indeed be represented in this way. Any random variable Y of the form $Z \mathbb{1}_{\{\tau > T\}}$ is referred to as a *survival claim* with maturity date T, and a random variable Z is said to be the *promised payoff* associated with Y.

It is known (see, e.g., Bielecki and Rutkowski (2004)) that

$$\mathbb{E}_{\widetilde{\mathbb{Q}}}(Y \,|\, \mathcal{G}_t) = \mathbb{E}_{\widetilde{\mathbb{Q}}}(Z \mathbb{1}_{\{\tau > T\}} \,|\, \mathcal{G}_t) = \mathbb{E}_{\widetilde{\mathbb{Q}}}(Z \mathbb{1}_{\{\tau > T\}} \,|\, \mathcal{G}_t^1)$$

$$= \mathbb{1}_{\{\tau > t\}} e^{\Gamma_t} \mathbb{E}_{\widetilde{\mathbb{Q}}}(Z e^{-\Gamma_T} \,|\, \mathcal{F}_t) = L_t m_t^Z,$$

where $L_t := \mathbb{1}_{\{\tau > t\}} e^{\Gamma_t}$ is a $\mathbb{G}$-martingale and $m_t^Z = \mathbb{E}_{\widetilde{\mathbb{Q}}}(Z e^{-\Gamma_T} \,|\, \mathcal{F}_t)$ is an $\mathbb{F}$-martingale. From the predictable representation theorem for a Brownian motion (or since the default-free market is complete), it follows that there exists an $\mathbb{F}$-adapted process μ^Z such that

$$m_t^Z = m_0^Z + \int_0^t \mu_u^Z \, dZ_u^1. \tag{103}$$

In Proposition 21, we have already described the structure of the process $\psi^{Y,\mathbb{P}}$ that specifies the projection of Y on $\mathcal{P}^0$. In the next two results, we shall give more explicit formulae for $\psi^{Y,\widetilde{\mathbb{Q}}}$ and $N^{Y,\widetilde{\mathbb{Q}}}$ within the present setup.

Lemma 16. *Consider a survival claim* $Y = Z\mathbb{1}_{\{\tau > T\}}$ *with the promised payoff* $Z \in L^2(\Omega, \mathcal{F}_T, \widetilde{\mathbb{Q}})$. *It holds that* $\psi_t^{Y, \widetilde{\mathbb{Q}}} = L_{t-}\mu_t^Z$ *for every* $t \in [0, T]$, *where by convention* $L_{0-} = 0$.

Proof. It is easy to check that $dL_t = -L_{t-}dM_t$. Since Γ is increasing, the process L is of finite variation, and thus

$$d(L_t m_t^Z) = L_{t-} \, dm_t^Z + m_t^Z \, dL_t = L_{t-}\mu_t^Z \, dZ_t^1 + m_t^Z \, dL_t,$$

and thus we obtain

$$d\langle \mathcal{Y}, Z^1\rangle_t = L_{t-}\mu_t^Z \, d\langle Z^1\rangle_t$$

and $\psi_t^{Y, \widetilde{\mathbb{Q}}} = L_{t-}\mu_t^Z$, which proves the result. $\qquad\square$

For the proof of the next auxiliary result, the reader is referred, for instance, to Jeanblanc and Rutkowski (2000) or Bielecki and Rutkowski (2004).

Lemma 17. *Consider a survival claim* $Y = Z\mathbb{1}_{\{\tau > T\}}$ *with the promised payoff* $Z \in L^2(\Omega, \mathcal{F}_T, \widetilde{\mathbb{Q}})$. *The process* $N^{Y, \widetilde{\mathbb{Q}}}$ *in the Galtchouk-Kunita-Watanabe decomposition of* Y *with respect to* Z^1 *under* $\widetilde{\mathbb{Q}}$ *is given by the expression*

$$N_t^{Y, \widetilde{\mathbb{Q}}} = \int_{[0, t)} n_u^Z \, dM_u,$$

where the process $M_t = H_t - \Gamma_{t \wedge \tau}$ *is a* $\mathbb{G}$*-martingale, strongly orthogonal in the martingale sense to* $\widetilde{W}$ *under* $\widetilde{\mathbb{Q}}$, *and where*

$$n_t^Z = -\mathbb{E}_{\widetilde{\mathbb{Q}}}\big(Ze^{\Gamma_t - \Gamma_T} \mid \mathcal{F}_t\big). \tag{104}$$

By combining Proposition 21 with the last two result, we obtain the following corollary, which furnishes an almost explicit representation for the process $\widetilde{\psi}^{Y, \mathbb{P}}$ associated with the projection on $\mathcal{P}^0$ of a survival claim.

Corollary 11. *Let* $Y = Z\mathbb{1}_{\{\tau > T\}}$ *be a survival claim, where* Z *belongs to* $L^2(\Omega, \mathcal{F}_T, \widetilde{\mathbb{Q}})$. *Then* $\Pi_{\mathbb{P}}^0(Y)$ *is given by the following expression*

$$\Pi_{\mathbb{P}}^0(Y) = \int_0^T \widetilde{\psi}_t^{Y, \mathbb{P}} \, dZ_t^1,$$

where for every $t \in [0, T]$

$$\widetilde{\psi}_t^{Y, \mathbb{P}} = L_{t-}\mu_t^Z - \widetilde{\psi}_t\Big(\widetilde{\eta}_0^{-1}\mathbb{E}_{\widetilde{\mathbb{Q}}}Y + \int_0^t \widetilde{\eta}_u^{-1} n_u^Z \, dM_u\Big) \tag{105}$$

where in turn $L_t = \mathbb{1}_{\{\tau > t\}}e^{\Gamma_t}$ *and the processes* $\widetilde{\psi}, \widetilde{\eta}, \mu^Z$ *and* n^Z *are given by (91), (92), (103) and (104), respectively.*

Projection of a Defaultable Bond

According to the adopted convention regarding the recovery scheme, the terminal payoff at time T of a defaultable bond equals $X = L\mathbb{1}_{\{\tau>T\}} + \delta L\mathbb{1}_{\{\tau\leq T\}}$ for some $L > 0$ and $\delta \in [0,1)$. Notice that the payoff X can be represented as follows $X = \delta L + (1 - \delta)LY$, where $Y = \mathbb{1}_{\{\tau>T\}}$ is a simple survival claim, with the promised payoff $Z = 1$. Using the linearity of the projection $\Pi_{\mathbb{P}}^0$, we notice that $\Pi_{\mathbb{P}}^0(X)$ can be evaluated as follows

$$\Pi_{\mathbb{P}}^0(X) = \delta L\, \Pi_{\mathbb{P}}^0(1) + (1 - \delta)L\, \Pi_{\mathbb{P}}^0(Y).$$

By virtue of Corollary 11, we conclude that

$$\Pi_{\mathbb{P}}^0(X) = -\delta L e^{\theta^2 T}\, \Pi_{\widetilde{\mathbb{Q}}}^0(\eta_T) + (1 - \delta)L \int_0^T \psi_t\, dZ_t^1,$$

where (cf. (105))

$$\psi_t = \mathbb{1}_{\{\tau>t\}}e^{\Gamma_t}\mu_t - \widetilde{\psi}_t\left(e^{-\theta^2 T}\,\mathbb{E}_{\widetilde{\mathbb{Q}}}(e^{-\Gamma_T}) + \int_0^t \widetilde{\eta}_u^{-1}\, n_u\, dM_u\right), \qquad (106)$$

where in turn the process $\widetilde{\psi}$ is given by (91), n by $n_t = -\mathbb{E}_{\widetilde{\mathbb{Q}}}(e^{\Gamma_t - \Gamma_T} \mid \mathcal{F}_t)$, and the process μ is such that

$$\mathbb{E}_{\widetilde{\mathbb{Q}}}(e^{-\Gamma_T} \mid \mathcal{F}_t) = \mathbb{E}_{\widetilde{\mathbb{Q}}}Y + \int_0^t \mu_u\, dZ_u^1, \qquad \forall t \in [0,T].$$

Example 5. Consider the special case when Γ is deterministic. It is easily seen that we now have $\mu = 0$ and $n_t = -e^{\Gamma_t - \Gamma_T}$. Consequently, (106) becomes

$$\psi_t = -\widetilde{\psi}_t e^{-\Gamma_T}\left(e^{-\theta^2 T} - \int_0^t \widetilde{\eta}_u^{-1}\, e^{\Gamma_t}\, dM_u\right),$$

and thus

$$\Pi_{\mathbb{P}}^0(X) = -\delta L e^{\theta^2 T}\, \Pi_{\widetilde{\mathbb{Q}}}^0(\eta_T)$$

$$- (1 - \delta)L \int_0^T \widetilde{\psi}_t e^{-\Gamma_T}\left(e^{-\theta^2 T} - \int_0^t \widetilde{\eta}_u^{-1}\, e^{\Gamma_t}\, dM_u\right)dZ_t^1,$$

where the processes $\widetilde{\psi}$ and $\widetilde{\eta}$ are given by (91) and (92), respectively.

10 Risk-Return Portfolio Selection

In the preceding sections, we have examined the Markowitz-type mean-variance hedging problem from the particular perspective of valuation of non-attainable contingent claims. In view of the dependence of the mean-variance price obtained through this procedure on agent's preferences, (formally reflected, among others, by the values of parameters d and v), this specific application of Markowitz-type methodology suffers from deficiencies, which may undermine its practical implementations.

In this section, we shall take a totally different perspective, and we shall assume that a given claim X can be purchased by an agent (an asset management fund, say) for some pre-specified price. For instance, the price of X can be given by an investment bank that is able to hedge this claim using some arbitrage-free model, or it can be simply given by the OTC market.

Let us emphasize that an agent is now assumed to be a pricetaker, so that the issue of preference-based valuation of a non-attainable claim will not be considered in this section.

We postulate that an agent would like to invest in X, but will not be able (or willing) to hedge this claim using the underlying primary assets (if any such assets are available). As a consequence, an agent will only have in its portfolio standard instruments that are widely available for trading. The two important issues we would like to address in this section are:

- What proportion of the initial endowment v should an agent invest in the claim X if the goal is to lower the standard deviation (or, equivalently, the variance) of return, and to keep the expected rate of return at the desired level.

- How much should an agent invest in X in order to enhance the expected rate of return, and to preserve at the same time the pre-specified level of risk, as measured by the standard deviation of the rate of return.

We shall argue that mathematical tools and results presented in the previous sections are sufficient to solve both these problems. It seems to us that this alternative application of the mean-variance methodology can be of practical importance as well.

For the sake of simplicity, we shall solve the optimization problems formulated above in the class $\Phi(\mathbb{F})$ of $\mathbb{F}$-admissible trading strategies. A similar study can be conducted for the case of $\mathbb{G}$-admissible strategies. For any $v > 0$ and any trading strategy $\phi \in \Phi(\mathbb{F})$, let $r(\phi)$ be the simple rate of return, defined as

$$r(\phi) = \frac{V_T^v(\phi) - v}{v}.$$

The minimization of the standard deviation of the rate of return, which equals

$$\sigma(r(\phi)) = \sqrt{\mathbb{V}_{\mathbb{P}}\left(\frac{V_T^v(\phi) - v}{v}\right)} = v^{-1}\sqrt{\mathbb{V}_{\mathbb{P}}(V_T^v(\phi))},$$

is, of course, equivalent to the minimization of the variance $\mathbb{V}_{\mathbb{P}}(V_T^v(\phi))$. Within the present context, it is natural to introduce the constraint

$$\mathbb{E}_{\mathbb{P}}(v^{-1}V_T^v(\phi)) \geq d = 1 + d_r,$$

where $d_r > 0$ represents the desired minimal level of the expected rate of return.

10.1 Auxiliary Problems

The following auxiliary problem $MV(dv, v)$ is merely a version of the previously considered problem $MV(d, v)$:

Problem $MV(dv, v)$:   For a fixed $v > 0$ and $d > 1$, minimize the variance $\mathbb{V}_\mathbb{P}(V_T^v(\phi))$ over all strategies $\phi \in \Phi(\mathbb{F})$, subject to $\mathbb{E}_\mathbb{P} V_T^v(\phi) \geq dv$.

We assume from now on that $\theta \neq 0$, and we denote by Θ the constant

$$\Theta = (e^{\theta^2 T} - 1)^{-1} > 0.$$

Recall that for the problem $MV(dv, v)$, the risk-return trade-off can be summarized by the minimal variance curve $\mathbf{v}^*(dv, v)$. By virtue of Proposition 16, we have

$$\mathbf{v}^*(dv, v) = \Theta v^2 (d - 1)^2 = \Theta v^2 d_r^2. \qquad (107)$$

Equivalently, the minimal standard deviation of the rate of return satisfies

$$\sigma_r^* = \sigma(r(\phi^*(dv, v))) = \sqrt{\mathbf{v}^*(dv, v)} = \sqrt{\Theta} d_r,$$

so, as expected, it is independent of the value of the initial endowment v.

Suppose now that a claim X is available for some price $p_X \neq 0$, referred to as the *market price*. It is convenient to introduce the normalized claim $\bar{X} = X p_X^{-1}$. Under this convention, by the postulated linearity property of the market price, the price $p_{\bar{X}}$ of one unit of $\bar{X}$ is manifestly equal to 1.

The next auxiliary problem we wish to solve reads: find $p \in \mathbb{R}$ such that the solution to the problem $MV(dv, v, p, p\bar{X})$ has the minimal variance. This means, of course, that we are looking for $p \in \mathbb{R}$ for which $\mathbf{v}^*(dv, v, p, p\bar{X})$ is minimal. Notice that the constraint on the expected rate of return becomes

$$\mathbb{E}_\mathbb{P}(v^{-1} V_T^{v-p}(\phi) + p\bar{X}) \geq d = 1 + d_r,$$

where $d_r > 0$. It is clear that the curve $\mathbf{v}^*(dv, v, p, p\bar{X})$ can be derived from the general expression for $\mathbf{v}^*(d, v, p, X)$, which was established in Proposition 17. Let us denote

$$\gamma_{\bar{X}} = \mathbb{V}_\mathbb{P}(\bar{X} - \mathbb{E}_\mathbb{P}(\bar{X} \mid \mathcal{F}_T))$$

and

$$\nu_{\bar{X}} = \mathbb{E}_\mathbb{Q} \bar{X} - 1.$$

Let us notice that the condition $d - v + p - \mathbb{E}_\mathbb{Q} X > 0$, which was imposed in part (i) of Proposition 17, now corresponds to the following inequality: $v d_r > p \nu_{\bar{X}}$. We shall assume from now on that $\bar{X} \neq 1$ (this assumption means simply that the claim X does not represent the savings account). Recall that $v > 0$ and $d_r = d - 1 > 1$.

Proposition 24. (i) *If we assume that $\gamma_{\bar{X}} > 0$ and that $\nu_{\bar{X}} \neq 0$, then the problem $MV(dv, v, p, p\bar{X})$ has a solution with the minimal variance with respect to p. The minimal variance equals*

$$\mathbf{v}^*(dv, v, p^*, p^*\bar{X}) = \Theta v^2 d_r^2 \left(1 - \frac{\nu_{\bar{X}}^2}{\Theta^{-1}\gamma_{\bar{X}} + \nu_{\bar{X}}^2} \right) \tag{108}$$

and the optimal value of p equals

$$p^* = \frac{v d_r \nu_{\bar{X}}}{\Theta^{-1}\gamma_{\bar{X}} + \nu_{\bar{X}}^2}. \tag{109}$$

(ii) *Let $\gamma_{\bar{X}} > 0$ and $\nu_{\bar{X}} = 0$. Then we have $p^* = 0$ and the minimal variance equals*

$$\mathbf{v}^*(dv, v, p^*, p^*\bar{X}) = \Theta v^2 d_r^2.$$

(iii) *Let $\gamma_{\bar{X}} = 0$ and $\nu_{\bar{X}} \neq 0$. If the inequality $\nu_{\bar{X}} > 0$ ($\nu_{\bar{X}} < 0$, respectively) holds then for any $p \geq v d_r \nu_{\bar{X}}^{-1}$ ($p \leq v d_r \nu_{\bar{X}}^{-1}$, respectively) the minimal variance $\mathbf{v}^*(dv, v, p, p\bar{X})$ is minimal with respect to p and it equals 0.*
(iv) *Let $\gamma_{\bar{X}} = \nu_{\bar{X}} = 0$. Then $\bar{X}$ is an attainable claim and $\mathbb{E}_{\mathbb{Q}}\bar{X} = 1$. In this case, for any $p \in \mathbb{R}$ the minimal variance equals*

$$\mathbf{v}^*(dv, v, p, p\bar{X}) = \Theta v^2 d_r^2.$$

Proof. Let us first prove parts (i)-(ii). It suffices to observe that, by virtue of Proposition 17, the minimal variance for the problem $MV(dv, v, p, p\bar{X})$ is given by the expression:

$$\mathbf{v}^*(dv, v, p, p\bar{X}) = \Theta(d_r v - p\nu_{\bar{X}})^2 + p^2 \gamma_{\bar{X}} \tag{110}$$

provided that $v d_r > p\nu_{\bar{X}}$. A simple argument shows that the minimal value for the right-hand side in (110) is obtained by setting $p = p^*$, where p^* is given by (109), and the minimal variance is given by (108). Moreover, it is easily seen that for p^* given by (109) the inequality $v d_r > p^* \nu_{\bar{X}}$ is indeed satisfied, provided that $\gamma_{\bar{X}} > 0$. Notice also that if $\mathbb{E}_{\mathbb{Q}}\bar{X} = 1$, we obviously have $v d_r > p\nu_{\bar{X}} = 0$ for any $p \in \mathbb{R}_+$, and thus we obtain the following optimal values:

$$p^* = 0, \quad \mathbf{v}^*(dv, v, p^*, p^*\bar{X}) = \Theta v^2 d_r^2.$$

Assume now that $v d_r \leq p\nu_{\bar{X}}$, so that the case $\nu_{\bar{X}} = 0$ (i.e., the case $\mathbb{E}_{\mathbb{Q}}\bar{X} = 1$) is excluded. Then, by virtue of part (ii) in Proposition 17, the minimal variance equals $p^2 \gamma_{\bar{X}}$ (notice that the assumption that $\gamma_{\bar{X}}$ is strictly positive is not needed here). Assume first that $\mathbb{E}_{\mathbb{Q}}\bar{X} < 1$, so that $\nu_{\bar{X}} < 0$. Then the condition $v d_r \leq p\nu_{\bar{X}}$ becomes $p \leq v d_r \nu_{\bar{X}}^{-1}$, and thus p is necessarily negative. The minimal variance corresponds to $p^* = v d_r \nu_{\bar{X}}^{-1}$, and it equals

$$\mathbf{v}^*(dv, v, p^*, p^*\bar{X}) = (p^*)^2 \gamma_{\bar{X}} = v^2 d_r^2 \nu_{\bar{X}}^{-2} \gamma_{\bar{X}}. \tag{111}$$

In, on the contrary, $\mathbb{E}_\mathbb{Q}\bar{X} > 1$, then $\nu_{\bar{X}} > 0$ and we obtain $p \geq vd_r\nu_{\bar{X}}^{-1}$, so that p is strictly positive. Again, the minimal variance corresponds to $p^* = vd_r\nu_{\bar{X}}^{-1}$, and it is given by (111). It is easy to check that the following inequality holds:

$$\Theta v^2 d_r^2 \left(1 - \frac{\nu_{\bar{X}}^2}{\Theta^{-1}\gamma_{\bar{X}} + \nu_{\bar{X}}^2}\right) < v^2 d_r^2 \nu_{\bar{X}}^{-2}\gamma_{\bar{X}}.$$

By combining the considerations above, we conclude that statements (i)-(ii) are valid. The proof of part (iii) is also based on the analysis above. We thus proceed to the proof of the last statement.

Notice that $\Theta^{-1}\gamma_{\bar{X}} + \nu_{\bar{X}}^2 = 0$ if and only if $\gamma_{\bar{X}} = 0$ and $\nu_{\bar{X}} = 0$. This means that $\bar{X}$ is $\mathcal{F}_T$-adapted (and thus $\mathbb{F}$-attainable) and $\mathbb{E}_\mathbb{Q}\bar{X} = 1$ (so that the arbitrage price of $\bar{X}$ coincides with its market price $p_{\bar{X}}$). Condition $vd_r - p\nu_{\bar{X}} > 0$ is now satisfied, and thus the minimal variance is given by (110), which now becomes

$$\mathbf{v}^*(dv, v, p, p\bar{X}) = \Theta v^2 d_r^2, \quad \forall p \in \mathbb{R}_+.$$

Obviously, the result does not depend on p. This proves part (iv).   □

In the last proposition, no a priori restriction on the value of the parameter p was imposed. Of course, one can also consider a related constrained problem by postulating, for instance, that the price p belongs to the interval $[0, v]$.

10.2 Minimization of Risk

We are in a position to examine the first question, which reads: how much to invest in the new opportunity in order to minimize the risk and to preserve at the same time the pre-specified level $d_r > 0$ of the expected rate of return.

Case of an attainable claim. Assume first that $\bar{X}$ is an $\mathbb{F}$-attainable contingent claim, so that $\mathbb{E}_\mathbb{P}(\bar{X} \mid \mathcal{F}_T) = \bar{X}$, and thus $\gamma_{\bar{X}} = 0$. If the claim $\bar{X}$ is correctly priced by the market, i.e., if $\mathbb{E}_\mathbb{Q}\bar{X} = p_{\bar{X}} = 1$ then, by virtue of part (iv) in Proposition 24, for any choice of p the minimal variance is the same as in the problem $MV(dv, v)$. Hence, as expected, the possibility of investing in the claim $\bar{X}$ has no bearing on the efficiency of trading.

Let us now consider the case where $\mathbb{E}_\mathbb{Q}\bar{X} \neq 1$, that is, the market price $p_{\bar{X}}$ does not coincide with the arbitrage price $\pi_0(\bar{X})$. Suppose first that $\mathbb{E}_\mathbb{Q}\bar{X} > 1$, that is, $\bar{X}$ is underpriced by the market. Then, in view of part (iii) in Proposition 24, the variance of the rate of return can be reduced to 0 by choosing p which satisfies

$$p \geq vd_r(\mathbb{E}_\mathbb{Q}\bar{X} - 1)^{-1} > 0.$$

Similarly, if $\mathbb{E}_\mathbb{Q}\bar{X} < 1$ then for any p such that

$$p \leq vd_r(\mathbb{E}_\mathbb{Q}\bar{X} - 1)^{-1} < 0$$

the variance equals 0. Off course, this feature is due to the presence of arbitrage opportunities in the market. We conclude that, as expected, in the case of an attainable claim the solution to the problem considered is this section is rather trivial, and thus it has no practical appeal.

Case of a non-attainable claim. We now assume that $\gamma_{\bar{X}} > 0$. Suppose first that $\mathbb{E}_\mathbb{Q} \bar{X} = 1$. By virtue of part (ii) in Proposition 24, under this assumption it is optimal not to invest in $\bar{X}$. To better appreciate this result, notice that for the conditional expectation $\widetilde{X} = \mathbb{E}_\mathbb{P}(\bar{X} \mid \mathcal{F}_T)$ we have $\mathbb{E}_\mathbb{Q} \widetilde{X} = \mathbb{E}_\mathbb{Q} \bar{X} = 1$ and $\mathbb{E}_\mathbb{P} \widetilde{X} = \mathbb{E}_\mathbb{P} \bar{X}$ (cf. Section 8.2). Therefore, trading in $\bar{X}$ is essentially equivalent to trading in an attainable claim $\widetilde{X}$, but trading in $\bar{X}$ results in the residual variance $p^2 \gamma_{\bar{X}}$. This observations explains why the solution $p^* = 0$ is optimal.

Suppose now that $\mathbb{E}_\mathbb{Q} \bar{X} \neq 1$. Then part (i) of Proposition 24 shows that the variance of the rate of return can always be reduced by trading in $\bar{X}$. Specifically, p^* is strictly positive provided that $\mathbb{E}_\mathbb{Q} \bar{X} > 1 = p_{\bar{X}}$, that is, the expected value of $\bar{X}$ under the martingale measure $\mathbb{Q}$ for the underlying market is greater than its market price.

Case of an independent claim. Assume that the claim $\bar{X}$ is independent of $\mathcal{F}_T$, so that $\gamma_{\bar{X}} > 0$ is the variance of $\bar{X}$. In this case $\mathbb{E}_\mathbb{Q} \bar{X} = \mathbb{E}_\mathbb{P} \bar{X}$ and thus (108) becomes

$$\mathbf{v}^* = \Theta v^2 d_r^2 \left(1 - \frac{(\mathbb{E}_\mathbb{P} \bar{X} - 1)^2}{\Theta^{-1} \mathbb{V}_\mathbb{P}(\bar{X}) + (\mathbb{E}_\mathbb{P} \bar{X} - 1)^2} \right).$$

From the last formula, it is clear that an agent should always to invest either a positive or negative amount of initial endowment v in an independent claim X, except for the case where $\mathbb{E}_\mathbb{P} \bar{X} = 1$. If $\mathbb{E}_\mathbb{P} \widetilde{X} \neq 1$ then the optimal value of p equals (cf. (109))

$$p^* = \frac{v d_r (\mathbb{E}_\mathbb{P} \bar{X} - 1)}{\Theta^{-1} \mathbb{V}_\mathbb{P}(\bar{X}) + (\mathbb{E}_\mathbb{P} \bar{X} - 1)^2}$$

so that it is positive if and only if $\mathbb{E}_\mathbb{P} \bar{X} > 1$.

Case of a claim with zero market price. The case when the market price of X is zero (that is, the equality $p_X = 0$ holds) is also of practical interest, since such a feature is typical for forward contracts. It should be stressed that this particular case is not covered by Proposition 24, however.

In fact, we deal here with the following variant of the mean-variance problem:

Find $\alpha \in \mathbb{R}$ such that the solution to the problem $\mathrm{MV}(dv, v, 0, \alpha X)$ has the minimal variance.

Under the assumption that $v d_r > \alpha \mathbb{E}_\mathbb{Q} X$, we have

$$\mathbf{v}^*(dv, v, 0, \alpha X) = \Theta(v d_r - \alpha \mathbb{E}_\mathbb{Q} X)^2 + \alpha^2 \gamma_X.$$

If, on the contrary, the inequality $v d_r \leq \alpha \mathbb{E}_\mathbb{Q} X$ is valid, then the minimal variance equals $\alpha^2 \gamma_X$. Of course, we necessarily have $\alpha \neq 0$ here (since $v d_r > 0$).

10.3 Maximization of Expected Return

Let us focus on part (i) in Proposition 24, that is, let us assume that $\gamma_{\bar{X}} > 0$ and $\nu_{\bar{X}} \neq 0$ (as was explained above, other cases examined in Proposition 24 are of minor practical interest). The question of maximization of the expected rate return for a pre-specified level of risk, can be easily solved by comparing (107) with (108). Indeed, for a given level d_r of the expected rate of return, and thus a given level $v^*(dv, v)$ of the minimal variance, it suffices to find a number $\widehat{d}_r$ which solves the following equation

$$\Theta v^2 d_r^2 = \Theta v^2 \widehat{d}_r^2 \left(1 - \frac{\nu_{\bar{X}}^2}{\Theta^{-1}\gamma_{\bar{X}} + \nu_{\bar{X}}^2} \right).$$

It is obvious that the last equation has the unique solution

$$\widehat{d}_r = d_r \sqrt{1 + \frac{\nu_{\bar{X}}^2}{\Theta^{-1}\gamma_{\bar{X}}}} > d_r.$$

The corresponding value of p^* is given by (109) with d_r substituted with $\widehat{d}_r$. It is thus clear that, under the present assumptions, a new investment opportunity can be used to enhance the expected rate of return. If we insist, in addition, that $p > 0$, then the latter statement remains valid, provided that $\mathbb{E}_{\mathbb{Q}} \bar{X} > 1$.

Part III. Indifference Pricing

In this part, we present a few alternative ways of pricing defaultable claims in the situation when perfect hedging is not possible. In the previous part, we have presented the mean-variance hedging framework. Now, we study the indifference price approach that was initiated by Hodges and Neuberger (1989). We shall refer to this approach as the "Hodges price" approach. This will lead us to solving portfolio optimization problems in incomplete market, and we shall use the dynamic programming (DP) approach.

We also present the Hamilton-Jacobi-Bellman (HJB) equations, when appropriate, even though this method typically requires strong assumptions to give closed-form solutions. In particular, when dealing with the general DP approach, we need not make any Markovian assumption about the underlying processes; such assumptions are fundamental for the HJB methodology to work.

In Section 11, we define the Hodges indifference price associated to strategies adapted with the reference filtration $\mathbb{F}$, and we solve the problem for exponential preferences and for some particular defaultable claims. We shall use results obtained here to provide basis for a comparison between the historical spread and the risk-neutral one.

In Section 12, using backward stochastic differential equations (BSDEs), we work with $\mathbb{G}$-adapted strategies, and we solve portfolio optimization problems for exponential utility functions. Our method relies on the ideas of Rouge and El Karoui (2000) and Musiela and Zariphopoulou (2004). The reader can refer to El Karoui and Mazliak (1997), El Karoui and Quenez (1997), El Karoui et al. (1997), or to the survey by Buckdahn (2000) for an introduction to the theory of backward stochastic differential equations and its applications in finance.

Section 13 is devoted to the study of a particular indifference price, based on the quadratic criterion; we call such a price the *quadratic hedging price* (see the introduction to Part II). In particular, we compare the indifference prices obtained using strategies adapted to the reference filtration $\mathbb{F}$ to the indifference prices obtained using strategies based on the enlarged filtration $\mathbb{G}$. It is worthwhile to stress, though, that the quadratic utility alone is not quite adequate for the pricing purposes, although it represents a good criterion for hedging purposes. This is one of the reasons we presented the mean-variance approach to pricing and hedging of defaultable claims in Part II.

In the last section, we present a very particular case of the duality approach for exponential utilities.

As in the previous part, we emphasize that a very important aspect of our analysis is the distinction between the case when admissible portfolios are adapted to the filtration $\mathbb{F}$, and the case when admissible portfolios are adapted to the filtration $\mathbb{G}$.

11 Hedging in Incomplete Markets

We recall briefly the probabilistic setting of Part II. The default-free asset is Z^1 with the dynamics

$$dZ_t^1 = Z_t^1(\nu dt + \sigma dW_t), \quad Z_0^1 > 0,$$

and the price process of the money market account has the dynamics

$$dZ_t^2 = rZ_t^2\, dt, \quad Z_0^2 = 1,$$

where r is the constant interest rate. The default-free market is complete and arbitrage free: one can hedge perfectly any square-integrable contingent claim $X \in \mathcal{F}_T$. The default time is some random time τ, and the default process is denoted as $H_t = \mathbb{1}_{\{\tau \le t\}}$. The reference filtration is the Brownian filtration $\mathcal{F}_t = \sigma(W_u, u \le t)$ and the enlarged filtration is $\mathcal{G}_t = \mathcal{F}_t \vee \mathcal{H}_t$ where $\mathcal{H}_t = \sigma(H_u, u \le t)$.

We assume that the hazard process $F_t = \mathbb{P}\{\tau \le t \,|\, \mathcal{F}_t\}$ is absolutely continuous with respect to Lebesgue measure, so that $F_t = \int_0^t f_u\, du$ (hence, it is an increasing process). Therefore, the process

$$M_t = H_t - \int_0^{t \wedge \tau} \gamma_u\, du = H_t - \int_0^{t \wedge \tau} \frac{f_u}{1 - F_u}\, du$$

is a $\mathbb{G}$-martingale, where γ is the *default intensity*. Note that the stochastic intensity γ is the intensity of the default time τ with respect to the reference filtration $\mathbb{F}$ generated by the Brownian motion W.

For a fixed $T > 0$, we introduce a *risk-neutral probability* $\mathbb{Q}$ for the market model (Z^1, Z^2) by setting $d\mathbb{Q}|_{\mathcal{G}_t} = \eta_t\, d\mathbb{P}|_{\mathcal{G}_t}$ for $t \in [0, T]$, where the Radon-Nikodym density η is the $\mathbb{F}$-martingale defined as

$$d\eta_t = -\theta\eta_t\, dW_t, \quad \eta_0 = 1,$$

where $\theta = (\nu - r)/\sigma$. Under $\mathbb{Q}$, the discounted process $\widetilde{Z}_t^1 = e^{-rt}Z_t^1$ is a martingale. It should be emphasized that $\mathbb{Q}$ is not necessarily a martingale measure for defaultable assets. Let us recall, however, that if $\widetilde{\mathbb{Q}}$ is any equivalent martingale measure on $\mathbb{G}$ for the default-free and defaultable market, then the restriction of $\widetilde{\mathbb{Q}}$ to $\mathbb{F}$ is equal to the restriction of $\mathbb{Q}$ to $\mathbb{F}$. A *defaultable claim* is simply any random variable X, which is $\mathcal{G}_T$-measurable. Hence, default-free claims are formally considered as special cases of defaultable claims.

11.1 Hodges Indifference Price

We present a general framework of the Hodges and Neuberger (1989) approach with some strictly increasing, strictly concave and continuously differentiable mapping u, defined on $\mathbb{R}$. We solve explicitly the problem in the case of exponential utility for portfolios adapted to the reference filtration.

The Hodges approach to pricing of unhedgeable claims is a utility-based approach and can be summarized as follows: the issue at hand is to assess the value of some (defaultable) claim X as seen from the perspective of an economic agent who optimizes his behavior relative to some utility function, say u. In order to provide such an assessment one can argue that one should first consider the following possible modes of agent's behavior and the associated optimization problems:

Problem $(\mathcal{P})$: Optimization in the default-free market.

The agent invests his initial wealth $v > 0$ in the default-free financial market using a self-financing strategy. The associated optimization problem is,

$$(\mathcal{P}) : \mathcal{V}(v) := \sup_{\phi \in \Phi(\mathbb{F})} \mathbb{E}_{\mathbb{P}}\{u(V_T^v(\phi))\},$$

where the wealth process $V_t = V_t^v(\phi)$, $t \in \mathbb{R}_+$, is solution of

$$dV_t = rV_t\, dt + \phi_t(dZ_t^1 - rZ_t^1 dt), \quad V_0 = v. \tag{112}$$

Recall that $\Phi(\mathbb{F})$ is the class of all admissible, $\mathbb{F}$-adapted, self-financing trading strategies (for the definition of this class, see Part II).

Problem $(\mathcal{P}_{\mathbb{F}}^X)$: Optimization in the default-free market using $\mathbb{F}$-adapted strategies and buying the defaultable claim.

The agent buys the contingent claim X at price p, and invests the remaining wealth $v - p$ in the financial market, using a trading strategy $\phi \in \Phi(\mathbb{F})$. The resulting *global terminal wealth* will be

$$V_T^{v-p,X}(\phi) = V_T^{v-p}(\phi) + X.$$

The associated optimization problem is

$$(\mathcal{P}_{\mathbb{F}}^X) : \mathcal{V}_X(v - p) := \sup_{\phi \in \Phi(\mathbb{F})} \mathbb{E}_{\mathbb{P}}\{u(V_T^{v-p}(\phi) + X)\},$$

where the process $V^{v-p}(\phi)$ is a solution of (112) with the initial condition $V_0^{v-p}(\phi) = v - p$. We emphasize that the class $\Phi(\mathbb{F})$ of admissible strategies is the same as in the problem $(\mathcal{P})$, that is, we restrict here our attention to trading strategies that are adapted to the reference filtration $\mathbb{F}$.

Problem $(\mathcal{P}_{\mathbb{G}}^X)$: Optimization in the default-free market using $\mathbb{G}$-adapted strategies and buying the defaultable claim.

The agent buys the contingent claim X at price p, and invests the remaining wealth $v - p$ in the financial market, using a strategy adapted to the enlarged filtration $\mathbb{G}$. The associated optimization problem is

$$(\mathcal{P}_{\mathbb{G}}^X) : \mathcal{V}_X^{\mathbb{G}}(v - p) := \sup_{\phi \in \Phi(\mathbb{G})} \mathbb{E}_{\mathbb{P}}\{u(V_T^{v-p}(\phi) + X)\},$$

where $\Phi(\mathbb{G})$ is the class of all $\mathbb{G}$-admissible trading strategies (for the definition of the class $\Phi(\mathbb{G})$, see Part II). Next, the utility based assessment of the value (price) of the claim X, as seen from the agent's perspective, is given in terms of the following definition.

Definition 10. *For a given initial endowment v, the $\mathbb{F}$-Hodges buying price of a defaultable claim X is the real number $p_{\mathbb{F}}^*(v)$ such that $\mathcal{V}(v) = \mathcal{V}_X(v - p_{\mathbb{F}}^*(v))$. Similarly, the $\mathbb{G}$-Hodges buying price of X is the real number $p_{\mathbb{G}}^*(v)$ such that $\mathcal{V}(v) = \mathcal{V}_X^{\mathbb{G}}(v - p_{\mathbb{G}}^*(v))$.*

Remark. We can define the $\mathbb{F}$-*Hodges selling price* $p_*^{\mathbb{F}}(v)$ of X by considering $-p$, where p is the buying price of $-X$, as specified in Definition 10.

If the contingent claim X is $\mathcal{F}_T$-measurable, then the $\mathbb{F}$- and the $\mathbb{G}$-Hodges prices coincide with the hedging price of X, i.e., $p_*^{\mathbb{F}}(v) = p_{\mathbb{G}}^*(v) = \pi_0(X) = \mathbb{E}_{\mathbb{P}}(\zeta_T X)$, where we denote $\zeta_t = \eta_t R_t$ with $R_t = (Z_t^2)^{-1} = e^{-rt}$. Indeed, assume that there exists a self-financing portfolio $\widehat{\phi}$ such that $X = V_T^{\pi_0(X)}(\widehat{\phi})$, and let h be the $\mathbb{F}$-Hodges buying price. Suppose first that $h < \pi_0(X)$. Then for any ϕ we obtain

$$V_T^{v-h}(\phi) + X = V_T^{v-h}(\phi) + V_T^{\pi_0(X)}(\widehat{\phi}) = V_T^{v-h+\pi_0(X)}(\psi),$$

where we denote $\psi = \widehat{\phi} + \phi \in \Phi(\mathbb{F})$. Hence

$$
\begin{aligned}
\mathcal{V}_X(v - h) &= \sup_{\phi \in \Phi(\mathbb{F})} \mathbb{E}_{\mathbb{P}}\{u(V_T^{v-h}(\phi) + X)\} \\
&= \sup_{\psi \in \Phi(\mathbb{F})} \mathbb{E}_{\mathbb{P}}\{u(V_T^{v-h+\pi_0(X)}(\psi))\} \geq \mathcal{V}(v),
\end{aligned}
$$

where the last inequality (which is a strict inequality) follows from $v < v - h + \pi_0(X)$ and the arbitrage principle. Therefore, the supremum over $\phi \in \Phi(\mathbb{F})$ of $\mathbb{E}_{\mathbb{P}}(u(V_T^{v-h}(\phi) + X))$ is greater than $\mathcal{V}(v)$. We conclude that the $\mathbb{F}$-Hodges buying price can not be smaller than the hedging price. Arguing in a similar way, one can show that the $\mathbb{F}$-Hodges selling price of an $\mathcal{F}_T$-measurable claim can not be smaller than the hedging price. Finally, almost identical arguments show that the $\mathbb{G}$-Hodges buying and selling price of an $\mathcal{F}_T$-measurable claim are equal to the hedging price of X (see Section 12.2).

Remark. It can be shown (see Rouge and El Karoui (2000), or Collin-Dufresne and Hugonnier (2002)) that in the general case of non-hedgeable contingent claim, the Hodges price belongs to the open interval

$$\left(\inf_{\widetilde{\mathbb{Q}}} \mathbb{E}_{\widetilde{\mathbb{Q}}}(Xe^{-rT}), \sup_{\widetilde{\mathbb{Q}}} \mathbb{E}_{\widetilde{\mathbb{Q}}}(Xe^{-rT}) \right),$$

where $\widetilde{\mathbb{Q}}$ runs over the set of all equivalent martingale measures, and thus it can not induce arbitrage opportunities.

11.2 Solution of Problem $(\mathcal{P})$

We briefly recall one of the solution methods for the problem $(\mathcal{P})$. To this end, we first observe that in view of (112) the process $e^{-rt}V_t^{v-p}(\phi)$, $t \in \mathbb{R}_+$, is a martingale under any equivalent martingale measure, hence $\zeta_t V_t^{v-p}(\phi)$, $t \in \mathbb{R}_+$, is a $\mathbb{P}$-martingale and, in particular, $\mathbb{E}_{\mathbb{P}}(V_T^v(\phi)\zeta_T) = v$. It follows that in order to obtain a terminal wealth equal to, say V, the initial endowment v has to be greater or equal to $\mathbb{E}_{\mathbb{P}}(V\zeta_T)$; this condition is commonly referred to as the *budget constraint*.

Now, let us denote by I the inverse of the monotonic mapping u' (the first derivative of u). It is well known (see, e.g., Karatzas and Shreve (1998)) that the optimal terminal wealth in the problem $(\mathcal{P})$ is given by the formula

$$V_T^{v,*} = I(\mu\zeta_T), \quad \mathbb{P}\text{-a.s.,} \tag{113}$$

where μ is a real number such that the budget constraint is binding, that is,

$$v = \mathbb{E}_{\mathbb{P}}(\zeta_T V_T^{v,*}). \tag{114}$$

Consequently, the optimal value of the objective criterion for the problem $(\mathcal{P})$ is $\mathcal{V}(v) = \mathbb{E}_{\mathbb{P}}(u(V_T^{v,*}))$.

The above results are obtained by means of convex duality theory. The disadvantage of this approach, however, is the fact that it is typically very difficult to identify an optimal trading strategy. Thus, in general, using the convex duality approach we can only partially solve the problem $(\mathcal{P})$. Specifically, we can compute the optimal value of the objective criterion, but we can't identify the optimal strategy. Later in this part, we shall use the BSDE approach in a more general setting. It will be seen that this approach will allow us to identify (at least in principle) an optimal trading strategy.

11.3 Solution of Problem $(\mathcal{P}_{\mathbb{F}}^X)$

In this subsection, we shall examine the problem $(\mathcal{P}_{\mathbb{F}}^X)$ for a defaultable claim of a particular form. First, we shall provide a solution $\mathcal{V}_X(v - p)$ to the related optimization problem. Next, we shall establish a quasi-explicit representation for the Hodges price of X in the case of exponential utility. Finally, we shall compare the spread obtained via the risk-neutral valuation with the spread determined by the Hodges price of a defaultable zero-coupon bond. The reader can refer to Bernis and Jeanblanc (2003) for other comments.

Particular Form of a Defaultable Claim

We restrict our attention to the case when X is of the form

$$X = X_1 \mathbb{1}_{\{\tau > T\}} + X_2 \mathbb{1}_{\{\tau \le T\}}, \tag{115}$$

where $X_i, i = 1, 2$ are $\mathbb{P}$-square-integrable and $\mathcal{F}_T$-measurable random variables. In this case, we have

$$V_T^{v-p,X}(\phi) = V_T^{v-p}(\phi) + X_1$$

if the default did not occur before maturity date T, that is, on the set $\{\tau > T\}$, and

$$V_T^{v-p,X}(\phi) = V_T^{v-p}(\phi) + X_2$$

otherwise. In other words,

$$V_T^{v-p,X}(\phi) = \mathbb{1}_{\{\tau > T\}}(V_T^{v-p}(\phi) + X_1) + \mathbb{1}_{\{\tau \le T\}}(V_T^{v-p}(\phi) + X_2).$$

Observe that the pay-off X_2 is not paid at time of default τ, but at the terminal time T.

Since the trading strategies are $\mathbb{F}$-adapted, the terminal wealth $V_T^{v-p}(\phi)$ is an $\mathcal{F}_T$-measurable random variable. Consequently, it holds that

$$\mathbb{E}_{\mathbb{P}}\{u(V_T^{v-p,X}(\phi))\}$$
$$= \mathbb{E}_{\mathbb{P}}\{u(V_T^{v-p}(\phi) + X_1)\mathbb{1}_{\{\tau > T\}} + u(V_T^{v-p}(\phi) + X_2)\mathbb{1}_{\{\tau \le T\}}\}$$
$$= \mathbb{E}_{\mathbb{P}}\left\{\mathbb{E}_{\mathbb{P}}\left(u\left(V_T^{v-p}(\phi) + X_1\right)\mathbb{1}_{\{\tau > T\}} + u\left(V_T^{v-p}(\phi) + X_2\right)\mathbb{1}_{\{\tau \le T\}} | \mathcal{F}_T\right)\right\}$$
$$= \mathbb{E}_{\mathbb{P}}\{u(V_T^{v-p}(\phi) + X_1)(1 - F_T) + u(V_T^{v-p}(\phi) + X_2)F_T\},$$

where $F_T = \mathbb{P}\{\tau \le T \mid \mathcal{F}_T\}$. Define, for every $\omega \in \Omega$ and $y \in \mathbb{R}$,

$$J_X(y, \omega) = u(y + X_1(\omega))(1 - F_T(\omega)) + u(y + X_2(\omega))F_T(\omega).$$

Notice that under the present assumptions, the problem $(\mathcal{P}_{\mathbb{F}}^X)$ is equivalent to the following problem:

$$(\mathcal{P}_{\mathbb{F}}^X) : \mathcal{V}_X(v - p) := \sup_{\phi \in \Phi(\mathbb{F})} \mathbb{E}_{\mathbb{P}}\{J_X(V_T^{v-p}(\phi), \omega)\}.$$

The mapping $J_X(\cdot, \omega)$ is a strictly concave and increasing real-valued mapping. Consequently, for any $\omega \in \Omega$ we can define the mapping $I_X(z, \omega)$ by setting $I_X(z, \omega) = (J_X'(\cdot, \omega))^{-1}(z)$ for $z \in \mathbb{R}$, where $(J_X'(\cdot, \omega))^{-1}$ denotes the inverse mapping of the derivative of J_X with respect to the first variable. To simplify the notation, we shall usually suppress the second variable, and we shall write $I_X(\cdot)$ in place of $I_X(\cdot, \omega)$.

The following lemma provides the form of the optimal solution.

Lemma 18. *The optimal terminal wealth for the problem* $(\mathcal{P}_{\mathbb{F}}^X)$ *is given by* $V_T^{v-p,*} = I_X(\lambda^*\zeta_T)$, $\mathbb{P}$-a.s., *for some* λ^* *such that*

$$v - p = \mathbb{E}_{\mathbb{P}}(\zeta_T V_T^{v-p,*}). \tag{116}$$

Thus the optimal global wealth equals $V_T^{v-p,X,*} = V_T^{v-p,*} + X = I_X(\lambda^*\zeta_T) + X$ *and the optimal value of the objective criterion for the problem* $(\mathcal{P}_{\mathbb{F}}^X)$ *is*

$$\mathcal{V}_X(v - p) = \mathbb{E}_{\mathbb{P}}(u(V_T^{v-p,X,*})) = \mathbb{E}_{\mathbb{P}}(u(I_X(\lambda^*\zeta_T) + X)). \tag{117}$$

Proof. As a consequence of predictable representation property (see, e.g., Karatzas and Shreve (1991)), one knows that in order to find the optimal wealth it is enough to maximize $u(\Delta)$ over the set of square-integrable and $\mathcal{F}_T$-measurable random variables Δ, subject to the budget constraint, given by

$$\mathbb{E}_{\mathbb{P}}(\zeta_T \Delta) \le v - p.$$

The associated Lagrange multiplier, say λ^*, is non-negative. Moreover, by the strict monotonicity of u, we know that, at optimum, the constraint is binding, and thus $\lambda^* > 0$. We check that $I_X(\lambda^*\zeta_T)$ is the optimal wealth.

The mapping $J_X(\cdot)$ is strictly concave (for all ω). Hence, for every wealth process $V^{v-p}(\phi)$, starting from $v - p$, by tangent inequality, we have

$$\mathbb{E}_{\mathbb{P}}\{J_X(V_T^{v-p}(\phi)) - J_X(V_T^{v-p,*})\} \leq \mathbb{E}_{\mathbb{P}}\{(V_T^{v-p}(\phi) - V_T^{v-p,*})J_X'(V_T^{v-p,*})\}.$$

Replacing $V^{v-p,*}$ by its expression given in Lemma 18 yields for any $\phi \in \Phi(\mathbb{F})$

$$\mathbb{E}_{\mathbb{P}}\{J_X(V_T^{v-p}(\phi)) - J_X(V_T^{v-p,*})\} \leq \lambda^* \mathbb{E}_{\mathbb{P}}\{\zeta_T(V_T^{v-p}(\phi) - V_T^{v-p,*})\} \leq 0,$$

where the last inequality follows from (116) and the budget constraint. To end the proof, it remains to observe that the first order conditions are also sufficient in the case of a concave criterion. Moreover, by virtue of strict concavity of the function J_X, the optimum is unique. □

Exponential Utility: Explicit Computation of the Hodges Price

For the sake of simplicity, we assume here that $r = 0$. Let us state the following result, the proof of which stems from Lemma 18, by direct computations.

Proposition 25. Let $u(x) = 1 - \exp(-\varrho x)$ for some $\varrho > 0$. Assume that for $i = 1, 2$ the random variable $\zeta_T e^{-\varrho X^i}$ is $\mathbb{P}$-integrable. Then we have

$$p_{\mathbb{F}}^*(v) = -\frac{1}{\varrho} \mathbb{E}_{\mathbb{P}}\left(\zeta_T \ln\left((1 - F_T)e^{-\varrho X_1} + F_T e^{-\varrho X_2}\right)\right) = \mathbb{E}_{\mathbb{P}}(\zeta_T \Psi),$$

where the $\mathcal{F}_T$-measurable random variable Ψ equals

$$\Psi = -\frac{1}{\varrho} \ln\left((1 - F_T)e^{-\varrho X_1} + F_T e^{-\varrho X_2}\right). \tag{118}$$

Thus, the $\mathbb{F}$-Hodges buying price $p_{\mathbb{F}}^*(v)$ is the arbitrage price of the associated claim Ψ. In addition, the claim Ψ enjoys the following meaningful property

$$\mathbb{E}_{\mathbb{P}}\{u(X - \Psi) \,|\, \mathcal{F}_T\} = 0. \tag{119}$$

Proof. In view of the form of the solution to the problem $(\mathcal{P})$, we obtain (cf. (113))

$$V_T^{v,*} = -\frac{1}{\varrho} \ln\left(\frac{\mu^* \zeta_T}{\varrho}\right).$$

The budget constraint $\mathbb{E}_{\mathbb{P}}(\zeta_T V_T^{v,*}) = v$ implies that the Lagrange multiplier μ^* satisfies

$$\frac{1}{\varrho} \ln\left(\frac{\mu^*}{\varrho}\right) = -\frac{1}{\varrho}\mathbb{E}_{\mathbb{P}}(\zeta_T \ln \zeta_T) - v. \tag{120}$$

In the case of an exponential utility, we have (recall that the variable ω is suppressed)

$$J_X(y) = (1 - e^{-\varrho(y+X_1)})(1 - F_T) + (1 - e^{-\varrho(y+X_2)})F_T,$$

so that

$$J'_X(y) = \varrho e^{-\varrho y} (e^{-\varrho X_1}(1 - F_T) + e^{-\varrho X_2} F_T).$$

Thus, setting

$$A = e^{-\varrho X_1}(1 - F_T) + e^{-\varrho X_2} F_T = e^{-\varrho \Psi},$$

we obtain

$$I_X(z) = -\frac{1}{\varrho} \ln\left(\frac{z}{A\varrho}\right) = -\frac{1}{\varrho} \ln\left(\frac{z}{\varrho}\right) - \Psi.$$

It follows that the optimal terminal wealth for the initial endowment $v - p$ is

$$V_T^{v-p,*} = -\frac{1}{\varrho} \ln\left(\frac{\lambda^* \zeta_T}{A\varrho}\right) = -\frac{1}{\varrho} \ln\left(\frac{\lambda^*}{\varrho}\right) - \frac{1}{\varrho} \ln \zeta_T - \Psi,$$

where the Lagrange multiplier λ^* is chosen so that:

$$\frac{1}{\varrho} \ln\left(\frac{\lambda^*}{\varrho}\right) = -\frac{1}{\varrho} \mathbb{E}_{\mathbb{P}}(\zeta_T \ln \zeta_T) - \mathbb{E}_{\mathbb{P}}(\zeta_T \Psi) - v + p, \tag{121}$$

which guarantees that the budget constraint $\mathbb{E}_{\mathbb{P}}(\zeta_T V_T^{v-p,*}) = v - p$ is satisfied. The $\mathbb{F}$-Hodges buying price is a real number $p^* = p_{\mathbb{F}}^*(v)$ such that

$$\mathbb{E}_{\mathbb{P}}\big(\exp(-\varrho V_T^{v,*})\big) = \mathbb{E}_{\mathbb{P}}\big(\exp(-\varrho(V_T^{v-p^*,*} + X))\big),$$

where μ^* and λ^* are given by (120) and (121), respectively. After substitution and simplifications, we arrive at the following equality

$$\mathbb{E}_{\mathbb{P}}\Big\{\exp\Big(-\varrho\big(\mathbb{E}_{\mathbb{P}}(\zeta_T \Psi) - p^* + X - \Psi\big)\Big)\Big\} = 1. \tag{122}$$

Using (115), it is easy to check that

$$\mathbb{E}_{\mathbb{P}}\big(e^{-\varrho(X-\Psi)} \mid \mathcal{F}_T\big) = 1 \tag{123}$$

so that equality (119) holds, and $\mathbb{E}_{\mathbb{P}}\big(e^{-\varrho(X-\Psi)}\big) = 1$. Combining (122) and (123), we conclude that $p_{\mathbb{F}}^*(v) = \mathbb{E}_{\mathbb{P}}(\zeta_T \Psi)$.   □

We briefly provide the analog of (118) for the $\mathbb{F}$-Hodges selling price of X. We have $p_*^{\mathbb{F}}(v) = \mathbb{E}_{\mathbb{P}}(\zeta_T \widetilde{\Psi})$, where

$$\widetilde{\Psi} = \frac{1}{\varrho} \ln\big((1 - F_T)e^{\varrho X_1} + F_T e^{\varrho X_2}\big). \tag{124}$$

Remark. It is important to notice that the $\mathbb{F}$-Hodges prices $p_{\mathbb{F}}^*(v)$ and $p_*^{\mathbb{F}}(v)$ do not depend on the initial endowment v. This is an interesting property of the exponential utility function. In view of (119), the random variable Ψ will be called the *indifference conditional hedge*.

Comparison with the Davis price. Let us present the results derived from the marginal utility pricing approach. The *Davis price* (see Davis (1997)) is given by

$$d^*(v) = \frac{\mathbb{E}_\mathbb{P}\{u'(V_T^{v,*})X\}}{\mathcal{V}'(v)}.$$

In our context, this yields

$$d^*(v) = \mathbb{E}_\mathbb{P}\{\zeta_T(X_1 F_T + X_2(1 - F_T))\}.$$

In this case, the risk aversion ϱ has no influence on the pricing of the contingent claim. In particular, when F is deterministic, the Davis price reduces to the arbitrage price of each (default-free) financial asset X^i, $i = 1, 2$, weighted by the corresponding probabilities F_T and $1 - F_T$.

Risk-Neutral Spread Versus Hodges Spreads

Let us consider the case of a defaultable bond with zero recovery, so that $X_1 = 1$ and $X_2 = 0$. It follows from (124) that the $\mathbb{F}$-Hodges buying and selling prices of the bond are (it will be convenient here to indicate the dependence of the Hodges price on maturity T)

$$D_\mathbb{F}^*(0, T) = -\frac{1}{\varrho} \mathbb{F}_\mathbb{P}\{\zeta_T \ln(e^{-\varrho}(1 - F_T) + F_T)\}$$

and

$$D_*^\mathbb{F}(0, T) = \frac{1}{\varrho} \mathbb{E}_\mathbb{P}\{\zeta_T \ln(e^\varrho(1 - F_T) + F_T)\},$$

respectively. Let $\widetilde{\mathbb{Q}}$ be a risk-neutral probability for the filtration $\mathbb{G}$, that is, for the enlarged market. The "market" price at time $t = 0$ of defaultable bond, denoted as $D^0(0, T)$, is thus equal to the expectation under $\widetilde{\mathbb{Q}}$ of its discounted pay-off, that is,

$$D^0(0, T) = \mathbb{E}_{\widetilde{\mathbb{Q}}}(\mathbb{1}_{\{\tau > T\}} R_T) = \mathbb{E}_{\widetilde{\mathbb{Q}}}((1 - \widetilde{F}_T) R_T),$$

where $\widetilde{F}_t = \widetilde{\mathbb{Q}}\{\tau \le t \mid \mathcal{F}_t\}$ for every $t \in [0, T]$. Let us emphasize that the risk-neutral probability $\widetilde{\mathbb{Q}}$ is chosen by the market, via the price of the defaultable asset. Hence, it should not be confused with the probability measure $\mathbb{Q}$, which combines, in a sense, the risk-neutral probability for the default-free market (Z^1, Z^2) with the real-life intensity of default.

Let us recall that in our setting the price process of the T-maturity unit discount Treasury (default-free) bond is $B(t, T) = e^{-r(T-t)}$. The Hodges buying and selling spreads at time $t = 0$ are defined as

$$S^*(0, T) = -\frac{1}{T} \ln \frac{D_\mathbb{F}^*(0, T)}{B(0, T)}$$

and

$$S_*(0, T) = -\frac{1}{T} \ln \frac{D_*^{\mathbb{F}}(0, T)}{B(0, T)},$$

respectively. Likewise, the *risk-neutral spread* at time $t = 0$ is given as

$$S^0(0, T) = -\frac{1}{T} \ln \frac{D^0(0, T)}{B(0, T)}.$$

Since $D_{\mathbb{F}}^*(0, 0) = D_*^{\mathbb{F}}(0, 0) = D^0(0, 0) = 1$, the respective *backward short spreads* at time $t = 0$ are given by the following limits (provided the limits exist)

$$s^*(0) = \lim_{T \downarrow 0} S^*(0, T) = -\frac{d^+ \ln D_{\mathbb{F}}^*(0, T)}{dT}\bigg|_{T=0} - r$$

and

$$s_*(0) = \lim_{T \downarrow 0} S_*(0, T) = -\frac{d^+ \ln D_*^{\mathbb{F}}(0, T)}{dT}\bigg|_{T=0} - r,$$

respectively. We also set

$$s^0(0) = \lim_{T \downarrow 0} S^0(0, T) = -\frac{d^+ \ln D^0(0, T)}{dT}\bigg|_{T=0} - r.$$

Assuming, as we do, that the processes $\widetilde{F}_T$ and F_T are absolutely continuous with respect to the Lebesgue measure, and using the observation that the restriction of $\widetilde{\mathbb{Q}}$ to $\mathcal{F}_T$ is equal to $\mathbb{Q}$, we find out that

$$\frac{D_{\mathbb{F}}^*(0, T)}{B(0, T)} = -\frac{1}{\varrho} \mathbb{E}_{\mathbb{Q}}\big\{ \ln \big(e^{-\varrho}(1 - F_T) + F_T \big) \big\}$$

$$= -\frac{1}{\varrho} \mathbb{E}_{\mathbb{Q}}\Big\{ \ln \Big(e^{-\varrho}\Big(1 - \int_0^T f_t \, dt\Big) + \int_0^T f_t \, dt \Big) \Big\},$$

and

$$\frac{D_*^{\mathbb{F}}(0, T)}{B(0, T)} = \frac{1}{\varrho} \mathbb{E}_{\mathbb{Q}}\big\{ \ln \big(e^{\varrho}(1 - F_T) + F_T \big) \big\}$$

$$= \frac{1}{\varrho} \mathbb{E}_{\mathbb{Q}}\Big\{ \ln \Big(e^{\varrho}\Big(1 - \int_0^T f_t \, dt\Big) + \int_0^T f_t \, dt \Big) \Big\}.$$

Furthermore,

$$\frac{D^0(0, T)}{B(0, T)} = \mathbb{E}_{\mathbb{Q}}(1 - \widetilde{F}_T) = \mathbb{E}_{\mathbb{Q}}\Big(1 - \int_0^T \widetilde{f}_t \, dt\Big).$$

Consequently,

$$s^*(0) = \frac{1}{\varrho}\big(e^{\varrho} - 1\big)f_0, \quad s_*(0) = \frac{1}{\varrho}\big(1 - e^{-\varrho}\big)f_0,$$

and $s^0(0) = \tilde{f}_0$. Now, if we postulate, for instance, that $s_*(0) = s^0(0)$ (it would be the case if the market price is the selling Hodges price), then we must have

$$\tilde{\gamma}_0 = \tilde{f}_0 = \frac{1}{\varrho}\left(1 - e^{-\varrho}\right)f_0 = \frac{1}{\varrho}\left(1 - e^{-\varrho}\right)\gamma_0$$

so that $\tilde{\gamma}_0 < \gamma_0$. Observe, however, that the case when the market price were equal to the buying Hodges price, that is $s^*(0) = s^0(0)$ would necessitate that $\tilde{\gamma}_0 > \gamma_0$. Similar calculations can be made for any $t \in [0, T)$.

12 Optimization Problems and BSDEs

The major distinction between this section and the previous one is that here we consider strategies ϕ that are predictable with respect to the full filtration $\mathbb{G}$. Unless explicitly stated otherwise, the underlying probability measure is the real-world probability $\mathbb{P}$. We consider the following dynamics for the risky asset Z^1

$$dZ_t^1 = Z_{t-}^1\left(\nu dt + \sigma dW_t + \varphi dM_t\right), \tag{125}$$

where $M_t = H_t - \int_0^{t \wedge \tau} \gamma_s \, ds$, and where we impose the condition $\varphi > -1$, which ensures that the price Z_t^1 remains strictly positive.

In order to simplify notation, we shall denote by ζ the process such that $dM_t = dH_t - \xi_t \, dt$ is a $\mathbb{G}$-martingale, i.e., $\xi_t = \gamma_t(1 - H_t)$. We assume that the hypothesis (H) holds, that is, any $\mathbb{F}$-martingale is a $\mathbb{G}$-martingale as well.

Throughout most of the section, we shall deal with the same market model as in the previous section, that is, we shall set $\varphi = 0$. Only in Section 14 we generalize the dynamics of the risky asset to the case when $\varphi \neq 0$, so that the dynamics of the risky asset Z^1 are sensitive to the default risk. In particular, the limit case $\varphi = -1$ corresponds to the case where the underlying risky asset has value 0 after the default.

We assume for simplicity that $r = 0$, and we change the notational convention for an admissible portfolio to the one that will be more suitable for problems considered here: instead of using the number of shares ϕ as before, we set $\pi = \phi Z^1$, so that π represents the value invested in the risky asset. The portfolio process π_t should not be confused with the arbitrage price process $\pi_t(X)$. In addition, we adopt here the following relaxed definition of admissibility of a self-financing trading strategy.

Definition 11. *The class $\Pi(\mathbb{F})$ (respectively $\Pi(\mathbb{G})$) of $\mathbb{F}$-admissible (respectively $\mathbb{G}$-admissible) trading strategies is the set of all $\mathbb{F}$-predictable (respectively $\mathbb{G}$-predictable) processes π such that $\int_0^T \pi_t^2 \, dt < \infty$, $\mathbb{P}$-a.s.*

The wealth process of a strategy π satisfies

$$dV_t(\pi) = \pi_t\left(\nu dt + \sigma dW_t + \varphi dM_t\right). \tag{126}$$

Note that with the present definition of admissible strategies the "martingale part" of the wealth process is a local martingale, in general.

Let X be a given contingent claim, represented by a $\mathcal{G}_T$-measurable random variable. We shall study the following problem:

$$\sup_{\pi \in \Pi(\mathbb{G})} \mathbb{E}_{\mathbb{P}}\{u(V_T^v(\pi) + X)\}.$$

12.1 Exponential Utility

In this section, we shall examine the problem introduced above in the case of the exponential utility, and setting $\varphi = 0$ in dynamics (125). First, we examine the existence and the form of a solution to the optimization problem, under additional technical assumptions. Subsequently, we shall derive the expression for the Hodges buying price.

Optimization Problem

Let $X \in \mathcal{G}_T$ be a given non-negative contingent claim, and let v be the initial endowment of an agent. Our first goal is to solve an optimization problem for an agent who buys a claim X. To this end, it suffices to find a strategy $\pi \in \Pi(\mathbb{G})$ that maximizes $\mathbb{E}_{\mathbb{P}}(u(V_T^v(\pi) + X))$, where the wealth process $V_t = V_t^v(\pi)$ (for simplicity, we shall frequently skip v and π from the notation) satisfies

$$dV_t = \phi_t \, dZ_t^1 = \pi_t(\nu dt + \sigma dW_t), \ V_0 = v.$$

We consider the exponential utility function $u(x) = 1 - e^{-\varrho x}$, with $\varrho > 0$. Therefore, we deal with the following problem:

$$\sup_{\pi \in \Pi(\mathbb{G})} \mathbb{E}_{\mathbb{P}}\{u(V_T^v(\pi) + X)\} = 1 - \inf_{\pi \in \Pi(\mathbb{G})} \mathbb{E}_{\mathbb{P}}(e^{-\varrho V_T^v(\pi)} e^{-\varrho X}).$$

Let us describe the idea of a solution. Suppose that we can find a process Z with $Z_T = e^{-\varrho X}$, which depends only on the claim X and parameters ϱ, σ, ν, and such that the process $e^{-\varrho V_t^v(\pi)} Z_t$ is a $\mathbb{G}$-submartingale under $\mathbb{P}$ for any admissible strategy π and is a martingale under $\mathbb{P}$ for some admissible strategy $\pi^* \in \Pi(\mathbb{G})$. Then, we would have

$$\mathbb{E}_{\mathbb{P}}(e^{-\varrho V_T^v(\pi)} Z_T) \geq e^{-\varrho V_0^v(\pi)} Z_0 = e^{-\varrho v} Z_0$$

for any $\pi \in \Pi(\mathbb{G})$, with equality for some strategy $\pi^* \in \Pi(\mathbb{G})$. Consequently, we would obtain

$$\inf_{\pi \in \Pi(\mathbb{G})} \mathbb{E}_{\mathbb{P}}(e^{-\varrho V_T^v(\pi)} e^{-\varrho X}) = \mathbb{E}_{\mathbb{P}}(e^{-\varrho V_T^v(\pi^*)} e^{-\varrho X}) = e^{-\varrho v} Z_0, \qquad (127)$$

and thus we would be in a position to conclude that π^* is an optimal strategy. In fact, it will turn out that in order to implement the above idea we shall need to restrict further the class of $\mathbb{G}$-admissible trading strategies.

We shall search for an auxiliary process Z in the class of all processes satisfying the following backward stochastic differential equation (BSDE)

$$dZ_t = f_t\, dt + \widehat{z}_t\, dW_t + \widetilde{z}_t\, dM_t,\ t \in [0, T),\ Z_T = e^{-\varrho X}, \tag{128}$$

where the process f will be determined later (see equation (130) below). By applying Itô's formula, we obtain

$$d(e^{-\varrho V_t}) = e^{-\varrho V_t}\left(\left(\tfrac{1}{2}\varrho^2 \pi_t^2 \sigma^2 - \varrho \pi_t \nu\right) dt - \varrho \pi_t \sigma\, dW_t\right),$$

so that

$$d(e^{-\varrho V_t} Z_t) = e^{-\varrho V_t}\left(f_t + Z_t(\tfrac{1}{2}\varrho^2 \pi_t^2 \sigma^2 - \varrho \pi_t \nu) - \varrho \pi_t \sigma \widehat{z}_t\right) dt$$
$$+ e^{-\varrho V_t}\left((\widehat{z}_t - \varrho \pi_t \sigma Z_t)\, dW_t + \widetilde{z}_t\, dM_t\right).$$

Let us choose π^* such that it minimizes, for every t, the following expression

$$Z_t\left(\tfrac{1}{2}\varrho^2 \pi_t^2 \sigma^2 - \varrho \pi_t \nu\right) - \varrho \pi_t \sigma \widehat{z}_t = -\varrho \pi_t(\nu Z_t + \sigma \widehat{z}_t) + \tfrac{1}{2}\varrho^2 \pi_t^2 \sigma^2 Z_t.$$

It is easily seen that

$$\pi_t^* = \frac{\nu Z_t + \sigma \widehat{z}_t}{\varrho \sigma^2 Z_t} = \frac{1}{\varrho \sigma}\left(\theta + \frac{\widehat{z}_t}{Z_t}\right), \tag{129}$$

Now, let us choose the process f, by postulating that

$$f_t = f(Z_t, \widehat{z}_t) = Z_t\left(\varrho \pi_t^* \nu - \tfrac{1}{2}\varrho^2 (\pi_t^*)^2 \sigma^2\right) + \varrho \pi_t^* \sigma \widehat{z}_t$$
$$= \varrho \pi_t^*(Z_t \nu + \sigma \widehat{z}_t) - \tfrac{1}{2}\varrho^2 (\pi_t^*)^2 \sigma^2 Z_t = \frac{(\nu Z_t + \sigma \widehat{z}_t)^2}{2\sigma^2 Z_t}. \tag{130}$$

In other words, we shall focus on the following BSDE:

$$dZ_t = \frac{(\nu Z_t + \sigma \widehat{z}_t)^2}{2\sigma^2 Z_t}\, dt + \widehat{z}_t\, dW_t + \widetilde{z}_t\, dM_t,\ t \in [0, T[,\ Z_T = e^{-\varrho X}. \tag{131}$$

Recall that W is a Brownian motion under $\mathbb{P}$, and that the risk-neutral probability $\mathbb{Q}$ is given by $d\mathbb{Q}|_{\mathcal{F}_t} = \eta_t\, d\mathbb{P}|_{\mathcal{F}_t}$, where $d\eta_t = -\eta_t \theta\, dW_t$ with $\theta = \nu/\sigma$ and $\eta_0 = 1$. Thus the process $W_t^{\mathbb{Q}} = W_t + \theta t$, $t \in [0, T]$, is a Brownian motion under $\mathbb{Q}$. It will be convenient to write equation (131) as

$$dZ_t = \left(\tfrac{1}{2}\theta^2 Z_t + \theta \widehat{z}_t + \tfrac{1}{2}Z_t^{-1}\widehat{z}_t^2\right) dt + \widehat{z}_t\, dW_t + \widetilde{z}_t\, dM_t,\ t \in [0, T[,\ Z_T = e^{-\varrho X}.$$

Equivalently,

$$dZ_t = \left(\tfrac{1}{2}\theta^2 Z_t + \tfrac{1}{2}Z_t^{-1}\widehat{z}_t^2\right) dt + \widehat{z}_t\, dW_t^{\mathbb{Q}} + \widetilde{z}_t\, dM_t,\ t \in [0, T[,\ Z_T = e^{-\varrho X}. \tag{132}$$

Remark. To the best of out knowledge, no general theorem, which would establish the existence of a solution to equation (132), is available. The comparison theorem works for BSDEs driven by a jump process when the drift satisfies some Lipschitz condition (see Royer (2003)). Hence, the proofs of Lepeltier and San-Martin (1997) and Kobylanski (2000), which rely on comparison results, may not be directly carried to the case of quadratic BSDEs driven by a jump process. We shall solve the BSDE (132) under rather restrictive assumptions on X. Hence, the general case remains an open problem.

Lemma 19. *Assume that there exists $\mathbb{G}$-predictable processes $\widehat{k}$, $\widetilde{k} > -1$ and a constant c such that*

$$\exp(K_T)\mathcal{E}_T(\widetilde{M}) = e^{-\varrho X}, \tag{133}$$

where

$$K_t = c + \int_0^t \widehat{k}_u \, dW_u^{\mathbb{Q}}, \quad \widetilde{M}_t = \int_0^t \widetilde{k}_u \, dM_u,$$

and $\mathcal{E}(\widetilde{M})$ is the Doléans exponential of $\widetilde{M}$. Then $U_t = \exp(K_t)\mathcal{E}_t(\widetilde{M})$ solves the following BSDE

$$dU_t = \tfrac{1}{2}U_t^{-1}\widehat{u}_t^2 \, dt + \widehat{u}_t \, dW_t^{\mathbb{Q}} + \widetilde{u}_t \, dM_t, \ t \in [0, T[, \ U_T = e^{-\varrho X}. \tag{134}$$

Proof. Since $d\mathcal{E}_t(\widetilde{M}) = \mathcal{E}_{t-}(\widetilde{M}) \, d\widetilde{M}_t$, the process U defined above satisfies

$$dU_t = \tfrac{1}{2}U_t\widehat{k}_t^2 \, dt + U_t\widehat{k}_t \, dW_t^{\mathbb{Q}} + U_{t-}\widetilde{k}_t \, dM_t$$

and thus

$$dU_t = \tfrac{1}{2}U_t^{-1}\widehat{u}_t^2 \, dt + \widehat{u}_t \, dW_t^{\mathbb{Q}} + \widetilde{u}_t \, dM_t$$

where we denote $\widehat{u}_t = U_t\widehat{k}_t$ and $\widetilde{u}_t = U_{t-}\widetilde{k}_t$. Since obviously $U_T = e^{-\varrho X}$, this ends the proof. □

Corollary 12. *Let X be a $\mathcal{G}_T$-measurable claim such that (133) holds for some $\mathbb{G}$-predictable processes $\widehat{k}$, $\widetilde{k} > -1$ and some constant c. Then there exists a solution $(Z, \widehat{z}, \widetilde{z})$ of the BSDE (132). Moreover, the process Z is strictly positive.*

Proof. Let us set $Y_t = e^{-(T-t)\theta^2/2}$ and let U be the process introduced in Lemma 19. Then the process $Z_t = U_tY_t$ satisfies

$$\begin{aligned}
dZ_t &= Y_t \, dU_t + \tfrac{1}{2}\theta^2 Y_t U_t \, dt \\
&= \tfrac{1}{2}\theta^2 Y_t U_t \, dt + \tfrac{1}{2}Y_t U_t^{-1}\widehat{u}_t^2 \, dt + Y_t\widehat{u}_t \, dW_t^{\mathbb{Q}} + Y_t\widetilde{u}_t \, dM_t \\
&= \tfrac{1}{2}\theta^2 Z_t \, dt + \tfrac{1}{2}Z_t^{-1}Y_t^2\widehat{u}_t^2 \, dt + Y_t\widehat{u}_t \, dW_t^{\mathbb{Q}} + Y_t\widetilde{u}_t \, dM_t \\
&= \tfrac{1}{2}\theta^2 Z_t \, dt + \tfrac{1}{2}Z_t^{-1}\widehat{z}_t^2 \, dt + \widehat{z}_t \, dW_t^{\mathbb{Q}} + \widetilde{z}_t \, dM_t
\end{aligned}$$

where we set $\widehat{z}_t = Y_t\widehat{u}_t$ and $\widetilde{z}_t = Y_t\widetilde{u}_t$. It is also clear that $Z_T = U_T = e^{-\varrho X}$ and Z is strictly positive. □

Recall that the process Z depends on the choice of a contingent claim X, as well as on the model's parameters ϱ, σ and ν. The next lemma shows that the processes Z and π^* introduced above have indeed the desired properties that were described at the beginning of this section. To achieve our goal, we need to restrict the class of admissible trading strategies, however. We say that an admissible strategy π is *regular with respect to X* if the martingale part of the process $e^{-\varrho V_t^v(\pi)} Z_t$ is a martingale under $\mathbb{P}$, rather than a local martingale. We denote by $\Pi_X(\mathbb{G})$ the class of all admissible trading strategies, which are regular with respect to X.

Lemma 20. *Let X be a $\mathcal{G}_T$-measurable claim such that (133) holds for some $\mathbb{G}$-predictable processes $\widehat{k}, \widetilde{k}$ and some constant c. Assume that the default intensity γ and the processes $\widehat{k}, \widetilde{k}$ are bounded. Suppose that the process $Z = Z(X, \varrho, \sigma, \nu)$ is a solution to the BSDE (131) given in Corollary 12. Then:*
(i) The process $e^{-\varrho V_t^v(\pi)} Z_t$ is a submartingale for any strategy $\pi \in \Pi_X(\mathbb{G})$.
(ii) The process $e^{-\varrho V_t^v(\pi^*)} Z_t$ is a martingale for the process π^* given by expression (129).
(iii) The process π^* belongs to the class $\Pi_X(\mathbb{G})$ of admissible trading strategies regular with respect to X.

Proof. In view of the definition of π^* and the choice of the process f (see formula (130)), the validity of part (i) is rather clear. To establish (ii), we shall first check that the process $e^{-\varrho V_t^*} Z_t$ is a martingale (and not only a local martingale) under $\mathbb{P}$, where $V_t^* = V_t^v(\pi^*)$. From the choice of π^*, we obtain

$$d(e^{-\varrho V_t^*} Z_t) = e^{-\varrho V_t^*}\left((\widehat{z}_t - \varrho \pi_t^* \sigma Z_t)\, dW_t + \widetilde{z}_t\, dM_t\right)$$
$$= -\theta e^{-\varrho V_t^*} Z_t\, dW_t + e^{-\varrho V_t^*} \widetilde{z}_t\, dM_t.$$

This means that

$$e^{-\varrho V_t^*} Z_t = e^{-\varrho v} Z_0 \exp\left(-\theta W_t - \tfrac{1}{2}\theta^2 t\right) \exp\left(-\int_0^t \frac{\widetilde{z}_s}{Z_s} \xi_s\, ds\right)\left(1 + \frac{\widetilde{z}_{\tau-}}{Z_{\tau-}} H_t\right).$$

The quantity $e^{-\varrho v} Z_0 \exp\left(-\theta W_t - \tfrac{1}{2}\theta^2 t\right)$ is clearly a continuous martingale under $\mathbb{P}$. Recall that

$$\widetilde{z}_t = Y_t \widetilde{u}_t = \widetilde{k}_t Z_t.$$

and thus $\widetilde{z}_t / Z_t = \widetilde{k}_t$ is a bounded process. We conclude that the process

$$\exp\left(-\int_0^t \frac{\widetilde{z}_s}{Z_s} \xi_s\, ds\right)\left(1 + \frac{\widetilde{z}_{\tau-}}{Z_{\tau-}} H_t\right)$$

is a bounded, purely discontinuous martingale under $\mathbb{P}$. To complete the proof, it remains to check that the process π^* given by (129) is $\mathbb{G}$-admissible, in the sense of Definition 11. To this end, it suffices to check that

$$\int_0^T \widehat{z}_t^2 Z_t^{-2}\, dt < \infty, \quad \mathbb{P}\text{-a.s.}$$

This is clear since the process $\widehat{z}_t/Z_t = \widehat{k}_t$ is bounded. We conclude that the strategy π^* belongs to the class $\Pi_X(\mathbb{G})$.   □

Recall now that in this section we examine the following problem:

$$\sup_{\pi \in \Pi_X(\mathbb{G})} \mathbb{E}_{\mathbb{P}}\big(u(V_T^v(\pi) + X)\big) = 1 - \inf_{\pi \in \Pi_X(\mathbb{G})} \mathbb{E}_{\mathbb{P}}\big(e^{-\varrho V_T^v(\pi)} e^{-\varrho X}\big).$$

We are in a position to state the following result.

Proposition 26. *Let X be a $\mathcal{G}_T$-measurable claim such that (133) holds for some $\mathbb{G}$-predictable processes $\widehat{k}$, $\widetilde{k}$ and some constant c. Assume that the default intensity γ and the processes $\widetilde{k}$, $\widehat{k}$ are bounded. Then*

$$\inf_{\pi \in \Pi_X(\mathbb{G})} \mathbb{E}_{\mathbb{P}}\big(e^{-\varrho V_T^v(\pi)} e^{-\varrho X}\big) = \mathbb{E}_{\mathbb{P}}\big(e^{-\varrho V_T^v(\pi^*)} e^{-\varrho X}\big) = e^{-\varrho v} Z_0^X,$$

where the optimal strategy $\pi^* \in \Pi_X(\mathbb{G})$ is given by the formula, for every $t \in [0, T]$,

$$\pi_t^* = \frac{1}{\varrho \sigma}\left(\theta + \frac{\widehat{z}_t^X}{Z_t^X}\right) = \frac{\theta + \widehat{k}_t}{\varrho \sigma},$$

where $Z_t^X = Z_t$ and $\widehat{z}_t^X = \widehat{z}_t$ are the two first components of a solution $(Z_t, \widehat{z}_t, \widetilde{z})$ of the BSDE

$$dZ_t = \frac{(\nu Z_t + \sigma \widehat{z}_t)^2}{2\sigma^2 Z_t}\, dt + \widehat{z}_t\, dW_t + \widetilde{z}_t\, dM_t, \ Z_T = e^{-\varrho X}. \tag{135}$$

More explicitly (see Corollary 12), we have $\widehat{z}_t = \widehat{k}_t Z_t$ and

$$Z_t = e^{-(T-t)\theta^2/2} \exp(K_t)\mathcal{E}_t(\widetilde{M}).$$

Proof. The proof is rather straightforward. We know that the process Z which solves (135) is such that: (i) the process $Z_t e^{-\varrho V_t^v(\pi^*)}$ is a martingale, and (ii) for any strategy $\pi \in \Pi_X(\mathbb{G})$ the process $Z_t e^{-\varrho V_t^v(\pi)}$ is equal to a martingale minus an increasing process (since the drift term is non-positive), and thus it is a submartingale. This shows that (127) holds with $\Pi(\mathbb{G})$ substituted with $\Pi_X(\mathbb{G})$.   □

It should be acknowledged that the assumptions of Proposition 26 are restrictive, so that it covers only a very special case of a claim X. Let us now comment briefly on the case of a general claim; we do not pretend here to give strict results, our aim is merely to give some hints how one can deal with the general case.

Recall that our aim is to find a solution $(Z, \widehat{z}, \widetilde{z})$ of the following BSDE

$$dZ_t = \big(\tfrac{1}{2}\theta^2 Z_t + \tfrac{1}{2}Z_t^{-1}\widehat{z}_t^2\big)dt + \widehat{z}_t\, dW_t^{\mathbb{Q}} + \widetilde{z}_t\, dM_t, \ t \in [0, T[, \ Z_T = e^{-\varrho X},$$

or equivalently, of the equation

$$dU_t = \tfrac{1}{2} U_t^{-1} \widehat{u}_t^2 \, dt + \widehat{u}_t \, dW_t^{\mathbb{Q}} + \widetilde{u}_t \, dM_t, \; t \in [0, T[, \; U_T = e^{-\varrho X},$$

Assume that the process U is strictly positive and set $X_t = \ln U_t$. Then, denoting $\widehat{x}_t = \widehat{u}_t U_t^{-1}, \widetilde{x}_t = \widetilde{u}_t U_{t-}^{-1}$ and applying Itô's formula, we obtain (recall that we denote $\xi_t = \gamma_t \mathbb{1}_{\{\tau > t\}}$)

$$
\begin{aligned}
dX_t &= \widehat{x}_t \, dW_t^{\mathbb{Q}} + \widetilde{x}_t \, dM_t + (\ln(1 + \widetilde{x}_t) - \widetilde{x}_t) \, dH_t \\
&= \widehat{x}_t \, dW_t^{\mathbb{Q}} + \widetilde{x}_t \, dM_t + (\ln(1 + \widetilde{x}_t) - \widetilde{x}_t)(dM_t + \xi_t \, dt) \\
&= \widehat{x}_t \, dW_t^{\mathbb{Q}} + \ln(1 + \widetilde{x}_t) \, dM_t + (\ln(1 + \widetilde{x}_t) - \widetilde{x}_t)\xi_t \, dt \\
&= \widehat{x}_t \, dW_t^{\mathbb{Q}} + x_t^* \, dM_t + (1 - e^{x_t^*} + x_t^*)\xi_t \, dt \\
&= \widehat{x}_t \, dW_t^{\mathbb{Q}} + x_t^* \, dH_t + (1 - e^{x_t^*})\xi_t \, dt,
\end{aligned}
$$

where $x_t^* = \ln(1 + \widetilde{x}_t)$ and the terminal condition is $X_T = -\varrho X$. It thus suffices to solve the following BSDE

$$dX_t = \widehat{x}_t \, dW_t^{\mathbb{Q}} + x_t^* \, dH_t + (1 - e^{x_t^*})\xi_t \, dt, \; t \in [0, T[, \; X_T = -\varrho X. \quad (136)$$

Assume first that $X \in \mathcal{F}_T$. In that case, it is obvious that we may take $\widehat{x} = \widetilde{x}^* = 0$ and thus $X_t = -\mathbb{E}_{\mathbb{Q}}(\varrho X \mid \mathcal{G}_t) = -\mathbb{E}_{\mathbb{Q}}(\varrho X \mid \mathcal{F}_t)$ is a solution. In the general case, we note that the continuous $\mathbb{G}$-martingales are stochastic integrals with respect to the Brownian motion $W^{\mathbb{Q}}$. We may thus transform the problem: it suffices to find a process x^* such that the process R, defined through the formula

$$R_t = \mathbb{E}_{\mathbb{Q}}\left(-\varrho X + \int_0^T (e^{x_s^*} - 1)\xi_s \, ds - x_\tau^* \mathbb{1}_{\{\tau \le T\}} \, \Big| \, \mathcal{G}_t \right),$$

is a continuous $\mathbb{G}$-martingale, so that $dR_t = \widehat{x}_t \, dW_t^{\mathbb{Q}}$ for some $\mathbb{G}$-predictable process $\widehat{x}$. Suppose that we can find a process x^* for which the last property is valid. Then, by setting

$$
\begin{aligned}
X_t &= R_t - \int_0^t (e^{x_s^*} - 1)\xi_s \, ds - x_\tau^* \mathbb{1}_{\{\tau \le t\}} \\
&= \mathbb{E}_{\mathbb{Q}}\left(-\varrho X + \int_t^T (e^{x_s^*} - 1)\xi_s \, ds - x_\tau^* \mathbb{1}_{\{t < \tau \le T\}} \, \Big| \, \mathcal{G}_t \right)
\end{aligned}
$$

we obtain a solution $(X, \widehat{x}, x^*)$ to (136).

Case of a survival claim. From now on, we shall focus on a survival claim $X = Y \mathbb{1}_{\{\tau > T\}}$, where Y is an $\mathcal{F}_T$-measurable random variable. Let us fix $t \in [0, T]$. On the set $\{t \le \tau\}$ we obtain

$$\mathbb{E}_{\mathbb{Q}}(\varrho Y \mathbb{1}_{\{\tau > T\}} \mid \mathcal{G}_t) = e^{\Gamma_t} \, \mathbb{E}_{\mathbb{Q}}(e^{-\Gamma_T} \varrho Y \mid \mathcal{F}_t)$$

and on the set $\{\tau < t\}$, we have $\mathbb{E}_{\mathbb{Q}}(\varrho Y \mathbb{1}_{\{\tau > T\}} \mid \mathcal{G}_t) = 0$. The jump of the term A_t, defined as

$$A_t = \mathbb{E}_\mathbb{Q}\left(\int_t^T (e^{x_s^*} - 1)\xi_s \, ds - x_\tau^* \mathbb{1}_{\{\tau \leq T\}} \,\Big|\, \mathcal{G}_t\right),$$

can be computed as follows. On the set $\{t \leq \tau\}$, we obtain

$$A_t = \int_t^T \mathbb{E}_\mathbb{Q}\big((e^{x_s^*} - 1)\gamma_s \mathbb{1}_{\{\tau > s\}} \,|\, \mathcal{G}_t\big) \, ds - \mathbb{E}_\mathbb{Q}\big(x_\tau^* \mathbb{1}_{\{\tau \leq T\}} \,|\, \mathcal{G}_t\big)$$

$$= \mathbb{1}_{\{\tau > t\}} e^{\Gamma_t} \mathbb{E}_\mathbb{Q}\left(\int_t^T (e^{x_s^*} - 1 - x_s^*) e^{-\Gamma_s} \gamma_s \, ds \,\Big|\, \mathcal{F}_t\right).$$

On the set $\{\tau < t\}$ for A_t we have

$$\mathbb{E}_\mathbb{Q}\left(\int_t^T (e^{x_s^*} - 1)\gamma_s \mathbb{1}_{\{\tau > s\}} \, ds - x_\tau^* \mathbb{1}_{\{\tau \leq T\}} \,\Big|\, \mathcal{G}_t\right) = -\mathbb{E}_\mathbb{Q}\big(x_\tau^* \mathbb{1}_{\{\tau \leq t\}} \,|\, \mathcal{G}_t\big) = -x_\tau^*.$$

We conclude that our problem is to find a process x^* such that

$$-\mathbb{E}_\mathbb{Q}\big(e^{-\Gamma_T} \varrho Y | \mathcal{F}_t\big) = -e^{-\Gamma_t} x_t^* - \mathbb{E}_\mathbb{Q}\left(\int_t^T (e^{x_s^*} - 1 - x_s^*) e^{-\Gamma_s} \gamma_s \, ds \,\Big|\, \mathcal{F}_t\right).$$

In other words, we need to solve the following BSDE with $\mathbb{F}$-adapted processes x^* and κ

$$d\big(x_t^* e^{-\Gamma_t}\big) = \big(e^{x_t^*} - 1 - x_t^*\big) e^{-\Gamma_t} \gamma_t \, dt + \kappa_t \, dW_t^\mathbb{Q}, \; t \in [0, T[, \; x_T^* = \varrho Y.$$

From integration by parts, this BSDE can be written

$$dx_t^* = \big(e^{x_t^*} - 1\big) e^{-\Gamma_t} \gamma_t \, dt + \kappa_t \, dW_t^\mathbb{Q}, \; t \in [0, T[, \; x_T^* = \varrho X.$$

Unfortunately, the standard results for existence of solutions to BSDEs do not apply here because the drift term is not of a linear growth with respect to x^*.

12.2 Hodges Buying and Selling Prices

Particular case. Assume, as before, that $r = 0$ and let us check that the Hodges buying price is the hedging price in case of attainable claims. Assume that a claim X is $\mathcal{F}_T$-measurable. By virtue of the predictable representation theorem, there exists a pair $(x, \hat{x})$, where x is a constant and $\hat{x}_t$ is an $\mathbb{F}$-adapted process, such that $X = x + \int_0^T \hat{x}_u \, dW_u^\mathbb{Q}$, where $W_t^\mathbb{Q} = W_t + \theta t$. Here $x = \mathbb{E}_\mathbb{Q} X$ is the arbitrage price $\pi_0(X)$ of X and the replicating portfolio is obtained through $\hat{x}$. Hence, the time t value of X is $X_t = x + \int_0^t \hat{x}_u \, dW_u^\mathbb{Q}$. Then $dX_t = \hat{x}_t \, dW_t^\mathbb{Q}$ and the process

$$Z_t = e^{-\theta^2(T-t)/2} e^{-\varrho X_t}$$

satisfies

$$dZ_t = Z_t\left(\left(\tfrac{1}{2}\theta^2 + \tfrac{1}{2}\varrho^2\widehat{x}_t^2\right)dt + \varrho\widehat{x}_t\,dW_t^{\mathbb{Q}}\right)$$

$$= \frac{1}{2\sigma^2 Z_t}(\nu Z_t + \sigma\varrho Z_t\widehat{x}_t)^2\,dt + \varrho Z_t\widehat{x}_t\,dW_t.$$

Hence $(Z_t, \varrho Z_t\widehat{x}_t, 0)$ is the solution of (135) with the terminal condition $e^{-\varrho X}$, and

$$Z_0 = e^{-\theta^2 T/2}e^{-\varrho x}.$$

Note that, for $X = 0$, we get $Z_0 = e^{-\theta^2 T/2}$, therefore

$$\inf_{\pi\in\Pi(\mathbb{G})} \mathbb{E}_{\mathbb{P}}(e^{-\varrho V_T^v(\pi)}) = e^{-\varrho v}e^{-\theta^2 T/2}.$$

The $\mathbb{G}$-Hodges buying price of X is the value of p such that

$$\inf_{\pi\in\Pi(\mathbb{G})} \mathbb{E}_{\mathbb{P}}\left(e^{-\varrho V_T^v(\pi)}\right) = \inf_{\pi\in\Pi(\mathbb{G})} \mathbb{E}_{\mathbb{P}}\left(e^{-\varrho(V_T^{v-p}(\pi)+X)}\right),$$

that is,

$$e^{-\varrho v}e^{-\theta^2 T/2} = e^{-\varrho(v-p+\pi_0(X))}e^{-\theta^2 T/2}.$$

We conclude easily that $p_*^{\mathbb{G}}(X) = \pi_0(X) = \mathbb{E}_{\mathbb{Q}}X$. Similar arguments show that $p_{\mathbb{G}}^*(X) = \pi_0(X)$.

General case. Assume now that a claim X is $\mathcal{G}_T$-measurable and the assumptions of Proposition 26 are satisfied. Since the process $\widetilde{z}$ introduced in Corollary 12 is strictly positive, we can use its logarithm. Let us assume that the processes $\widehat{k}$ and $\widetilde{k}$ are strictly positive, and let us denote $\widehat{\psi}_t = Z_t/\widehat{z}_t = \widehat{k}_t^{-1}$, $\widetilde{\psi}_t = Z_t/\widetilde{z}_t = \widetilde{k}_t^{-1}$ and

$$\kappa_t = \frac{\widetilde{\psi}_t}{\ln(1 + \widetilde{\psi}_t)} \geq 0.$$

Then we get

$$d(\ln Z_t) = \tfrac{1}{2}\theta^2\,dt + \widehat{\psi}_t\,dW_t^{\mathbb{Q}} + \ln(1 + \widetilde{\psi}_t)\big(dM_t + \xi_t(1 - \kappa_t)\,dt\big),$$

and thus

$$d(\ln Z_t) = \tfrac{1}{2}\theta^2\,dt + \widehat{\psi}_t\,dW_t^{\mathbb{Q}} + \ln(1 + \widetilde{\psi}_t)\,d\widehat{M}_t,$$

where

$$d\widehat{M}_t = dM_t + \xi_t(1 - \kappa_t)\,dt = dH_t - \xi_t\kappa_t\,dt.$$

The process $\widehat{M}$ is a martingale under the probability measure $\widehat{\mathbb{Q}}$ defined as $d\widehat{\mathbb{Q}}|_{\mathcal{G}_t} = \widehat{\eta}_t\,d\mathbb{P}|_{\mathcal{G}_t}$, where $\widehat{\eta}$ satisfies

$$d\widehat{\eta}_t = -\widehat{\eta}_{t-}\big(\theta\,dW_t + \xi_t(1 - \kappa_t)\,dM_t\big)$$

with $\widehat{\eta}_0 = 1$.

Proposition 27. *The $\mathbb{G}$-Hodges buying price of X with respect to the exponential utility is the real number p such that $e^{-\varrho(v-p)}Z_0^X = e^{-\varrho v}Z_0^0$, that is, $p_{\mathbb{G}}^*(X) = \varrho^{-1}\ln(Z_0^0/Z_0^X)$ or, equivalently, $p_{\mathbb{G}}^*(X) = \mathbb{E}_{\widehat{\mathbb{Q}}}X$.*

Our previous study establishes that the dynamic hedging price of a claim X is the process $X_t = \mathbb{E}_{\widehat{\mathbb{Q}}}(X \mid \mathcal{G}_t)$. This price is the expectation of the payoff, under some martingale measure, as is any price in the range of no-arbitrage prices.

13 Quadratic Hedging

We assume here that the wealth process follows

$$dV_t^v(\pi) = \pi_t\big(\nu\,dt + \sigma\,dW_t\big), \quad V_0^v(\pi) = v,$$

where we assume that $\pi \in \Pi(\mathbb{F})$ or $\pi \in \Pi(\mathbb{G})$, depending on the case studied below. The more general case

$$dV_t^v(\pi) = \pi_t\big(\nu\,dt + \sigma\,dW_t + \varphi\,dM_t\big), \quad V_0^v(\pi) = v$$

is too long to be presented here. In this section, we examine the issue of the quadratic pricing and hedging, specifically, for a given $\mathbb{P}$-square-integrable claim X we solve the following minimization problems:

• For a given initial endowment v, solve the minimization problem:

$$\min_{\pi} \mathbb{E}_{\mathbb{P}}\big((V_T^v(\pi) - X)^2\big).$$

A solution to this problem provides the portfolio which, among the portfolios with a given initial wealth, has the closest terminal wealth to a given claim X, in sense of L^2-norm under $\mathbb{P}$.

• Solve the minimization problem:

$$\min_{\pi,v} \mathbb{E}_{\mathbb{P}}\big((V_T^v(\pi) - X)^2\big).$$

The minimal value of v is called the *quadratic hedging price* and the optimal π the *quadratic hedging strategy*.

The mean-variance hedging problem was examined in a fairly general framework of incomplete markets by means of BSDEs in several papers; see, for example, Mania (2000), Mania and Tevzadze (2003), Bobrovnytska and Schweizer (2004), Hu and Zhou (2004) or Lim (2004). Since this list is by no means exhaustive, the interested reader is referred to the references quoted in the above-mentioned papers.

13.1 Quadratic Hedging with $\mathbb{F}$-Adapted Strategies

We shall first solve, for a given initial endowment v, the following minimization problem

$$\min_{\pi \in \Pi(\mathbb{F})} \mathbb{E}_\mathbb{P}\big((V_T^v(\pi) - X)^2\big),$$

where the claim $X \in \mathcal{G}_T$ is given as

$$X = X_1 \mathbb{1}_{\{\tau > T\}} + X_2 \mathbb{1}_{\{\tau \leq T\}}$$

for some $\mathcal{F}_T$-measurable and $\mathbb{P}$-square-integrable random variables X_1 and X_2. Using the same approach as in the previous section, we define the auxiliary function J_X by setting

$$J_X(y) = (y - X_1)^2(1 - F_T) + (y - X_2)^2 F_T,$$

so that its derivative equals

$$J_X'(y) = 2\big(y - X_1(1 - F_T) - X_2 F_T\big).$$

Hence

$$I_X(z) = \tfrac{1}{2}z + X_1(1 - F_T) + X_2 F_T,$$

and thus the optimal terminal wealth equals

$$V_T^{v,*} = \tfrac{1}{2}\lambda^* \zeta_T + X_1(1 - F_T) + X_2 F_T,$$

where λ^* is specified through the budget constraint:

$$\mathbb{E}_\mathbb{P}(\zeta_T V_T^{v,*}) = \tfrac{1}{2}\lambda^* \mathbb{E}_\mathbb{P}(\zeta_T^2) + \mathbb{E}_\mathbb{P}(\zeta_T X_1(1 - F_T)) + \mathbb{E}_\mathbb{P}(\zeta_T X_2 F_T) = v.$$

We deduce that

$$
\begin{aligned}
\min_\pi &\, \mathbb{E}_\mathbb{P}\big((V_T^v - X)^2\big) \\
&= \mathbb{E}_\mathbb{P}\left[\big(\tfrac{1}{2}\lambda^* \zeta_T + X_1(1 - F_T) + X_2 F_T - X_1\big)^2 (1 - F_T)\right] \\
&\quad + \mathbb{E}_\mathbb{P}\left[\big(\tfrac{1}{2}\lambda^* \zeta_T + X_1(1 - F_T) + X_2 F_T\big) - X_2\big)^2 F_T\right] \\
&= \tfrac{1}{4}(\lambda^*)^2\, \mathbb{E}_\mathbb{P}(\zeta_T^2) + \mathbb{E}_\mathbb{P}\big((X_1 - X_2)^2 F_T(1 - F_T)\big) \\
&= \frac{1}{2\mathbb{E}_\mathbb{P}(\zeta_T^2)}\big(v - \mathbb{E}_\mathbb{P}(\zeta_T(X_1 + F_T(X_2 - X_1)))\big)^2 \\
&\quad + \mathbb{E}_\mathbb{P}\big((X_1 - X_2)^2 F_T(1 - F_T)\big).
\end{aligned}
$$

Therefore, we obtain the following result.

Proposition 28. *If we restrict our attention to* $\mathbb{F}$*-adapted strategies, the quadratic hedging price of the claim* $X = X_1 \mathbb{1}_{\{\tau > T\}} + X_2 \mathbb{1}_{\{\tau \leq T\}}$ *equals*

$$\mathbb{E}_{\mathbb{P}}\big(\zeta_T(X_1 + F_T(X_2 - X_1))\big) = \mathbb{E}_{\mathbb{Q}}\big(X_1(1 - F_T) + F_T X_2\big).$$

The optimal quadratic hedging of X *is the strategy which duplicates the* $\mathcal{F}_T$*-measurable contingent claim* $X_1(1 - F_T) + F_T X_2$.

Let us now examine the case of a generic $\mathcal{G}_T$-measurable random variable X. In this case, we shall only examine the solution of the second problem introduced above, that is,

$$\min_{v,\pi} \mathbb{E}_{\mathbb{P}}\big((V_T^v(\pi) - X)^2\big).$$

As we have explained in the previous part, this problem is essentially equivalent to a problem where we restrict our attention to the terminal wealth. From the properties of conditional expectations, we have

$$\min_{V \in \mathcal{F}_T} \mathbb{E}_{\mathbb{P}}\big((V - X)^2\big) = \mathbb{E}_{\mathbb{P}}\big((\mathbb{E}_{\mathbb{P}}(X \mid \mathcal{F}_T) - X)^2\big)$$

and the initial value of the strategy with terminal value $\mathbb{E}_{\mathbb{P}}(X \mid \mathcal{F}_T)$ is

$$\mathbb{E}_{\mathbb{P}}\big(\zeta_T \mathbb{E}_{\mathbb{P}}(X \mid \mathcal{F}_T)\big) = \mathbb{E}_{\mathbb{P}}(\zeta_T X).$$

In essence, the latter statement is a consequence of the completeness of the default-free market model. In conclusion, the quadratic hedging price equals $\mathbb{E}_{\mathbb{P}}(\zeta_T X) = \mathbb{E}_{\mathbb{Q}} X$ and the quadratic hedging strategy is the replicating strategy of the attainable claim $\mathbb{E}_{\mathbb{P}}(X \mid \mathcal{F}_T)$ associated with X.

13.2 Quadratic Hedging with $\mathbb{G}$-Adapted Strategies

Our next goal is to solve, for a given initial endowment v, the following minimization problem

$$\min_{\pi \in \Pi(\mathbb{G})} \mathbb{E}_{\mathbb{P}}\big((V_T^v(\pi) - X)^2\big).$$

We have seen in Part II that one way of solving this problem is to project the random variable X on the set of stochastic integrals. Here, we present an alternative approach.

We are looking for $\mathbb{G}$-adapted processes X, Θ and Ψ such that the process

$$J_t(\pi) = \big(V_t^v(\pi) - X_t\big)^2 \Theta_t + \Psi_t, \quad \forall t \in [0, T], \tag{137}$$

is a $\mathbb{G}$-submartingale for any $\mathbb{G}$-adapted trading strategy π and a $\mathbb{G}$-martingale for some strategy π^*. In addition, we require that $X_T = X$, $\Theta_T = 1$, $\Phi_T = 0$. Let us assume that the dynamics of these processes are of the form

$$dX_t = x_t \cdot dt + \widehat{x}_t \, dW_t + \widetilde{x}_t \, dM_t, \qquad (138)$$

$$d\Theta_t = \Theta_{t-}\big(\vartheta_t \, dt + \widehat{\vartheta}_t \, dW_t + \widetilde{\vartheta}_t \, dM_t\big), \qquad (139)$$

$$d\Psi_t = \psi_t \, dt + \widehat{\psi}_t \, dW_t + \widetilde{\psi}_t \, dM_t, \qquad (140)$$

where the drifts x_t, ϑ_t and ψ_t are yet to be determined. From Itô's formula, we obtain (recall that $\xi_t = \gamma_t \mathbb{1}_{\{\tau > t\}}$)

$$
\begin{aligned}
d(V_t - X_t)^2 &= 2(V_t - X_t)(\pi_t \sigma - \widehat{x}_t) \, dW_t - 2(V_t - X_{t-})\widetilde{x}_t \, dM_t \\
&\quad + \big[(V_t - X_{t-} - \widetilde{x}_t)^2 - (V_t - X_{t-})^2\big] dM_t \\
&\quad + \Big(2(V_t - X_t)(\pi_t \nu - x_t) + (\pi_t \sigma - \widehat{x}_t)^2 \\
&\quad + \xi_t\big[(V_t - X_t - \widetilde{x}_t)^2 - (V_t - X_t)^2\big]\Big) dt,
\end{aligned}
$$

where we denote $V_t = V_t^v(\pi)$. The process $J(\pi)$ is a martingale if and only if its drift term $k(t, \pi_t, x_t, \vartheta_t, \psi_t) = 0$ for every $t \in [0, T]$.

Straightforward calculations show that

$$
\begin{aligned}
k(t, \pi_t, \vartheta_t, x_t, \psi_t) &= \psi_t + \Theta_t\Big[\vartheta_t(V_t - X_t)^2 \\
&\quad + 2(V_t - X_t)\big[(\pi_t \nu - x_t) + \widehat{\vartheta}_t(\pi_t \sigma - \widehat{x}_t) + \xi_t \widetilde{x}_t\big] \\
&\quad + (\pi_t \sigma - \widehat{x}_t)^2 + \xi_t(\widetilde{\vartheta}_t + 1)\big[(V_t - X_t - \widetilde{x}_t)^2 - (V_t - X_t)^2\big]\Big].
\end{aligned}
$$

In the first step, for any $t \subset [0, T]$ we shall find π_t^* such that the minimum of $k(t, \pi_t, x_t, \vartheta_t, \psi_t)$ is attained. Subsequently, we shall choose the auxiliary processes $x = x^*$, $\vartheta = \vartheta^*$ and $\psi = \psi^*$ in such a way that $k(t, \pi_t^*, x_t^*, \vartheta_t^*, \psi_t^*) = 0$. This choice will imply that $k(t, \pi_t, x_t^*, \vartheta_t^*, \psi_t^*) \geq 0$ for any trading strategy π and any $t \in [0, T]$.

The strategy π^*, which minimizes $k(t, \pi_t, x_t, \vartheta_t, \psi_t)$, is the solution of the following equation:

$$(V_t^v(\pi) - X_t)(\nu + \widehat{\vartheta}_t \sigma) + \sigma(\pi_t \sigma - \widehat{x}_t) = 0, \quad \forall t \in [0, T].$$

Hence, the strategy π^* is implicitly given by

$$\pi_t^* = \sigma^{-1}\widehat{x}_t - \sigma^{-2}(\nu + \widehat{\vartheta}_t \nu)(V_t^v(\pi^*) - X_t) = A_t - B_t(V_t^v(\pi^*) - X_t),$$

where we denote

$$A_t = \sigma^{-1}\widehat{x}_t, \quad B_t = \sigma^{-2}(\nu + \widehat{\vartheta}_t \sigma).$$

After some computations, we see that the drift term of the process J admits the following representation:

$$
\begin{aligned}
k(t, \pi_t, \vartheta_t, x_t, \psi_t) &= \psi_t + \Theta_t(V_t - X_t)^2(\vartheta_t - \sigma^2 B_t^2) \\
&\quad + 2\Theta_t(V_t - X_t)\big(\sigma^2 A_t B_t - \widehat{\vartheta}_t \widehat{x}_t - \widetilde{\vartheta}_t \widetilde{x}_t \xi_t - x_t\big) + \Theta_t \xi_t(\widetilde{\vartheta}_t + 1)\widetilde{x}_t^2.
\end{aligned}
$$

From now on, we shall assume that the auxiliary processes ϑ, x and ψ are chosen as follows:

$$\vartheta_t = \vartheta_t^* = \sigma^2 B_t^2,$$
$$x_t = x_t^* = \sigma^2 A_t B_t - \widehat{\vartheta}_t \widehat{x}_t - \widetilde{\vartheta}_t \widetilde{x}_t \xi_t,$$
$$\psi_t = \psi_t^* = -\Theta_t \xi_t (\widetilde{\vartheta}_t + 1) \widetilde{x}_t^2.$$

It is rather clear that if the drift coefficients ϑ, x, ψ in (138)-(140) are chosen as above, then the drift term in dynamics of J is always non-negative, and it is equal to 0 for the strategy π^*, where $\pi_t^* = A_t - B_t(V_t^v(\pi^*) - X_t)$.

Our next goal is to solve equations (138)-(140). Let us first consider equation (139). Since $\vartheta_t = \sigma^2 B_t^2$, it suffices to find the three-dimensional process $(\Theta, \widehat{\vartheta}, \widetilde{\vartheta})$ which is a solution to the following BSDE:

$$d\Theta_t = \Theta_t \left(\sigma^{-2} (\nu + \widehat{\vartheta}_t \sigma)^2 \, dt + \widehat{\vartheta}_t \, dW_t + \widetilde{\vartheta}_t \, dM_t \right), \quad \Theta_T = 1.$$

It is obvious that the processes $\widehat{\vartheta} = 0$, $\widetilde{\vartheta} = 0$ and Θ, given as

$$\Theta_t = \exp(-\theta^2(T - t)), \quad \forall t \in [0, T], \tag{141}$$

solve this equation.

In the next step, we search for a three-dimensional process $(X, \widehat{x}, \widetilde{x})$, which solves equation (138) with $x_t = x_t^* = \sigma^2 A_t(\nu/\sigma^2) = \theta \widehat{x}_t$. It is clear that $(X, \widehat{x}, \widetilde{x})$ is the unique solution to the linear BSDE

$$dX_t = \theta \widehat{x}_t \, dt + \widehat{x}_t \, dW_t + \widetilde{x}_t \, dM_t, \quad X_T = X.$$

The unique solution to this equation is $X_t = \mathbb{E}_{\mathbb{Q}}(X \mid \mathcal{G}_t)$, where $\mathbb{Q}$ is the risk-neutral probability measure, so that $d\mathbb{Q} = \eta_t \, d\mathbb{P}$, where

$$d\eta_t = -\theta \eta_t \, dW_t, \quad \eta_0 = 1.$$

The components $\widehat{x}$ and $\widetilde{x}$ are given by the integral representation of the $\mathbb{G}$-martingale X with respect to $W^{\mathbb{Q}}$ and M. Notice also that since $\widehat{\vartheta} = 0$, the optimal portfolio π^* is given by the feedback formula

$$\pi_t^* = \sigma^{-1} \left(\widehat{x}_t - \theta(V_t^v(\pi^*) - X_t) \right).$$

Finally, since $\widetilde{\vartheta} = 0$, we have $\psi_t = -\xi_t \widetilde{x}_t^2 \Theta_t$. Therefore, we can solve explicitly the BSDE (140) for the process Ψ. Indeed, we are now looking for a three-dimensional process $(\Psi, \widehat{\psi}, \widetilde{\psi})$, which is the unique solution of the BSDE

$$d\Psi_t = -\Theta_t \xi_t \widetilde{x}_t^2 \, dt + \widehat{\psi}_t \, dW_t + \widetilde{\psi}_t \, dM_t, \quad \Psi_T = 0.$$

Noting that the process

$$\Psi_t + \int_0^t \Theta_s \xi_s \tilde{x}_s^2 \, ds$$

is a $\mathbb{G}$-martingale under $\mathbb{P}$, we obtain the value of Ψ in a closed form:

$$\Psi_t = \mathbb{E}_{\mathbb{P}} \left(\int_t^T \Theta_s \xi_s \tilde{x}_s^2 \, ds \,\Big|\, \mathcal{G}_t \right). \tag{142}$$

Substituting (141) and (142) in (137), we conclude that the value function for our problem is $J_t^* = J_t(\pi^*)$, where in turn

$$J_t(\pi^*) = (V_t^v(\pi^*) - X_t)^2 e^{-\theta^2(T-t)} + \mathbb{E}_{\mathbb{P}} \left(\int_t^T \Theta_s \xi_s \tilde{x}_s^2 \, ds \,\Big|\, \mathcal{G}_t \right)$$

$$= (V_t^v(\pi^*) - X_t)^2 e^{-\theta^2(T-t)} + \int_t^T e^{-\theta^2(T-s)} \mathbb{E}_{\mathbb{P}} \left(\gamma_s \tilde{x}_s^2 \mathbb{1}_{\{\tau > s\}} \,\big|\, \mathcal{G}_t \right) ds$$

$$= (V_t^v(\pi^*) - X_t)^2 e^{-\theta^2(T-t)}$$

$$+ \mathbb{1}_{\{\tau > t\}} \int_t^T e^{-\theta^2(T-s)} \mathbb{E}_{\mathbb{P}} \left(\gamma_s \tilde{x}_s^2 e^{\Gamma_t - \Gamma_s} \,\big|\, \mathcal{F}_t \right) ds,$$

where we have identified the process $\tilde{x}$ with its $\mathbb{F}$-adapted version (recall that any $\mathbb{G}$-predictable process is equal, prior to default, to an $\mathbb{F}$-predictable process). In particular,

$$J_0^* = e^{-\theta^2 T} \left((v - X_0)^2 + \mathbb{E}_{\mathbb{P}} \left(\int_0^T e^{\theta^2 s} \gamma_s \tilde{x}_s^2 c^{-\Gamma_s} \, ds \right) \right).$$

From the last formula, it is obvious that the quadratic hedging price is $X_0 = \mathbb{E}_{\mathbb{Q}} X$. We are in a position to formulate the main result of this section. A corresponding theorem for a default-free financial model was established by Kohlmann and Zhou (2000).

Proposition 29. *Let a claim X be $\mathcal{G}_T$-measurable and $\mathbb{P}$-square-integrable. The optimal trading strategy π^*, which solves the quadratic problem*

$$\min_{\pi \in \Pi(\mathbb{G})} \mathbb{E}_{\mathbb{P}}((V_T^v(\pi) - X)^2),$$

is given by the feedback formula

$$\pi_t^* = \sigma^{-1} \big(\hat{x}_t - \theta(V_t^v(\pi^*) - X_t) \big),$$

where $X_t = \mathbb{E}_{\mathbb{Q}}(X \,|\, \mathcal{G}_t)$ for every $t \in [0, T]$, and the process $\hat{x}_t$ is specified by

$$dX_t = \hat{x}_t \, dW_t^{\mathbb{Q}} + \tilde{x}_t \, dM_t.$$

The quadratic hedging price of X is equal to $\mathbb{E}_{\mathbb{Q}} X$.

Survival Claim

Let us consider a simple survival claim $X = \mathbb{1}_{\{\tau > T\}}$, and let us assume that Γ is deterministic, specifically, $\Gamma(t) = \int_0^t \gamma(s)\, ds$. In that case, from the well-known representation theorem (see Bielecki and Rutkowski (2004), Page 159), we have $dX_t = \tilde{x}_t\, dM_t$ with $\tilde{x}_t = -e^{\Gamma(t)-\Gamma(T)}$. Hence

$$
\begin{aligned}
\Psi_t &= \mathbb{E}_{\mathbb{P}}\left(\int_t^T \Theta_s \xi_s \tilde{x}_s^2\, ds \,\Big|\, \mathcal{G}_t \right) \\
&= \mathbb{E}_{\mathbb{P}}\left(\int_t^T \Theta_s \gamma(s) \mathbb{1}_{\{\tau > s\}} e^{2\Gamma(s) - 2\Gamma(T)}\, ds \,\Big|\, \mathcal{G}_t \right) \\
&= \mathbb{1}_{\{\tau > t\}}\, e^{\Gamma(t) - 2\Gamma(T)}\, \mathbb{E}_{\mathbb{P}}\left(\int_t^T e^{-\theta^2(T-s)} \gamma(s) e^{\Gamma(s)}\, ds \,\Big|\, \mathcal{F}_t \right) \\
&= \mathbb{1}_{\{\tau > t\}}\, e^{\Gamma(t) - 2\Gamma(T)} \int_t^T e^{-\theta^2(T-s)} \gamma(s) e^{\Gamma(s)}\, ds.
\end{aligned}
$$

One can check that, at time 0, the value function is indeed smaller that the one obtained with $\mathbb{F}$-adapted portfolios.

Case of an Attainable Claim

Assume now that a claim X is $\mathcal{F}_T$-measurable. Then $X_t = \mathbb{E}_{\mathbb{Q}}(X \,|\, \mathcal{G}_t)$ is the price of X, and it satisfies $dX_t = \hat{x}_t\, dW_t^{\mathbb{Q}}$. The optimal strategy is, in a feedback form,

$$
\pi_t^* = \sigma^{-1}\big(\hat{x}_t - \theta(V_t - X_t)\big)
$$

and the associated wealth process satisfies

$$
dV_t = \pi_t^*(\nu dt + \sigma dW_t) = \pi_t^* \sigma\, dW_t^{\mathbb{Q}} = \sigma^{-1}\big(\sigma \hat{x}_t - \nu(V_t - X_t)\big)\, dW_t^{\mathbb{Q}}.
$$

Therefore,

$$
d(V_t - X_t) = -\theta(V_t - X_t)\, dW_t^{\mathbb{Q}}.
$$

Hence, if we start with an initial wealth equal to the arbitrage price $\pi_0(X)$ of X, then we obtain that $V_t = X_t$ for every $t \in [0, T]$, as expected.

Hodges Price

Let us emphasize that the Hodges price has no real meaning here, since the problem $\min \mathbb{E}_{\mathbb{P}}((V_T^v)^2)$ has no financial interpretation. We have studied in the preceding part a more pertinent problem, with a constraint on the expected value of V_T^v under $\mathbb{P}$. Nevertheless, from a mathematical point of view, the Hodges price would be the value of p such that

$$(v^2 - (v-p)^2) = \int_0^T e^{\theta^2 s} \mathbb{E}_{\mathbb{P}}(\gamma_s \tilde{x}_s^2 e^{-\Gamma_s}) \mathbb{1}_{\{\tau > t\}} ds$$

In the case of the example studied in Section 13.2, the Hodges price would be the non-negative value of p such that

$$2vp - p^2 = e^{-2\Gamma_T} \int_0^T e^{\theta^2 s} \gamma_s e^{\Gamma_s} ds.$$

Let us also mention that our results are different from results of Lim (2004). Indeed, Lim studies a model with Poisson component, and thus in his approach the intensity of this process does not vanish after the first jump.

14 Optimization in Incomplete Markets

In this last section, we shall briefly (and rather informally) examine a specific optimization problem associated with a defaultable claim. The interested reader is referred to Lukas (2001) for more details on the approach examined in this section. We now assume that the only risky asset available in the market is

$$dZ_t^1 = Z_t^1 (\nu \, dt + \sigma \, dW_t + \varphi \, dM_t),$$

and we assume that $r = 0$. We deal with the following problem:

$$\sup_\pi \mathbb{E}_{\mathbb{P}} \big(u(V_{\tau \wedge T}^v(\pi) + X) \big)$$

for the claim X of the form

$$X = \mathbb{1}_{\{\tau > T\}} g(Z_T^1) + \mathbb{1}_{\{\tau \leq T\}} h(Z_\tau^1)$$

for some functions $g, h : \mathbb{R} \to \mathbb{R}$. Note that here the recovery payment is paid at hit, that is, at the time of default. In addition, we assume that the default intensity γ under $\mathbb{P}$ is constant (hence, it is constant under any equivalent martingale measure as well). After time τ, the market reduces to a standard Black-Scholes model, and thus the solution to the corresponding optimization problem is well known.

In the particular case of the exponential utility $u(x) = 1 - \exp(-\varrho x)$, $\varrho > 0$, we are in a position to use the duality theory. This problem was studied by, among others, Rouge and El Karoui (2000), Delbaen et al. (2002) and Collin-Dufresne and Hugonnier (2002).

Let $H(\mathbb{Q} \,|\, \mathbb{P})$ stand for the relative entropy of $\mathbb{Q}$ with respect to $\mathbb{P}$. Recall that if a probability measure $\mathbb{Q}$ is absolutely continuous with respect to $\mathbb{P}$ then

$$H(\mathbb{Q} \,|\, \mathbb{P}) = \mathbb{E}_{\mathbb{P}} \left(\frac{d\mathbb{Q}}{d\mathbb{P}} \ln \frac{d\mathbb{Q}}{d\mathbb{P}} \right) = \mathbb{E}_{\mathbb{Q}} \left(\ln \frac{d\mathbb{Q}}{d\mathbb{P}} \right).$$

Otherwise, the relative entropy $H(\mathbb{Q} \,|\, \mathbb{P})$ equals ∞.

It is well known that, under suitable technical assumptions (see Rouge and El Karoui (2000) or Delbaen et al. (2002) for details), we have

$$\sup_{\pi} \mathbb{E}_{\mathbb{P}}\left(1 - e^{-\varrho(V_T^v(\pi)+X)}\right)$$
$$= 1 - \exp\left(- \inf_{\pi} \inf_{\mathbb{Q} \in \mathcal{Q}_T} \left(H(\mathbb{Q}\,|\,\mathbb{P}) + \varrho\,\mathbb{E}_{\mathbb{Q}}(V_T^v(\pi) + X)\right)\right),$$

where π runs over a suitable class of admissible portfolios, and $\mathcal{Q}_T$ stands for the set of equivalent martingale measures on the σ-field $\mathcal{G}_T$.

Since for any admissible portfolio π the expected value under any martingale measure $\mathbb{Q} \in \mathcal{Q}_T$ of the terminal wealth $V_T^v(\pi)$ equals v, we obtain

$$\sup_{\pi} \mathbb{E}_{\mathbb{P}}\left(1 - e^{-\varrho(V_T^v(\pi)+X)}\right) = 1 - \exp\left(- \inf_{\mathbb{Q} \in \mathcal{Q}_T} \left(H(\mathbb{Q}\,|\,\mathbb{P}) + \varrho\,\mathbb{E}_{\mathbb{Q}}X + \varrho v\right)\right).$$

Furthermore, since, without loss of generality, we may stop all the processes considered here at the default time τ, we end up with the following equality

$$\inf_{\pi} \mathbb{E}_{\mathbb{P}}\left(e^{-\varrho(V_{T\wedge\tau}^v(\pi)+X)}\right) = \exp\left(- \inf_{\mathbb{Q} \in \mathcal{Q}_{T\wedge\tau}} \left(H(\mathbb{Q}\,|\,\mathbb{P}) + \varrho\,\mathbb{E}_{\mathbb{Q}}X + \varrho v\right)\right),$$

where π runs over the class of all admissible trading strategies, and $\mathcal{Q}_{T\wedge\tau}$ stands the set of equivalent martingale measures on the σ-field $\mathcal{G}_{T\wedge\tau}$. The following result provides a description of the class $\mathcal{Q}_{T\wedge\tau}$.

Lemma 21. *The class $\mathcal{Q}_{T\wedge\tau}$ of all equivalent martingale measures on the space $(\Omega, \mathcal{G}_{T\wedge\tau})$ is the set of all probability measures $\mathbb{Q}_{k,h}$ of the form*

$$d\mathbb{Q}_{k,h}|_{\mathcal{G}_{T\wedge\tau}} = \eta_{T\wedge\tau}(k,h)\,d\mathbb{P},$$

where the Radon-Nikodym density process $\eta(k,h)$ is given by the formula

$$\eta_t(k,h) = \mathcal{E}_t(kM)\mathcal{E}_t(hW), \quad \forall t \in [0,T],$$

for some $\mathbb{F}$-adapted process k such that the inequality $k_t > -1$ holds for every $t \in [0,T]$, and for the associated process $h_t = -\theta - \varphi\gamma\sigma^{-1}(1+k_t)$, where $\theta = \nu/\sigma$. Under the martingale measure $\mathbb{Q} = \mathbb{Q}_{k,h}$ the process

$$W_{t\wedge\tau}^h = W_{t\wedge\tau} - \int_0^{t\wedge\tau} h_s\,ds, \quad \forall t \in [0,T],$$

is a stopped Brownian motion, and the process

$$M_{t\wedge\tau}^k = M_{t\wedge\tau} - \int_0^{t\wedge\tau} \gamma k_s\,ds, \quad \forall t \in [0,T],$$

is a martingale stopped at τ.

Straightforward calculations show that the relative entropy of a martingale measure $\mathbb{Q} = \mathbb{Q}_{k,h} \in \mathcal{Q}_{T \wedge \tau}$ with respect to $\mathbb{P}$ equals

$$\mathbb{E}_\mathbb{Q}\left(\int_0^{\tau \wedge T} h_s \, dW_s^h + \int_0^{\tau \wedge T} \left(\tfrac{1}{2}h_s^2 - \gamma k_o + \gamma(1 + k_s)\ln(1 + k_s) \right) ds \right)$$

$$+ \mathbb{E}_\mathbb{Q}\left(\int_0^{\tau \wedge T} \ln(1 + k_s) \, dM_s^k \right).$$

Consequently, the optimization problem

$$\inf_{\mathbb{Q} \in \mathcal{Q}_{T \wedge \tau}} \left(H(\mathbb{Q} \,|\, \mathbb{P}) + \varrho \, \mathbb{E}_\mathbb{Q} X \right)$$

can be reduced to the following problem

$$\inf_{k,h} \mathbb{E}_\mathbb{Q}\left(\int_0^{\tau \wedge T} \left(\tfrac{1}{2}h_s^2 - \gamma k_s + \gamma(1 + k_s)\ln(1 + k_s) \right) ds + \varrho X \right), \qquad (143)$$

where the processes k and h are as specified in the statement of Lemma 21. Let us set

$$\ell(k_s) = \tfrac{1}{2}h_s^2 - \gamma k_s + \gamma(1 + k_s)\ln(1 + k_s)$$

so that

$$\ell(k) = \tfrac{1}{2}\left(\theta + \varphi\gamma(1 + k) \right)^2 - \gamma k + \gamma(1 + k)\ln(1 + k). \qquad (144)$$

Consider a dynamic version of the minimization problem (143)

$$\inf_{k,h} \mathbb{E}_\mathbb{Q}\left(\int_t^{\tau \wedge T} \ell(k_s) \, ds + \varrho \mathbb{1}_{\{\tau \leq T\}} h(Z_\tau^1) + \varrho \mathbb{1}_{\{\tau > T\}} g(Z_T^1) \,\Big|\, \mathcal{G}_t \right).$$

Let us denote $K_s^t = e^{-\int_t^s \gamma(1 + k_u)\,du}$ for $t \leq s$. Then, on the pre-default event $\{\tau > t\}$, we obtain the following problem:

$$\inf_{k,h} \mathbb{E}_\mathbb{Q}\left(\int_t^T K_s^t\left(\ell(k_s) + \varrho\gamma(1 + k_s)h(Z_s^1(1 + \varphi)) \right) ds + \varrho K_T^t g(Z_T^1) \,\Big|\, \mathcal{F}_t \right).$$

The value function $J(t, x)$ of the latter problem satisfies the HJB equation

$$\partial_t J(t, x) + \tfrac{1}{2}\sigma^2 x^2 \partial_{xx} J(t, x)$$
$$+ \inf_{k > -1} \left(-\varphi\gamma(1 + k)x\partial_x J(t, x) - \gamma(1 + k)J(t, x) + \psi(k, x) \right) = 0$$

with the terminal condition $J(T, x) = \varrho g(x)$, where we denote

$$\psi(k, x) = \ell(k) + \varrho\gamma(1 + k)h(x(1 + \varphi))$$

and where the function ℓ is given by (144). The minimizer is given by $k = k^*(t, x)$, which is the unique root of the following equation:

$$\frac{\varphi}{\sigma^2}\left(\nu + \varphi\gamma(1 + k) \right) + \ln(1 + k) = J(t, x) + \varphi x \partial_x J(t, x) - \varrho h(x(1 + \varphi)),$$

and the optimal portfolio π^* is given by the formula

$$\pi_t^* = (\varrho\sigma^2)^{-1}\big(\nu + \varphi\gamma(1 + k^*(t, Z_t^1)) - \sigma^2 Z_{t-}^1 \partial_x J(t, Z_{t-}^1)\big).$$

Remark. Note that in the case $\varphi = 0$ this result is consistent with our result established in Section 12.1. When $\varphi = 0$, the process Z^1 is continuous, and thus we obtain

$$\pi_t^* = (\varrho\sigma)^{-1}\big(\theta - \sigma Z_t^1 \partial_x J(t, Z_t^1)\big),$$

where the value function $J(t, x)$ satisfies the simplified HJB equation

$$\partial_t J(t, x) + \tfrac{1}{2}\sigma^2 x^2 \partial_{xx} J(t, x)$$
$$+ \inf_{k > -1} \big(\ell(k) - \gamma(1 + k)J(t, x) + \varrho\gamma(1 + k)h(x)\big) = 0,$$

where in turn

$$\ell(k) = \tfrac{1}{2}\theta^2 - \gamma k + \gamma(1 + k)\ln(1 + k).$$

References

Arvanitis A. and Gregory, J. (2001) *Credit: The Complete Guide to Pricing, Hedging and Risk Management.* Risk Publications.

Arvanitis, A. and Laurent, J.-P. (1999) On the edge of completeness. *Risk*, October, 61–65.

Barles, G., Buckdahn, R. and Pardoux, E. (1997) Backward stochastic differential equations and integral-partial differential equations. *Stochastics and Stochastics Reports* **60**, 57–83.

Bélanger, A., Shreve, S. and Wong, D. (2001) A unified model for credit derivatives. To appear in *Mathematical Finance*.

Bernis, G. and Jeanblanc, M. (2003) Hedging defaultable derivatives *via* utility theory. Preprint, Evry University.

Bielecki, T.R. and Jeanblanc, M. (2003) Genuine mean-variance hedging of credit risk: A case study. Working paper.

Bielecki, T.R., Jeanblanc, M. and Rutkowski, M. (2004a) Modeling and valuation of credit risk. In: *CIME-EMS Summer School on Stochastic Methods in Finance, Bressanone, July 6-12, 2003*, Springer-Verlag, Berlin Heidelberg New York.

Bielecki, T.R. and Rutkowski, M. (2003) Dependent defaults and credit migrations. *Applicationes Mathematicae* **30**, 121–145.

Bielecki, T.R. and Rutkowski, M. (2004) *Credit Risk: Modeling, Valuation and Hedging.* Corrected 2nd printing. Springer-Verlag, Berlin Heidelberg New York.

Bielecki, T.R., Jin, H., Pliska, S.R. and Zhou, X.Y. (2004b) Dynamic mean-variance portfolio selection with bankruptcy prohibition. Forthcoming in *Mathematical Finance*.

Black, F. and Cox, J.C. (1976) Valuing corporate securities: Some effects of bond indenture provisions. *Journal of Finance* **31**, 351–367.

Blanchet-Scalliet, C. and Jeanblanc, M. (2004) Hazard rate for credit risk and hedging defaultable contingent claims. *Finance and Stochastics* **8**, 145–159.

Bobrovnytska, O. and Schweizer, M. (2004) Mean-variance hedging and stochastic control: Beyond the Brownian setting. *IEEE Transactions on Automatic Control* **49**, 396–408.

Buckdahn, R. (2001) Backward stochastic differential equations and viscosity solutions of semilinear parabolic deterministic and stochastic PDE of second order. In: *Stochastic Processes and Related Topics*, R. Buckdahn, H.-J. Engelbert and M. Yor, editors, Taylor and Francis, pp. 1–54.

Collin-Dufresne, P. and Hugonnier, J.-N. (2001) Event risk, contingent claims and the temporal resolution of uncertainty. Working paper, Carnegie Mellon University.

Collin-Dufresne, P. and Hugonnier, J.-N. (2002) On the pricing and hedging of contingent claims in the presence of extraneous risks. Preprint, Carnegie Mellon University.

Collin-Dufresne, P., Goldstein, R.S. and Hugonnier, J.-N. (2003) A general formula for valuing defaultable securities. Preprint.

Cossin, D. and Pirotte, H. (2000) *Advanced Credit Risk Analysis*. J. Wiley, Chichester.

Davis, M. (1997) Option pricing in incomplete markets. In: *Mathematics of Derivative Securities*, M.A.H. Dempster and S.R. Pliska, editors, Cambridge University Press, Cambridge, pp. 216–227

Delbaen, F., Grandits, P., Rheinländer, Th., Samperi, D., Schweizer, M. and Stricker, Ch. (2002) Exponential hedging and entropic penalties. *Mathematical Finance* **12**, 99–124.

Duffie, D. (2003) *Dynamic Asset Pricing Theory*. 3rd ed. Princeton University Press, Princeton.

Duffie, D. and Singleton, K. (2003) *Credit Risk: Pricing, Measurement and Management*. Princeton University Press, Princeton.

El Karoui, N. and Mazliak, L. (1997) Backward stochastic differential equations. *Pitman Research Notes in Mathematics*, Longman.

El Karoui, N., Peng, S. and Quenez, M.-C. (1997) Backward stochastic differential equations in finance. *Mathematical Finance* **7**, 1–71.

El Karoui, N. and Quenez, M.-C. (1995) Dynamic programming and pricing of contingent claims in an incomplete market. *SIAM Journal on Control and Optimization* **33**, 29–66.

El Karoui, N. and Quenez, M.-C. (1997) Imperfect markets and backward stochastic differential equations. In: *Numerical Methods in Finance*, L.C.G. Rogers, D. Talay, editors, Cambridge University Press, Cambridge, pp. 181–214.

Elliott, R.J., Jeanblanc, M. and Yor, M. (2000) On models of default risk. *Mathematical Finance* **10**, 179–195.

Gouriéroux, C., Laurent, J.-P. and Pham, N. (1998) Mean-variance hedging and numéraire. *Mathematical Finance* **8**, 179–200.

Greenfield, Y. (2000) Hedging of credit risk embedded in derivative transactions. Thesis, Carnegie Mellon University.

Gregory, J. and Laurent, J.-P. (2003) I will survive. *Risk*, June, 103–107.

Hodges, S.D. and Neuberger, A. (1989) Optimal replication of contingent claim under transaction costs. *Review of Futures Markets* **8**, 222–239.

Hu, Y. and Zhou, X.Y. (2004) Constrained stochastic LQ control with random coefficients and application to mean-variance portfolio selection. Preprint.

Hugonnier, J.-N., Kramkov, D. and Schachermayer, W. (2002) On the utility based pricing of contingent claims in incomplete markets. Preprint.

Jamshidian, F. (2002) Valuation of credit default swap and swaptions. To appear in *Finance and Stochastics*.

Jankunas, A. (2001) Optimal contingent claims. *Annals of Applied Probability* **11**, 735–749

Jarrow, R.A. and Yu, F. (2001) Counterparty risk and the pricing of defaultable securities. *Journal of Finance* **56**, 1756–1799.

Jeanblanc, M. and Rutkowski, M. (2000) Modelling of default risk: Mathematical tools. Working paper, Université d'Evry and Politechnika Warszawska.

Jeanblanc, M. and Rutkowski, M. (2002) Default risk and hazard processes. In: *Mathematical Finance – Bachelier Congress 2000*, H. Geman, D. Madan, S.R. Pliska and T. Vorst, editors, Springer-Verlag, Berlin Heidelberg New York, pp. 281–312.

Jeanblanc, M. and Rutkowski, M. (2003) Modelling and hedging of default risk. In: *Credit Derivatives: The Definitive Guide*, J. Gregory, editor, Risk Books, London, pp. 385–416.

Jeanblanc, M., Yor, M. and Chesney, M. (2004) *Mathematical Methods for Financial Markets*. Springer-Verlag. Forthcoming.

Karatzas, I. and Shreve, S.E. (1998) *Brownian Motion and Stochastic Calculus*. 2nd ed. Springer-Verlag, Berlin Heidelberg New York.

Karatzas, I. and Shreve, S.E. (1998) *Methods of Mathematical Finance*. Springer-Verlag, Berlin Heidelberg New York.

Kobylanski, M. (2000) Backward stochastic differential equations and partial differential equations with quadratic growth. *Annals of Probability* **28**, 558–602.

Kohlmann, M. and Zhou, X.Y. (2000) Relationship between backward stochastic differential equations and stochastic controls: A linear-quadratic approach. *SIAM Journal on Control and Optimization* **38**, 1392–1407.

Kramkov, D. (1996) Optional decomposition of supermartingales and hedging contingent claims in incomplete security markets. *Probability Theory and Related Fields* **105**, 459–479.

Kusuoka, S. (1999) A remark on default risk models. *Advances in Mathematical Economics* **1**, 69–82.

Lando, D. (1998) On Cox processes and credit-risky securities. *Review of Derivatives Research* **2**, 99–120.

Last, G. and Brandt, A. (1995) *Marked Point Processes on the Real Line: The Dynamic Approach*. Springer-Verlag, Berlin Heidelberg New York.

Laurent, J.-P. and Gregory, J. (2002) Basket defaults swaps, CDOs and factor copulas. Working paper.

Lepeltier, J.-P. and San-Martin, J. (1997) Existence for BSDE with superlinear-quadratic coefficient. *Stochastics and Stochastics Reports* **63**, 227–240.

Li, X., Zhou, X.Y. and Lim, A.E.B. (2001) Dynamic mean-variance portfolio selection with no-shorting constraints. *SIAM Journal on Control and Optimization* **40**, 1540–1555.

Lim, A.E.B. (2004) Hedging default risk. Preprint.

Lim, A.E.B. and Zhou, X.Y. (2002) Mean-variance portfolio selection with random parameters. *Mathematics of Operations Research* **27**, 101–120.

Lotz, C. (1998) Locally minimizing the credit risk. Working paper, University of Bonn.

Lukas, S. (2001) On pricing and hedging defaultable contingent claims. Thesis, Humboldt University.

Ma, J. and Yong, Y. (1999) *Forward-Backward Stochastic Differential Equations and Their Applications.* Springer-Verlag, Berlin Heidelberg New York.

Mania, M. (2000) A general problem of an optimal equivalent change of measure and contingent claim pricing in an incomplete market. *Stochastic Processes and their Applications* **90**, 19–42.

Mania, M. and Tevzadze, R. (2003) Backward stochastic PDE and imperfect hedging. *International Journal of Theoretical and Applied Finance* **6**, 663–692.

Merton, R. (1974) On the pricing of corporate debt: The risk structure of interest rates. *Journal of Finance* **29**, 449–470.

Musiela, M. and Rutkowski, M. (1997) *Martingale Methods in Financial Modelling.* Springer-Verlag, Berlin Heidelberg New York.

Musiela, M. and Zariphopoulou, T. (2004) An example of indifference prices under exponential preferences. *Finance and Stochastics* **8**, 229–239.

Pham, N., Rheinländer, Th. and Schweizer, M. (1998) Mean-variance hedging for continuous processes: New results and examples. *Finance and Stochastics* **2**, 173–98.

Pliska, S.R. (2001) Dynamic Markowitz problems with no bankruptcy. Presentation at JAFEE Conference, 2001.

Protter, P. (2003) *Stochastic Integration and Differential Equations.* 2nd ed. Springer-Verlag, Berlin Heidelberg New York.

Rheinländer, Th. (1999) Optimal martingale measures and their applications in mathematical finance. Ph.D. Thesis, TU-Berlin.

Rheinländer, Th. and Schweizer, M. (1997) On L^2-projections on a space of stochastic integrals. *Annals of Probability* **25**, 1810–1831.

Rong, S. (1997) On solutions of backward stochastic differential equations with jumps and applications. *Stochastic Processes and their Applications* **66**, 209–236.

Rouge, R. and El Karoui, N. (2000) Pricing via utility maximization and entropy. *Mathematical Finance* **10**, 259–276.

Royer, M. (2003) Equations différentielles stochastiques rétrogrades et martingales non-linéaires. Doctoral thesis.

Schönbucher, P.J. (1998) Pricing credit risk derivatives. Working paper, University of Bonn.

Schönbucher, P.J. (2003) *Credit Derivatives Pricing Models.* J.Wiley, Chichester.

Schweizer, M. (2001) A guided tour through quadratic hedging approaches. In: *Option Pricing, Interest Rates and Risk Management,* E. Jouini, J. Cvitanić and M. Musiela, editors, Cambridge University Press, pp. 538–574.

Vaillant, N. (2001) A beginner's guide to credit derivatives. Working paper, Nomura International.

Wong, D. (1998) A unifying credit model. Working paper, Research Advisory Services, Capital Markets Group, Scotia Capital Markets.

Yong, J. and Zhou, X.Y. (1999) *Stochastic Controls: Hamiltonian Systems and HJB Equations.* Springer-Verlag, Berlin Heidelberg New York.

Zhou, X.Y. (2003) Markowitz's world in continuous time, and beyond. In: *Stochastic Modeling and Optimization,* D.D. Yao et al., editors, Springer, New York, pp. 279–310.

Zhou, X.Y. and Li, D. (2000) Continuous time mean-variance portfolio selection: A stochastic LQ framework. *Applied Mathematics and Optimization* **42**, 19–33.

Zhou, X.Y. and Yin, G. (2004) Dynamic mean-variance portfolio selection with regime switching: A continuous-time model. *IEEE Transactions on Automatic Control* **49**, 349–360.

On the Geometry of Interest Rate Models

Tomas Björk

Department of Finance
Stockholm School of Economics
Box 6501
S-113 83 Stockholm SWEDEN
email: tomas.bjork@hhs.se

Summary. In this chapter, which is a substantial extension of an earlier essay [3], we give an overview of some recent work on the geometric properties of the evolution of the forward rate curve in an arbitrage free bond market. The main problems to be discussed are as follows.

- When is a given forward rate model consistent with a given family of forward rate curves?
- When can the inherently infinite dimensional forward rate process be realized by means of a Markovian finite dimensional state space model.

We consider interest rate models of Heath-Jarrow-Morton type, where the forward rates are driven by a multidimensional Wiener process, and where he volatility is allowed to be an arbitrary smooth functional of the present forward rate curve. Within this framework we give necessary and sufficient conditions for consistency, as well as for the existence of a finite dimensional realization, in terms of the forward rate volatilities. We also study stochastic volatility HJM models, and we provide a systematic method for the construction of concrete realizations.

Key words: HJM models, stochastic volatility, factor models, forward rates, state space models, Markovian realizations, infinite dimensional stochastic differential equations, invariant manifolds, geometry.
MSC 2000 subject classification. 91B28, 91B70

Acknowledgements: support from the Jan Wallander and Tom Hedelius Foundation is gratefully acknowledged. I am grateful to D. Filipovic, J. Teichmann and J. Zabzcyk for a number of very helpful discussions. A number of valuable comments from unknown referees on the underlying papers helped to improve these considerably. I am also highly indebted to B.J. Christensen, C.Landen and L. Svensson for their generosity in letting me use our joint results for this overview.

1 Introduction

We start by presenting the probabilistic framework and formulating the main problems to be studied.

1.1 Setup

We consider a bond market model (see [4], [33]) living on a filtered probability space $(\Omega, \mathcal{F}, \mathbb{F}, \mathbb{Q})$ where $\mathbb{F} = \{\mathcal{F}_t\}_{t\geq 0}$. The basis is assumed to carry a standard m-dimensional Wiener process W, and we also assume that the filtration $\mathbb{F}$ is the internal one generated by W.

By $p_t(x)$ we denote the price, at t, of a zero coupon bond maturing at $t + x$, and the forward rates $r_t(x)$ are defined by

$$r_t(x) = -\frac{\partial \log p_t(x)}{\partial x}.$$

Note that we use the Musiela parameterization, where x denotes the time **to** maturity. The short rate R is defined as $R_t = r_t(0)$, and the money account B is given by $B_t = \exp\left\{\int_0^t R_s ds\right\}$. The model is assumed to be free of arbitrage in the sense that the measure $\mathbb{Q}$ above is a martingale measure for the model. In other words, for every fixed time of maturity $T \geq 0$, the process $Z_t(T) = p_t(T - t)/B_t$ is a $\mathbb{Q}$-martingale.

Let us now consider a given forward rate model of the form

$$\begin{cases} dr_t(x) = \beta_t(x)dt + \sigma_t(x)dW, \\ r_0(x) = r_0^o(x). \end{cases} \tag{1}$$

where, for each x, $\beta(x)$ and $\sigma(x)$ are given optional processes. The initial curve $\{r_0^o(x); \ x \geq 0\}$ is taken as given. It is interpreted as the *observed forward rate curve*.

The standard Heath-Jarrow-Morton drift condition ([26]) can easily be transferred to the Musiela parameterization. The result (see [10], [32]) is as follows.

Proposition 1.1 (The Forward Rate Equation) *Under the martingale measure $\mathbb{Q}$ the r-dynamics are given by*

$$dr_t(x) = \left\{\frac{\partial}{\partial x}r_t(x) + \sigma_t(x)\int_0^x \sigma_t(u)^\star du\right\} dt + \sigma_t(x)dW_t, \tag{2}$$

$$r_0(x) = r_0^o(x). \tag{3}$$

where $\star$ denotes transpose.

1.2 The Geometric Point of View

At a first glance it is natural to view the forward rate equation (2) as an infinite system of stochastic differential equations (SDEs in what follows): We have one equation for each fixed x, so we are handling a continuum of SDEs.

An alternative, more geometrically oriented, view of (2) is to regard it as a **single equation** for an **infinite dimensional object**. The infinite dimensional object under study is the **forward rate curve**, i.e. the curve $x \longmapsto r_t(x)$. Denoting the forward rate curve at time t by r_t, and the entire forward rate curve process by r, we thus take the point of view that r is a process evolving in some infinite dimensional function space $\mathcal{H}$ of forward rate curves. For a fixed t we will thus view each outcome of r_t as a vector (or point) in $\mathcal{H}$. In the same way, the volatility process $\sigma = [\sigma_1, \ldots, \sigma_m]$ is viewed as a process evolving in $\mathcal{H}^m$, so that for each t the outcome of $\sigma_t = [\sigma_{1t}, \ldots, \sigma_{mt}]$ is regarded as a point in $\mathcal{H}^m$.

In order to avoid detailed technical discussions at this preliminary level we postpone the precise definition of $\mathcal{H}$, as well as the necessary technical conditions on σ, until Section 3.3. For the time being the reader is asked to think (loosely) of $\mathcal{H}$ as a space of C^∞ functions and to assume that all SDEs appearing in the text do admit unique strong solutions.

In order to emphasize the geometric point of view, we can now rewrite the forward rate equation (2) as

$$dr_t = \{\mathbf{F}r_t + \sigma_i \mathbf{H}\sigma_i\}\, dt + \sigma_t dW_t, \tag{4}$$

$$r_0 = r^0, \tag{5}$$

where the operators $\mathbf{F}$ and $\mathbf{H}$, both defined on $\mathcal{H}$, are given by

$$\mathbf{F} = \frac{\partial}{\partial x}, \tag{6}$$

$$[\mathbf{H}f](x) = \int_0^x f(s)ds, \quad \text{for } f \in \mathcal{H}. \tag{7}$$

and where we use the obvious interpretation $\mathbf{H}\sigma_t = [\mathbf{H}\sigma_{1t}, \ldots, \mathbf{H}\sigma_{mt}]$.

1.3 Main Problems

Suppose now that we are given a concrete model $\mathcal{M}$ within the above framework, i.e. suppose that we are given a concrete specification of the volatility process σ. We now formulate a couple of natural problems:

1. Take, in addition to $\mathcal{M}$, also as given a parameterized family $\mathcal{G}$ of forward rate curves. Under which conditions is the family $\mathcal{G}$ **consistent** with the dynamics of $\mathcal{M}$? Here consistency is interpreted in the sense that, given an initial forward rate curve in $\mathcal{G}$, the interest rate model $\mathcal{M}$ will only produce forward rate curves belonging to the given family $\mathcal{G}$.

2. When does the given, inherently infinite dimensional, interest rate model $\mathcal{M}$ admit a **finite dimensional realization**? More precisely, we seek conditions under which the forward rate process $r_t(x)$ induced by the model $\mathcal{M}$, can be realized by a system of the form

$$dZ_t = a\left(Z_t\right) dt + b\left(Z_t\right) dW_t, \tag{8}$$

$$r_t(x) = G\left(Z_t, x\right). \tag{9}$$

where Z (interpreted as the state vector process) is a finite dimensional diffusion, $a(z)$, $b(z)$ and $G(z, x)$ are deterministic functions and W is the same Wiener process as in (2). Expressed in other terms, we thus wish to investigate under what conditions the HJM model is generated by a finite dimensional Markovian state space model.

As will be seen below, these two problems are intimately connected, and the main purpose of this chapter, which is a substantial extension of a previous paper [3], is to give an overview of some recent work in this area. The text is based on [6], [5], [9], [7], and [8], but the presentation given below is more focused on geometric intuition than the original articles, where full proofs, technical details and further results can be found. In the analysis below we use ideas from systems and control theory (see [30]) as well as from nonlinear filtering theory (see [12]). References to the literature will sometimes be given in the text, but will also be summarized in the Notes at the end of each section.

It should be noted that the functional analytical framework of the entire theory has recently been improved in a quite remarkable way by Filipović and Teichmann. In a series of papers these authors have considerably extended the Hilbert space framework of the papers mentioned above. In doing so, they have also clarified many structural problems and derived a large number of concrete results. However: a full understanding of these extensions require a high degree of detailed technical knowledge in analysis on Frechet spaces so the scope of the present chapter prohibits us from doing complete justice to this beautiful part of the theory. The interested reader is referred to the original papers [23], [24], and [25].

The organization of the text is as follows. In Section 2 we treat the relatively simple case of linear realizations. Section 3 is devoted to a study of the general consistency problem, including a primer on the Filpovic inverse consistency theory, and in Section 4 we use the consistency results from Section 3 in order to give a fairly complete picture of the nonlinear realization problem. The problem of actually constructing a concrete realization is treated in Section 5, in Section 6 we discuss very briefly the Filipović–Teicmann extensions, and in Section 7 we extend the theory to include stochastic volatility models.

2 A Primer on Linear Realization Theory

In the general case, the forward rate equation (4) is a highly nonlinear infinite dimensional SDE but, as can be expected, the special case of linear dynamics is much easier to handle. In this section we therefore concentrate on linear forward rate models, and look for finite dimensional linear realizations. Almost all geometric ideas presented in this chapter will then be generalized to the nonlinear case studied later in the text.

2.1 Deterministic Forward Rate Volatilities

For the rest of the section we only consider the case when the volatility process $\sigma_t(x) = [\sigma_{1t}(x), \ldots, \sigma_{mt}(x)]$ is a deterministic function $\sigma(x)$ of x only.

Assumption 2.1 *Each component σ_i, of the volatility process σ is a deterministic vector in $\mathcal{H}$ for $i = 1, \ldots, m$. Equivalently, The volatility σ is a C^∞-mapping $\sigma : \mathbb{R}_+ \to \mathbb{R}^m$.*

Under this assumption, the forward rate equation (4) takes the form

$$dr_t = \{\mathbf{F}_t + D\}\, dt + \sigma dW_t, \tag{10}$$

where the function D, which we view as a vector $D \in \mathcal{H}$, is defined by

$$D(x) = \sigma(x) \int_0^x \sigma(s)^\star ds. \tag{11}$$

The point to note here is that, because of our choice of a deterministic volatility $\sigma(x)$, the forward rate equation (10) is a **linear** (or rather affine) SDE evolving in the infinite dimensional function space $\mathcal{H}$.

Because of the linear structure of the equation (albeit in infinite dimensions) we expect to be able to provide an explicit solution of (10). We now recall that a scalar equation of the form

$$dy_t = [ay_t + b]\, dt + cdW_t$$

has the solution

$$y_t = e^{at}y_0 + \int_0^t e^{a(t-s)}bds + \int_0^t e^{a(t-s)}cdW_s,$$

and we are thus led to conjecture that the solution to (10) is given by the formal expression

$$r_t = e^{\mathbf{F}t}r_0 + \int_0^t e^{\mathbf{F}(t-s)} Dds + \int_0^t e^{\mathbf{F}(t-s)}\sigma dW_s.$$

We now have to make precise mathematical sense of the formal exponent $e^{\mathbf{F}t}$. From the context it is clear that it acts on vectors in $\mathcal{H}$, i.e. on real valued C^∞ functions. It is in fact an operator $e^{\mathbf{F}t} : \mathcal{H} \to \mathcal{H}$ and we have to figure out how it acts. From the usual series expansion of the exponential function one is led to write

$$e^{\mathbf{F}t} f = \sum_{n=0}^{\infty} \frac{t^n}{n!} F^n f. \tag{12}$$

In our case $F^n = \frac{\partial^n}{\partial x^n}$, so we have

$$\left[e^{\mathbf{F}t} f \right] (x) = \sum_{n=0}^{\infty} \frac{t^n}{n!} \frac{\partial^n f}{\partial x^n} (x) \tag{13}$$

This is, however, just a Taylor series expansion of f around the point x, so for analytic f we have $\left[e^{\mathbf{F}t} f \right] (x) = f(x+t)$. We have in fact the following precise result (which can be proved rigorously).

Proposition 2.1 *The operator* $\mathbf{F}$ *is the infinitesimal generator of the semigroup of left translations, i.e. for any* $f \in \mathcal{H}$ *(and in fact for any continuous f) we have*

$$\left[e^{\mathbf{F}t} f \right] (x) = f(t+x). \tag{14}$$

Furthermore, the solution of the forward rate equation (10) is given by

$$r_t = e^{\mathbf{F}t} r^o + \int_0^t e^{\mathbf{F}(t-s)} D ds + \int_0^t e^{\mathbf{F}(t-s)} \sigma dW_s \tag{15}$$

or, written in component form, by

$$r_t(x) = r^o(x+t) + \int_0^t D(x+t-s) ds + \int_0^t \sigma(x+t-s) dW_s. \tag{16}$$

From (15) it is clear by inspection that we may write the solution of the forward rate equation (10) as

$$r_t = q_t + \delta_t, \tag{17}$$
$$dq_t = \mathbf{F} q_t dt + \sigma dW_t, \tag{18}$$
$$q_0 = 0 \tag{19}$$

where δ is given by

$$\delta_t = e^{\mathbf{F}t} r^o + \int_0^t e^{\mathbf{F}(t-s)} D ds, \tag{20}$$

or on component form

$$\delta_t(x) = r^o(x+t) + \int_0^t D(x+t-s) ds. \tag{21}$$

Since $\delta_t(x)$ is not affected by the input W, we see that the problem of finding a realization for the term structure system (10) is equivalent to that of finding a realization for (18)-(19). Since we have a linear dynamical system it seems natural to look for **linear** realizations and we are thus led to the following definition.

Definition 2.1 *A triple $[A, B, C(x)]$, where A is an $(n \times n)$-matrix, B is an $(n \times m)$-matrix and C is an n dimensional row vector function, is called an n-dimensional realization of the systems (18) if q has the representation*

$$q_t(x) = C(x)Z_t, \tag{22}$$
$$dZ_t = AZ_t dt + BdW_t, \tag{23}$$
$$Z_0 = 0. \tag{24}$$

Our main problems are now as follows.

- Consider an a priori given volatility structure $\sigma(x)$.

- When does there exist a finite dimensional realization?

- If there exists a finite dimensional realization, what is the minimal dimension?

- How do we construct a minimal realization from knowledge of σ?

2.2 Finite Dimensional Realizations

In this section we will investigate the existence of a finite dimensional realization (FDR) from a geometric point of view. There are also a number of other ways to attack the problem, but it is in fact the geometrical point of view which later in the text will be generalized to the nonlinear case. The discussion will be rather informal and some technical questions are sidestepped.

We recall the q-equation (18) as

$$dq_t = \mathbf{F}q_t dt + \sigma dW_t, \tag{25}$$
$$q_0 = 0.$$

Expressing the operator exponential $e^{F(t-s)}$ as a power series, and using Proposition 2.1, we may write the solution to (25) as

$$q(t) = \int_0^t \sum_0^\infty \frac{(t-s)^n}{n!} \mathbf{F}^n \sigma dW(s). \tag{26}$$

From this expression we see that, for each t, the random vector q_t is in fact given as a random infinite linear combination of the (deterministic) vectors $\sigma, \mathbf{F}\sigma, \mathbf{F}^2\sigma, \ldots$.

Thus we see that the q-process will in fact evolve in the (deterministic) subspace $\mathcal{R} \subseteq \mathcal{H}$ defined by

$$\mathcal{R} = span\left[\sigma, \mathbf{F}\sigma, \mathbf{F}^2\sigma, \ldots\right].\tag{27}$$

The subspace $\mathcal{R}$ is thus **invariant** under the action of the q process, and it is rather obvious that it is in fact the minimal (under inclusion) invariant subspace for q.

The obvious conjecture is that there exists an FDR if and only if $\mathcal{R}$ is finite dimensional. This conjecture is in fact correct and we have the following main result.

Proposition 2.2 *Consider a given volatility function*

$$\sigma = [\sigma_1, \cdots, \sigma_m].$$

Then there exists an FDR if and only if,

$$\dim(\mathcal{R}) < \infty,\tag{28}$$

with $\mathcal{R}$ defined by

$$\mathcal{R} = span\left[\mathbf{F}^k\sigma_i \; ; \; i = 1, \cdots, m; \; k = 0, 1, \cdots\right].\tag{29}$$

*Furthermore; the minimal dimension of an FDR, also known as the **McMillan degree**, is given by* $\dim(\mathcal{R})$.

Proof. For brevity we only give the proof for the case $m = 1$. The proof for general case is almost identical. Assume first that there exists an n-dimensional FDR of the form

$$q_t(x) = C(x)Z_t,\tag{30}$$

with Z-dynamics as in (24), and with

$$C(x) = [C_1(x), \ldots, C_n(x)].$$

Writing this as

$$q_t = CZ_t,\tag{31}$$

it is now obvious that the finite dimensional subspace of $\mathcal{H}$ spanned by the vectors $C_1, \ldots, C_n$ will in fact be invariant and thus contain $\mathcal{R}$ (since $\mathcal{R}$ is minimal invariant). Thus the existence of an FDR implies the finite dimensionality of $\mathcal{R}$.

Conversely, assume that $\mathcal{R}$ is n-dimensional. We now prove the existence of an FDR by actually constructing an explicit realization of the form (22)-(24). The finite dimensionality of $\mathcal{R}$ implies that (with n as above) there exists a linear relation of the form

$$\mathbf{F}^n\sigma = \sum_{i=0}^{n-1} \gamma_i \mathbf{F}^i\sigma\tag{32}$$

where $\gamma_0, \ldots, \gamma_{n-1}$ are real numbers. Thus, the vectors $\sigma, \mathbf{F}\sigma, \ldots \mathbf{F}^{n-1}\sigma$ are linearly independent and span $\mathcal{R}$.

Since $\sigma, \mathbf{F}\sigma, \ldots, \mathbf{F}^{n-1}\sigma$ is a basis for the invariant subspace $\mathcal{R}$ we can write

$$q_t = \sum_{i=0}^{n-1} Z_t^i \mathbf{F}^i \sigma. \tag{33}$$

where the processes $Z^0, \ldots, Z^{n-1}$ are the coordinate processes of q for the given basis. We now want to find the dynamics of $Z = \left[Z^0, \ldots, Z^{n-1}\right]$ so we make the **Ansatz**

$$dZ_t^i = \alpha_i\left(Z_t\right) dt + \beta_i\left(Z_t\right) dW_t, \quad i = 0, \ldots, n-1, \tag{34}$$

and the problem is to identify the unknown functions α and β.

Our strategy for finding α and β is as follows.

- Compute dq from (33)-(34).

- Compare the expression thus obtained with the original q dynamics given by

$$dq_t = \mathbf{F}q_t dt + \sigma dW_t. \tag{35}$$

- Identify the coefficients.

From (33)-(34) and using the notation

$$\sigma^{(i)} - \mathbf{F}^i \sigma,$$

we obtain

$$dq_t = \left(\sum_{i=0}^{n-1} \alpha_i\left(Z_t\right)\sigma^{(i)}\right) dt + \left(\sum_{i=0}^{n-1} \beta_i\left(Z_t\right)\sigma^{(i)}\right) dW_t. \tag{36}$$

We now want to compare this expression with the dynamics in (35). To do this we first use (33) to obtain

$$\mathbf{F}q_t = \sum_{i=0}^{n-1} Z_t^i \mathbf{F}^{i+1}\sigma = \sum_{i=0}^{n-1} Z_t^i \sigma^{i+1} = \sum_{i=1}^{n} Z_t^{i-1}\sigma^i. \tag{37}$$

After inserting (32) into (37) and collecting terms we have

$$\mathbf{F}q_t = \sum_{i=0}^{n-1} c_i\left(Z_t\right)\sigma^i, \tag{38}$$

where

$$c_i\left(Z\right) = Z^{i-1} + \gamma_i Z^{n-1}, \quad i = 0, \ldots, n-1, \tag{39}$$

with the convention $Z^{-1} = 0$. We may thus write the q dynamics in (35) as

$$dq_t = \left(\sum_{i=0}^{n-1} c_i \left(Z_t \right) \sigma^i \right) dt + \sigma^0 dW_t. \tag{40}$$

We may now identify coefficients by comparing (38) with (40) to obtain

$$\beta_0(Z) = 1,$$
$$\beta_i = 0, \quad i = 1, \ldots, n-1.$$

and

$$\alpha_i(Z) = c_i(Z), \quad i = 0, \ldots, n-1. \tag{41}$$

with c_i as in (39). We have thus derived the explicit realization

$$q_t = CZ_t, \tag{42}$$

where

$$C = \left[\sigma, \mathbf{F}\sigma, \ldots, \mathbf{F}^{n-1}\sigma \right], \tag{43}$$

and where the Z dynamics are given by

$$dZ_t^0 = \gamma_0 Z_t^{n-1} dt + dW_t, \tag{44}$$
$$dZ_t^i = \left(Z_t^{i-1} - \gamma_i Z_t^{n-1} \right) dt, \quad i = 1, \ldots, n-1. \tag{45}$$

$\square$

We note in passing that the proof above, apart from proving the existence result, also provides us with the concrete realization (43)-(45). In the proof this is only done for the case of a scalar Wiener process, but the method can easily be extended to the multi-dimensional case. See [7] for worked out examples.

We now go on to find a more explicit characterization of condition (28). Recalling that the operator $\mathbf{F}$ is defined as $\mathbf{F} = \partial/\partial x$, we see from Proposition 2.2 that the forward rate system admits an FDR if and only if the space spanned by the components of σ and all their derivatives is finite dimensional. In other words; there exists an FDR if and only if the components of σ satisfy a linear system of ordinary differential equations (ODEs in what follows) with constant coefficients. This leads us to the topic of quasi-exponential functions.

Definition 2.2 *A **quasi-exponential** (or QE) function is by definition any function of the form*

$$f(x) = \sum_i e^{\lambda_i x} + \sum_j e^{\alpha_i x} \left[p_j(x) \cos(\omega_j x) + q_j(x) \sin(\omega_j x) \right], \tag{46}$$

where $\lambda_i, \alpha_1, \omega_j$ are real numbers, whereas p_j and q_j are real polynomials.

QE functions will turn up over and over again, so we list some simple well known properties.

Lemma 2.1 *The following hold for the quasi-exponential functions.*

- *A function is QE if and only if it is a component of the solution of a vector valued linear ODE with constant coefficients.*

- *A function is QE if and only if it can be written as $f(x) = ce^{Ax}b$. where c is a row vector, A is a square matrix and b is a column vector.*

- *If f is QE, then f' is QE.*

- *If f is QE, then its anti-derivative is QE.*

- *If f and g are QE, then fg is QE.*

We can thus restate Proposition 2.2.

Proposition 2.3 *The forward rate equation admits a finite dimensional realization if and only of each component of σ is quasi-exponential.*

2.3 Economic Interpretation of the State Space

In general, the state space of a realization of a given system has no concrete (e.g. economic) interpretation. In the case of the forward rate equation, however, the states of the *minimal* realization turn out to have a simple economic interpretation.

Proposition 2.4 *Assume that $\mathcal{R}$ is n-dimensional, so that the existence of an FDR is guaranteed. Then, for any **minimal** realization, i.e. a realization with an n-dimensional state vector Z, there will exist an affine transformation mapping the state vector into a vector of benchmark forward rates.*

The moral of this is that, in a minimal realization, you can always choose your state variables as a fixed set of forward rates. It can also be shown that the maturities of the benchmark forward rates can be chosen without restrictions.

For precise statements, proofs and examples, see [6].

2.4 Connections to Systems and Control Theory

The geometric ideas of the previous section are in fact standard in the theory of mathematical systems and control. To see this, consider again the equation

$$dq_t = \mathbf{F}q_t dt + \sigma dW_t, \tag{47}$$

$$q_0 = 0. \tag{48}$$

Let us now formally "divide by dt", which gives us

$$\frac{dq_t}{dt} = \mathbf{F}q_t + \sigma\frac{dW_t}{dt},$$

where the formal time derivative dW_t/dt is interpreted as "white noise". We interpret this equation as an input-output system where the random input signal $t \longmapsto dW_t/dt$ is transformed into the infinite dimensional output signal $t \longmapsto q_t$. We thus view the equation as a stochastic version of the following controlled ODE

$$\frac{dq_t}{dt} = \mathbf{F}q_t + \sigma u_t, \tag{49}$$
$$q_0 = 0,$$

where u is a deterministic input signal (which in our case is replaced by white noise). Generally speaking, tricks like this does not work directly, since we are ignoring the difference between standard differential calculus, which is used to analyze (49), and Itô calculus which we use when dealing with SDEs. In this case, however, because of the linear structure, the second order Itô term will not come into play, so we are safe. (See the discussion in Section 3.4 around the Stratonovich integral for how to treat the nonlinear situation.)

The reader who is familiar with systems and control theory (see [11]) will now recognize the space $\mathcal{R}$ above as the **reachable subspace** of the control system (49). Not surprisingly, there is also a frequency domain approach to our realization problem. See [6] for details.

2.5 Examples

We now give some simple illustrations of the theory. We only consider the case of a scalar driving Wiener process.

Example 2.1 $\sigma(x) = \sigma e^{-ax}$
We consider a model driven by a one-dimensional Wiener process, having the forward rate volatility structure

$$\sigma(x) = \sigma e^{-ax},$$

where σ in the right hand side denotes a constant. (The reader will probably recognize this example as the Hull-White model.) We start by determining $\mathcal{R}$ which in this case is defined as

$$\mathcal{R} = span\left[\frac{d^k}{dx^k}\sigma e^{-ax} \; ; \; k \geq 0\right].$$

It is obvious that $\mathcal{R}$ is one-dimensional, so we can expect to find a one-dimensional realization. The existence of an FDR could of course also have been seen directly by observing that σ is quasi-exponential. Since we have $\mathbf{F}\sigma = -a\sigma$ we see that, in the notation of (32) we have $\gamma_0 = -a$. Thus denoting the single state variable Z^0 by Z, we may use (43)-(45) to obtain the realization

$$q_t(x) = e^{-ax} Z_t, \tag{50}$$

$$dZ_t = -aZ_t dt + dW_t. \tag{51}$$

A full realization of the forward rate process r_t is then obtained from (17) as

$$r_t(x) = \sigma e^{-ax} Z_t + r^0(x+t) + \frac{1}{2}\sigma^2 e^{-2ax}\left\{e^{-2at} - 1\right\}. \tag{52}$$

Example 2.2 $\sigma(x) = e^{-x^2}$
In this example $\mathcal{R}$ is given by

$$\mathcal{R} = span\left[\frac{d^k}{dx^k}e^{-x^2},\ k \geq 0\right] = span\left[x^k e^{-x^2},\ k \geq 0\right],$$

which is easily seen to be infinite dimensional. Thus we see that in this case there exists no finite dimensional *linear* realization. We will return to this example later and we will in fact prove that neither does there exists a non-linear FDR.

2.6 Notes

This section is to a large extent based on [6] where, however, the focus is more on the frequency domain approach. The first paper to appear in this area was to our knowledge the seminal preprint [32], where the Musiela parameterization is introduced and the space $\mathcal{R}$ is discussed in some detail. Because of the linear structure, the theory above is closely connected to (and in a sense inverse to) the theory of affine term structures developed in [17]. The standard reference on infinite dimensional SDEs is [16], where one also can find a presentation of the connections between control theory and infinite dimensional linear stochastic equations.

3 The Consistency Problem

We now turn to a more serious study of the geometric properties of the forward rate equation in the general nonlinear case. We begin by studying when a given submanifold of forward rate curves is **consistent** (a precise definition will be given below) with a given interest rate model. This problem is of interest from an applied as well as from a theoretical point of view. In particular we will use the results from this section to analyze problems about existence of finite dimensional factor realizations for interest rate models on forward rate form. Invariant manifolds are, however, also of interest in their own right, so we begin by discussing a concrete problem which naturally leads to the invariance concept.

3.1 Parameter Recalibration

A standard procedure when dealing with concrete interest rate models on a high frequency (say, daily) basis can be described as follows:

1. At time $t = 0$, use market data to fit (calibrate) the model to the observed bond prices.

2. Use the calibrated model to compute prices of various interest rate derivatives.

3. The following day ($t = 1$), repeat the procedure in 1. above in order to recalibrate the model, etc..

To carry out the calibration in step 1. above, the analyst typically has to produce a forward rate curve $\{r^o(x); x \geq 0\}$ from the observed data. However, since only a finite number of bonds actually trade in the market, the data consist of a discrete set of points, and a need to fit a curve to these points arises. This curve-fitting may be done in a variety of ways. One way is to use splines, but also a number of parameterized families of smooth forward rate curves have become popular in applications—the most well-known probably being the Nelson-Siegel (see [34]) family. Once the curve $\{r^o(x); x \geq 0\}$ has been obtained, the parameters of the interest rate model may be calibrated to this.

Now, from a purely logical point of view, the recalibration procedure in step 3. above is of course slightly nonsensical: If the interest rate model at hand is an exact picture of reality, then there should be no need to recalibrate. The reason that everyone insists on recalibrating is of course that any model in fact only is an approximate picture of the financial market under consideration, and recalibration allows incorporating newly arrived information in the approximation. Even so, the calibration procedure itself ought to take into account that it will be repeated. It appears that the optimal way to do so would involve a combination of time series and cross-section data, as opposed to the purely cross-sectional curve-fitting, where the information contained in previous curves is discarded in each recalibration.

The cross-sectional fitting of a forward curve and the repeated recalibration is thus, in a sense, a pragmatic and somewhat non-theoretical endeavor. Nonetheless, there are some nontrivial theoretical problems to be dealt with in this context, and the problem to be studied in this section concerns the **consistency** between, on the one hand, the dynamics of a given interest rate model, and, on the other hand, the forward curve family employed.

What, then, is meant by consistency in this context? Assume that a given interest rate model $\mathcal{M}$ (e.g. the Hull–White model) in fact *is* an exact picture of the financial market. Now consider a particular family $\mathcal{G}$ of forward rate curves (e.g. the Nelson-Siegel family) and assume that the interest rate model is calibrated using this family. We then say that the pair $(\mathcal{M}, \mathcal{G})$ is **consistent** (or, that $\mathcal{M}$ and $\mathcal{G}$ are consistent) if all forward curves which may be produced by the interest rate model $\mathcal{M}$ are contained within the family $\mathcal{G}$. Otherwise, the pair $(\mathcal{M}, \mathcal{G})$ is inconsistent.

Thus, if $\mathcal{M}$ and $\mathcal{G}$ are consistent, then the interest rate model actually produces forward curves which belong to the relevant family. In contrast, if $\mathcal{M}$ and $\mathcal{G}$ are inconsistent, then the interest rate model will produce forward curves outside the family used in the calibration step, and this will force the analyst to change the model parameters all the time—not because the model is an approximation to reality, but simply because the family does not go well with the model.

Put into more operational terms this can be rephrased as follows.

- Suppose that you are using a fixed interest rate model $\mathcal{M}$. If you want to do recalibration, then your family $\mathcal{G}$ of forward rate curves should be chosen is such a way as to be consistent with the model $\mathcal{M}$.

Note however that the argument also can be run backwards, yielding the following conclusion for empirical work.

- Suppose that a particular forward curve family $\mathcal{G}$ has been observed to provide a good fit, on a day-to-day basis, in a particular bond market. Then this gives you modeling information about the choice of an interest rate model in the sense that you should try to use/construct an interest rate model which is consistent with the family $\mathcal{G}$.

We now have a number of natural problems to study,

I   Given an interest rate model $\mathcal{M}$ and a family of forward curves $\mathcal{G}$, what are necessary and sufficient conditions for consistency?

II   Take as given a specific family $\mathcal{G}$ of forward curves (e.g. the Nelson-Siegel family). Does there exist any interest rate model $\mathcal{M}$ which is consistent with $\mathcal{G}$?

III   Take as given a specific interest rate model $\mathcal{M}$ (e.g. the Hull-White model). Does there exist any finitely parameterized family of forward curves $\mathcal{G}$ which is consistent with $\mathcal{M}$?

In this section we will mainly address problem (I) above. Problem II has been studied, for special cases, in [19], [20], whereas Problem III can be shown (see Proposition 4.1) to be equivalent to the problem of finding a finite dimensional factor realization of the model $\mathcal{M}$ and we provide a fairly complete solution in Section 4.

3.2 Invariant Manifolds

We now move on to give precise mathematical definition of the consistency property discussed above, and this leads us to the concept of an **invariant manifold**.

Definition 3.1 (Invariant Manifold) *Take as given the forward rate process dynamics (2). Consider also a fixed family (manifold) of forward rate curves $\mathcal{G}$. We say that*

$\mathcal{G}$ is locally **invariant** under the action of r if, for each point $(s, r) \in \mathbb{R}_+ \times \mathcal{G}$, the condition $r_s \in \mathcal{G}$ implies that $r_t \in \mathcal{G}$, on a (possibly random) time interval with positive length. If r stays forever on $\mathcal{G}$, we say that $\mathcal{G}$ is globally invariant.

The purpose of this section is to characterize invariance in terms of local characteristics of $\mathcal{G}$ and $\mathcal{M}$, and in this context local invariance is the best one can hope for. In order to save space, local invariance will therefore be referred to as invariance.

To get some intuitive feeling for the invariance concepts one can consider the following two-dimensional deterministic system

$$\frac{dy_1}{dt} = y_2,$$
$$\frac{dy_2}{dt} = -y_1.$$

For this system it is obvious that the unit circle $\mathcal{C} = \left\{ (y_1, y_2) : y_1^2 + y_2^2 = 1 \right\}$ is globally invariant, i.e. if we start the system on $\mathcal{C}$ it will stay forever on $\mathcal{C}$. The 'upper half' of the circle, $\mathcal{C}_u = \left\{ (y_1, y_2) : y_1^2 + y_2^2 = 1, y_2 > 0 \right\}$, is on the other hand only locally invariant, since the system will leave $\mathcal{C}_u$ at the point $(1, 0)$. This geometric situation is in fact the generic one also for our infinite dimensional stochastic case. The forward rate trajectory will never leave a locally invariant manifold at a point in the relative interior of the manifold. Exit from the manifold can only take place at the relative boundary points. We have no general method for determining whether a locally invariant manifold also is globally invariant or not. Problems of this kind have to be solved separately for each particular case.

3.3 The Space

In order to study the consistency problem we need (see Remark 3.1 below) a very regular space to work in.

Definition 3.2 *Consider a fixed real number* $\gamma > 0$. *The space* $\mathcal{H}_\gamma$ *is defined as the space of all infinitely differentiable functions*

$$r : \mathbb{R}_+ \to \mathbb{R}$$

satisfying the norm condition $\|r\|_\gamma < \infty$. *Here the norm is defined as*

$$\|r\|_\gamma^2 = \sum_{n=0}^\infty 2^{-n} \int_0^\infty \left(\frac{d^n r}{dx^n}(x) \right)^2 e^{-\gamma x} dx.$$

Note that $\mathcal{H}$ is not a space of distributions, but a space of functions. We will often suppress the subindex γ. With the obvious inner product $\mathcal{H}$ is a pre-Hilbert space, and in [9] the following result is proved.

Proposition 3.1 *The space $\mathcal{H}$ is a Hilbert space, i.e. it is complete. Furthermore, every function in the space is in fact real analytic, and can thus be uniquely extended to a holomorphic function in the entire complex plane.*

Remark 3.1 The reason for our choice of $\mathcal{H}$ as the underlying space, is that the linear operator $\mathbf{F} = d/dx$ is bounded in this space. Together with the assumptions above, this implies that both μ and σ are smooth vector fields on $\mathcal{H}$, thus ensuring the existence of a strong local solution to the forward rate equation for every initial point $r^o \in \mathcal{H}$.

The Forward Curve Manifold

We consider as given a mapping

$$G : \mathcal{Z} \to \mathcal{H}, \tag{53}$$

where the parameter space $\mathcal{Z}$ is an open connected subset of $\mathbb{R}^d$, i.e. for each parameter value $z \in \mathcal{Z} \subseteq \mathbb{R}^d$ we have a curve $G(z) \in \mathcal{H}$. The value of this curve at the point $x \in \mathbb{R}_+$ will be written as $G(z, x)$, so we see that G can also be viewed as a mapping

$$G : \mathcal{Z} \times \mathbb{R}_+ \to \mathbb{R}. \tag{54}$$

The mapping G is thus a formalization of the idea of a finitely parameterized family of forward rate curves, and we now define the forward curve manifold as the set of all forward rate curves produced by this family.

Definition 3.3 *The **forward curve manifold** $\mathcal{G} \subseteq \mathcal{H}$ is defined as*

$$\mathcal{G} = Im\,(G),$$

where we use the notation $Im\,(G)$ for the image (or range) of the mapping G.

The Interest Rate Model

We consider a given volatility σ of the form

$$\sigma : \mathcal{H} \times \mathbb{R}_+ \to \mathbb{R}^m.$$

In other words, $\sigma(r, x)$ is a functional of the infinite dimensional r-variable, and a function of the real variable x. An alternative, and more instructive, way of viewing a component σ_i is to see it as a mapping where point $r \in \mathcal{H}$ is mapped to the real valued function $\sigma_i(r, \cdot)$. We will in fact assume that this real valued function is a member of $\mathcal{H}$, which means that we can view each component σ_i as a **vector field** $\sigma_i : \mathcal{H} \to \mathcal{H}$ on the space $\mathcal{H}$. Denoting the forward rate curve at time t by r_t we then have the following forward rate equation.

$$dr_t(x) = \left\{ \frac{\partial}{\partial x} r_t(x) + \sigma(r_t, x) \int_0^x \sigma(r_t, u)^\star du \right\} dt + \sigma(r_t, x)dW_t. \qquad (55)$$

Remark 3.2 For notational simplicity we have assumed that the r-dynamics are time homogenous. The case when σ is of the form $\sigma(t, r, x)$ can be treated in exactly the same way. See [5].

We obviously need some regularity assumptions and these will be collected in Assumption 3.1 below. See [5] for further technical details.

The Problem

Our main problem is the following.

- Suppose that we are given
  - A volatility σ, specifying an interest rate model $\mathcal{M}$ as in (55)
  - A mapping G, specifying a forward curve manifold $\mathcal{G}$.
- Is $\mathcal{G}$ then invariant under the action of r?

3.4 The Invariance Conditions

In order to study the invariance problem from a geometrical point of view we introduce some compact notation.

Definition 3.4 *We define* $\mathbf{H}\sigma$ *by*

$$\mathbf{H}\sigma(r, x) = \int_0^x \sigma(r, s)ds$$

Suppressing the x-variable, the Itô dynamics for the forward rates are thus given by

$$dr_t = \left\{ \frac{\partial}{\partial x} r_t + \sigma(r_t)\mathbf{H}\sigma(r_t)^\star \right\} dt + \sigma(r_t)dW_t \qquad (56)$$

and we write this more compactly as

$$dr_t = \mu_0(r_t)dt + \sigma(r_t)dW_t. \qquad (57)$$

In this way we see clearly how (57) is an SDE on $\mathcal{H}$, specified by its diffusion vector fields $\sigma_1, \ldots, \sigma_m$ and drift vector field μ_0, where μ_0 is given by the bracket term in (56). To get some intuition we now formally "divide by dt" and obtain

$$\frac{dr}{dt} = \mu_0(r_t) + \sigma(r_t)\dot{W}_t, \qquad (58)$$

where the formal time derivative $\dot{W}_t$ is interpreted as an "input signal" chosen by chance. As in Section 2.4 we are thus led to study the associated deterministic control system

$$\frac{dr}{dt} = \mu_0(r_t) + \sigma(r_t)u_t. \tag{59}$$

The intuitive idea is now that $\mathcal{G}$ is invariant under (57) if and only if $\mathcal{G}$ is invariant under (59) for all choices of the input signal u. It is furthermore geometrically obvious that this happens if and only if the velocity vector $\mu(r) + \sigma(r)u$ is tangential to $\mathcal{G}$ for all points $r \in \mathcal{G}$ and all choices of $u \in \mathbb{R}^m$. Since the tangent space of $\mathcal{G}$ at a point $G(z)$ is given by $Im\,[G_z'(z)]$, where G_z' denotes the Frechet derivative (Jacobian), we are led to conjecture that $\mathcal{G}$ is invariant if and only if the condition

$$\mu_0(r) + \sigma(r)u \in Im\,[G_z'(z)]$$

is satisfied for all $u \in \mathbb{R}^m$. This can also be written

$$\mu_0(r) \in Im\,[G_z'(z)]\,,$$
$$\sigma(r) \in Im\,[G_z'(z)]\,,$$

where the last inclusion is interpreted component wise for σ.

This "result" is, however, not correct due to the fact that the argument above neglects the difference between ordinary calculus, which is used for (59), and Itô calculus, which governs (57). In order to bridge this gap we have to rewrite the analysis in terms of Stratonovich integrals instead of Itô integrals.

Definition 3.5 *For given semimartingales X and Y, the* **Stratonovich integral** *of X with respect to Y, $\int_0^t X(s) \circ dY(s)$, is defined as*

$$\int_0^t X_s \circ dY_s = \int_0^t X_s dY_s + \frac{1}{2}\langle X, Y\rangle_t\,. \tag{60}$$

The first term on the right hand side is the Itô integral. In the present case, with only Wiener processes as driving noise, we can define the 'quadratic variation process' $\langle X, Y\rangle$ in (60) by

$$d\langle X, Y\rangle_t = dX_t dY_t, \tag{61}$$

with the usual 'multiplication rules' $dW \cdot dt = dt \cdot dt = 0$, $dW \cdot dW = dt$. We now recall the main result and *raison d'être* for the Stratonovich integral.

Proposition 3.2 (Chain Rule) *Assume that the function $F(t, y)$ is smooth. Then we have*

$$dF(t, Y_t) = \frac{\partial F}{\partial t}(t, Y_t)dt + \frac{\partial F}{\partial y} \circ dY_t\,. \tag{62}$$

Thus, in the Stratonovich calculus, the Itô formula takes the form of the standard chain rule of ordinary calculus.

Returning to (57), the Stratonovich dynamics are given by

$$dr_t = \left\{ \frac{\partial}{\partial x} r_t + \sigma(r_t) \mathbf{H} \sigma(r_t)^\star \right\} dt - \frac{1}{2} d\langle \sigma(r_t), W_t \rangle \qquad (63)$$
$$+ \sigma(r_t) \circ dW_t.$$

In order to compute the Stratonovich correction term above we use the infinite dimensional Itô formula (see [16]) to obtain

$$d\sigma(r_t) = \{\dots\} dt + \sigma'_r(r_t)\sigma(r_t)dW_t, \qquad (64)$$

where σ'_r denotes the Frechet derivative of σ w.r.t. the infinite dimensional r-variable. From this we immediately obtain

$$d\langle \sigma(r_t), W_t \rangle = \sigma'_r(r_t)\sigma(r_t)dt. \qquad (65)$$

Remark 3.3 If the Wiener process W is multidimensional, then σ is a vector $\sigma = [\sigma_1, \dots, \sigma_m]$, and the right hand side of (65) should be interpreted as

$$\sigma'_r(r_t)\sigma(r_t, x) = \sum_{i=1}^{m} \sigma'_{ir}(r_t)\sigma_i(r_t)$$

Thus (63) becomes

$$dr_t = \left\{ \frac{\partial}{\partial x} r_t + \sigma(r_t) \mathbf{H} \sigma(r_t)^\star - \frac{1}{2} \sigma'_r(r_t)\sigma(r_t) \right\} dt \qquad (66)$$
$$+ \sigma(r_t) \circ dW_t$$

We now write (66) as
$$dr_t = \mu(r_t)dt + \sigma(r_t) \circ dW_t \qquad (67)$$

where

$$\mu(r, x) = \frac{\partial}{\partial x} r(x) + \sigma(r_t, x) \int_0^x \sigma(r_t, u)^\star du - \frac{1}{2} \left[\sigma'_r(r_t)\sigma(r_t) \right](x). \qquad (68)$$

For all these arguments to make sense, we need some formal regularity assumptions.

Assumption 3.1 *We assume the following .*

- *For each $i = 1, \dots, m$ the volatility vector field $\sigma_i : \mathcal{H} \to \mathcal{H}$ is smooth.*
- *The mapping*

$$r \longmapsto \sigma(r)\mathbf{H}\sigma(r)^\star - \frac{1}{2}\sigma'_r(r)\sigma(r)$$

  is a smooth map from $\mathcal{H}$ to $\mathcal{H}$.

- *The mapping $z \longmapsto G(z)$ is a smooth embedding, so in particular the Frechet derivative $G'_z(z)$ is injective for all $z \in \mathcal{Z}$.*

Given the heuristics above, our main result is not surprising. The formal proof, which is somewhat technical, is left out. See [5].

Theorem 3.1 (Invariance Theorem) *Under Assumption 3.1, the forward curve manifold $\mathcal{G}$ is locally invariant for the forward rate process $r_t(x)$ in $\mathcal{M}$ if and only if,*

$$G'_x(z) + \sigma(r) \mathbf{H}\sigma(r)^\star - \frac{1}{2}\sigma'_r(r)\sigma(r) \in Im[G'_z(z)] , \tag{69}$$

$$\sigma(r) \in Im[G'_z(z)] , \tag{70}$$

hold for all $z \in \mathcal{Z}$ with $r = G(z)$.

Here, G'_z and G'_x denote the Frechet derivative of G with respect to z and x, respectively. The condition (70) is interpreted component wise for σ. Condition (69) is called *the consistent drift condition*, and (70) is called *the consistent volatility condition*.

Remark 3.4 It is easily seen that if the family G is invariant under shifts in the x-variable, then we will automatically have the relation

$$G'_x(z) \subset Im[G'_z(z)],$$

so in this case the relation (69) can be replaced by

$$\sigma(r)\mathbf{H}\sigma(r)^\star - \frac{1}{2}\sigma'_r(r)\sigma(r) \in Im[G'_z(z)],$$

with $r = G(z)$ as usual.

3.5 Examples

The results above are extremely easy to apply in concrete situations. As a test case we consider the Nelson–Siegel (see [34]) family of forward rate curves. We analyze the consistency of this family with the Ho-Lee and Hull-White interest rate models. It should be emphasized that these examples are chosen only in order to illustrate the general methodology. For more examples and details, see [5].

The Nelson-Siegel Family

The Nelson–Siegel (henceforth NS) forward curve manifold $\mathcal{G}$ is parameterized by $z \in \mathbb{R}^4$, the curve $x \longmapsto G(z, x)$ as

$$G(z, x) = z_1 + z_2 e^{-z_4 x} + z_3 x e^{-z_4 x} . \tag{71}$$

For $z_4 \neq 0$, the Frechet derivatives are easily obtained as

$$G'_z(z, x) = \left[1, \ e^{-z_4 x}, \ x e^{-z_4 x}, \ -(z_2 + z_3 x) x e^{-z_4 x} \right] , \tag{72}$$

$$G'_x(z, x) = (z_3 - z_2 z_4 - z_3 z_4 x) e^{-z_4 x} . \tag{73}$$

In order for the image of this map to be included in $\mathcal{H}_\gamma$, we need to impose the condition $z_4 > -\gamma/2$. In this case, the natural parameter space is thus $\mathcal{Z} = \{ z \in \mathbb{R}^4 : z_4 \neq 0, z_4 > -\gamma/2 \}$. However, as we shall see below, the results are uniform w.r.t. γ. Note that the mapping G indeed is smooth, and for $z_4 \neq 0$, G and G'_z are also injective.

In the degenerate case $z_4 = 0$, we have

$$G(z, x) = z_1 + z_2 + z_3 x , \tag{74}$$

We return to this case below.

The Hull-White and Ho-Lee Models

As our test case, we analyze the Hull and White (henceforth HW) extension of the Vasiček model. On short rate form the model is given by

$$dR(t) = \{ \Phi(t) - aR(t) \} \, dt + \sigma dW(t), \tag{75}$$

where $a, \sigma > 0$. As is well known, the corresponding forward rate formulation is

$$dr_t(x) = \beta(t, x) dt + \sigma e^{-ax} dW_t. \tag{76}$$

Thus, the volatility function is given by $\sigma(x) = \sigma e^{-ax}$, and the conditions of Theorem 3.1 become

$$G'_x(z, x) + \frac{\sigma^2}{a} \left[e^{-ax} - e^{-2ax} \right] \in Im[G'_z(z, x)], \tag{77}$$

$$\sigma e^{-ax} \in Im[G'_z(z, x)]. \tag{78}$$

To investigate whether the NS manifold is invariant under HW dynamics, we start with (78) and fix a z-vector. We then look for constants (possibly depending on z) A, B, C, and D, such that for all $x \geq 0$ we have

$$\sigma e^{-ax} = A + B e^{-z_4 x} + C x e^{-z_4 x} - D(z_2 + z_3 x) x e^{-z_4 x}. \tag{79}$$

This is possible if and only if $z_4 = a$, and since (78) must hold for all choices of $z \in \mathcal{Z}$ we immediately see that HW is inconsistent with the full NS manifold (see also the Notes below).

Proposition 3.3 (Nelson-Siegel and Hull-White) *The Hull-White model is inconsistent with the NS family.*

We have thus obtained a negative result for the HW model. The NS manifold is 'too small' for HW, in the sense that if the initial forward rate curve is on the manifold, then the HW dynamics will force the term structure off the manifold within an arbitrarily short period of time. For more positive results see [5].

Remark 3.5 It is an easy exercise to see that the minimal manifold which is consistent with HW is given by

$$G(z, x) = z_1 e^{-ax} + z_2 e^{-2ax}.$$

In the same way, one may easily test the consistency between NS and the model obtained by setting $a = 0$ in (75). This is the continuous time limit of the Ho and Lee model [27], and is henceforth referred to as HL. Since we have a pedagogical point to make, we give the results on consistency, which are as follows.

Proposition 3.4 (Nelson-Siegel and Ho-Lee)

(a) The full NS family is inconsistent with the Ho-Lee model.

(b) The degenerate family $G(z, x) = z_1 + z_3 x$ is in fact consistent with Ho-Lee.

Remark 3.6 We see that the minimal invariant manifold provides information about the model. From the result above, the HL model is closely tied to the class of **affine** forward rate curves. Such curves are unrealistic from an economic point of view, implying that the HL model is overly simplistic.

3.6 The Filipović State Space Approach to Consistency

As we very easily detected above, neither the HW nor the HL model is consistent with the Nelson-Siegel family of forward rate curves. A much more difficult problem is to determine whether **any** interest rate model is. This is Problem II in Section 3.1 for the NS family, and in a very general setting, inverse consistency problems like this has been studied in great detail by Filipović in [19], [20], and [21]. In this section we will give an introduction to the Filipović state space approach to the (inverse) consistency problem, and we will also study a small laboratory example.

The study will be done within the framework of a factor model.

Definition 3.6 *A* **factor model** *for the forward rate process r consists of the following objects and relations.*

- A d-dimensional **factor** or **state** process Z with $\mathbb{Q}$-dynamics of the form

$$dZ_t = a(Z_t)dt + b(Z_t)dW_t, \tag{80}$$

where W is an m-dimensional Wiener process. We denote by a_i the $i:th$ component of the column vector a, and by b_i the $i:th$ row of the matrix b.

- A smooth **output mapping**

$$G : \mathbb{R}^d \to \mathcal{H}.$$

For each $z \in \mathbb{R}^d$, $G(z)$ is thus a real valued C^∞ function and it's value at the point $x \in \mathbb{R}$ is denoted by $G(z, x)$.

- The forward rate process is then defined by

$$r_t = G(Z_t), \tag{81}$$

or on component form

$$r_t(x) = G(Z_t, x). \tag{82}$$

Since we have given the Z dynamics under the *martingale measure* $\mathbb{Q}$, it is obvious that there has to be some consistency requirements on the relations between a, b and G in order for r in (81) to be a specification of the forward rate process under $\mathbb{Q}$. The obvious way of deriving the consistency requirements is to compute the r dynamics from (80)-(81) and then to compare the result with the general form of the forward rate equation in (4). For ease of notation we will use the shorthand notation

$$G_x = \frac{\partial G}{\partial x}, \qquad G_i = \frac{\partial G}{\partial z_i}, \qquad G_i = \frac{\partial^2 G}{\partial z_i \partial z_j} \tag{83}$$

From the Itô formula, (80), and (81) we obtain

$$dr_t = \left\{ \sum_{i=1}^d G_i(Z_t)a_i(Z_t)dt + \frac{1}{2} \sum_{i,j=1}^d G_{ij}(Z_t)b_i(Z_t)b_j^\star(Z_t) \right\} dt \tag{84}$$

$$+ \sum_{i=1}^d G_i(Z_t)b_i(Z_t)dW_t \tag{85}$$

where $\star$ denotes transpose. Going back to the forward rate equation (4) we can identify the volatility process as

$$\sigma_t = \sum_{i=1}^d G_i(Z_t)b_i(Z_t).$$

We now insert this into the drift part of (4). We then use (81) to deduce that $\mathbf{F}r_t = G_x(Z_t)$ and also insert this expression into the drift part of (4). Comparing the resulting equation with (84) gives us the required consistency conditions.

Proposition 3.5 (Filipović) *The following relation must hold identically in* (z, x).

$$G_x(z, x) + \sum_{i,j=1}^{d} b_i(z)b_j^\star(z)G_i(z, x) \int_0^x G_j(z, s)ds$$

$$= \sum_{i=1}^{d} G_i(z, x)a_i(z) + \frac{1}{2} \sum_{i,j=1}^{d} G_{ij}(z, x)b_i(z)b_j^\star(z) \tag{86}$$

We can view the consistency equation (86) in three different ways.

- We can check consistency for a given specification of G, a,b.

- We can specify a and b. Then (86) is a PDE for the determination of a consistent output function G.

- We can specify G, i.e. we can specify a finite dimensional manifold of forward rate curves, and then use (86) to investigate whether there exists an underlying consistent state vector process Z, and if so, to find a and b. This inverse problem is precisely Problem II in Section 3.1.

We will focus on the last inverse problem above, and to see how the consistency equation can be used, we now go on to study two simple laboratory examples.

Example 3.1 In this example we consider the 2-dimensional manifold of **linear** forward rate curves, i.e. the output function G defined by

$$G(z, x) = z_1 + z_2 x. \tag{87}$$

This is not a very natural example from a finance point of view, but it is a good illustration of technique. The question we ask is whether there exist some forward rate model consistent with the class of linear forward rate curves and if so what the factor dynamics look like. For simplicity we restrict ourselves to the case of a scalar driving Wiener process, but the reader is invited to analyze the (perhaps more natural) case with a two-dimensional W.

We thus model the factor dynamics as

$$dZ_{1,t} = a_1(Z_t)dt + b_1(Z_t)dW_t, \tag{88}$$
$$dZ_{2,t} = a_1(Z_t)dt + b_2(Z_t)dW_t. \tag{89}$$

In this case we have

$$G_x(z, x) = z_2, \quad G_1(z, x) = 1, \quad G_2(z, x) = x,$$
$$G_{11}(z, x) = 0, \quad G_{12}(z, x) = 0, \quad G_{22}(z, x) = 0,$$

and

$$\int_0^x G_1(z,s)ds = x, \qquad \int_0^x G_2(z,s)ds = \frac{1}{2}x^2,$$

so the consistency equation (86) becomes

$$z_2 + b_1^2(z)x + b_1(z)b_2(z)\frac{1}{2}x^2 + b_2(z)b_1(z)x^2 + b_2^2(z)\frac{1}{2}x^3 = a_1(z) + a_2(z)x \quad (90)$$

Identifying coefficients we see directly that $b_2 = 0$ so the equation reduces to

$$z_2 + b_1^2(z)x = a_1(z) + a_2(z)x \qquad (91)$$

which gives us the relations $a_1 = z_2$ and $a_2 = b_1^2$. Thus we see that for this choice of G there does indeed exist a class of consistent factor models, with factor dynamics given by

$$dZ_{1,t} = Z_{2,t}dt + b_1(Z_t)dW_t \qquad (92)$$
$$dZ_{2,t} = b_1^2(Z_t)dt. \qquad (93)$$

Here b_1 can be chosen completely freely (subject only to regularity conditions). Choosing $b_1(z) = 1$, we see that the factor Z_2 is essentially running time, and the model is then in fact a special case of the Ho-Lee model.

Example 3.2 We now go on to study the more complicated two-dimensional manifold of *exponential* forward rate curves, given by the output function

$$G(z,x) = z_1 e^{z_2 x}. \qquad (94)$$

This is a simplified version of the Nelson-Siegel manifold, so it will give us some insight into the consistency problem for the NS case. In this case we will assume two independent driving Wiener processes W^1, and W^2, and we will assume factor dynamics of the form

$$dZ_{1,t} = a_1(Z_t)dt + b_{11}(Z_t)dW_t^1, \qquad (95)$$
$$dZ_{2,t} = a_2(Z_t)dt + b_{22}(Z_t)dW_t^2. \qquad (96)$$

Note that the factors are being driven by independent Wiener processes. The reader is invited to study the general case when both W^1 and W^2 enters into both equations. In our case we have

$$b_1(z) = [b_{11}(z), 0], \tag{97}$$

$$b_2(z) = [0, b_{22}(z)], \tag{98}$$

$$G_x(z, x) = z_1 z_2 e^{z_2 x}, \tag{99}$$

$$G_1(z, x) = e^{z_2 x}, \tag{100}$$

$$\int_0^x G_1(z, s)ds = z_2^{-1}\left(e^{z_2 x} - 1\right), \tag{101}$$

$$G_{11}(z, x) = 0, \tag{102}$$

$$G_2(z, x) = z_1 x e^{z_2 x}, \tag{103}$$

$$G_{22}(z, x) = z_1 x^2 e^{z_2 x}, \tag{104}$$

$$\int_0^x G_2(z, s)ds = z_1 z_2^{-1} x e^{z_2 x} - z_1 z_2^{-2} e^{z_2 x} + z_2^{-2}. \tag{105}$$

The consistency equation thus becomes

$$z_1 z_2 e^{z_2 x}$$
$$+ b_{11}^2(z) z_2^{-1} e^{2 z_2 x} - b_{11}^2(z) z_2^{-1} e^{z_2 x}$$
$$+ b_{22}^2(z) z_1^2 z_2^{-1} x^2 e^{2 z_2 x} - b_{22}^2(z) z_1^2 z_2^{-2} x e^{2 z_2 x} + b_{22}^2(z) z_1 z_2^{-2} x e^{z_2 x}$$
$$= a_1(z) e^{z_2 x} + a_2(z) z_1 x e^{z_2 x} + \frac{1}{2} b_{22}^2(z) z_1 x^2 e^{z_2 x}$$

Rearranging terms we have

$$e^{z_2 x} \left[z_1 z_2 + b_{11}(z) z_2^{-1} - a_1(z) \right]$$
$$+ x e^{z_2 x} \left[b_{22}^2(z) z_1 z_2^{-2} - a_2(z) z_1 \right]$$
$$+ x^2 e^{z_2 x} \left[-\frac{1}{2} b_{22}^2(z) z_1 \right]$$
$$+ e^{2 z_2 x} \left[b_{11}^2(z) z_2^{-1} \right]$$
$$+ x e^{2 z_2 x} \left[-b_{22}^2(z) z_1^2 z_2^{-2} \right]$$
$$+ x^2 e^{2 z_2 x} \left[b_{22}^2(z) z_1^2 z_2^{-1} \right]$$
$$= 0$$

From the linear independence of the quasi-exponential functions, we immediately obtain

$$b_{11}(z) = 0, \quad b_{22}(z) = 0,$$
$$a_2(z) = 0, \quad a_1(z) = z_1 z_2.$$

The only consistent Z-dynamics are thus given by

$$dZ_{1,t} = z_{2,t} Z_{1,t}, dt \tag{106}$$

$$dZ_{2,t} = 0, \tag{107}$$

which implies that Z_2 is constant and that

$$Z_{1,t} = Z_{1,0}e^{Z_2 t}.$$

We thus see that, apart from allowing randomness in the initial values, both Z_1 and Z_2 evolve along deterministic paths, where in fact Z_2 stays constant whereas Z_1 grows exponentially at the rate Z_2. In other words, there exists no non-trivial factor model which is consistent with the class of exponential forward rate curves.

As we have seen, the calculations quickly become rather messy, and it is thus a formidable task to find the set of consistent factor models for a more complicated manifold like, say, the Nelson-Siegel family of forward rate curves. Since the NS family is four-dimensional we would need a four dimensional factor model with four independent Wiener processes (all of which would be driving each of the four equations).

In [19] the case of the NS family was indeed studied, and it was proved that **no nontrivial Wiener driven model is consistent with NS**. Thus, for a model to be consistent with Nelson-Siegel, it must be deterministic (apart from randomness in the initial conditions). In [20] (which is a technical tour de force) this result was then extended to a much larger exponential polynomial family than the NS family.

3.7 Notes

The section is largely based on [5] and [19]. In our presentation we have used strong solutions of the infinite dimensional forward rate SDE. This is of course restrictive. The invariance problem for weak solutions has been studied by Filipović in [22] and [21]. An alternative way of studying invariance is by using some version of the Stroock–Varadhan support theorem, and this line of thought is carried out in depth in [38].

4 The General Realization Problem

We now turn to Problem 2 in Section 1.3, i.e. the problem when a given forward rate model has a finite dimensional factor realization. For ease of exposition we mostly confine ourselves to time invariant forward rate dynamics. Time varying systems can be treated similarly (see [9]). We will use some ideas and concepts from differential geometry, and a general reference here is [37]. The section is based on [9].

4.1 Setup

We consider a given volatility structure $\sigma : \mathcal{H} \to \mathcal{H}^m$ and study the induced forward rate model (on Stratonovich form)

$$dr_t = \mu(r_t)dt + \sigma(r_t) \circ dW_t \tag{108}$$

where as before (see Section 3.4).

$$\mu(r) = \frac{\partial}{\partial x}r + \sigma(r)\mathbf{II}\upsilon(r)^* - \frac{1}{2}\sigma_r'(r)\sigma(r) \tag{109}$$

Throughout the rest of the section, Assumption 3.1 is in force.

4.2 The Geometric Problem

Given a specification of the volatility mapping σ, and an initial forward rate curve r^o we now investigate when (and how) the corresponding forward rate process possesses a finite, dimensional realization. We are thus looking for smooth d-dimensional vector fields a and b, an initial point $z_0 \in \mathbb{R}^d$, and a mapping $G : \mathbb{R}^d \rightarrow \mathcal{H}$ such that r, locally in time, has the representation

$$dZ_t = a(Z_t)\,dt + b(Z_t)\,dW_t, \ Z_0 = z_0 \tag{110}$$
$$r_t(x) = G(Z_t, x). \tag{111}$$

Remark 4.1 Let us clarify some points. Firstly, note that in principle it may well happen that, given a specification of σ, the r-model has a finite dimensional realization given a particular initial forward rate curve r^o, while being infinite dimensional for all other initial forward rate curves in a neighborhood of r^o. We say that such a model is a **non-generic** or **accidental** finite dimensional model. If, on the other hand, r has a finite dimensional realization for all initial points in a neighborhood of r^o, then we say that the model is a **generically** finite dimensional model. In this text we are solely concerned with the generic problem. Secondly, let us emphasize that we are looking for **local** (in time) realizations.

We can now connect the realization problem to our studies of invariant manifolds.

Proposition 4.1 *The forward rate process possesses a finite dimensional realization if and only if there exists an invariant finite dimensional submanifold $\mathcal{G}$ with $r^o \in \mathcal{G}$.*

Proof. See [5] for the full proof. The intuitive argument runs as follows. Suppose that there exists a finite dimensional invariant manifold $\mathcal{G}$ with $r^o \in \mathcal{G}$. Then $\mathcal{G}$ has a local coordinate system, and we may define the Z process as the local coordinate process for the r-process. On the other hand it is clear that if r has a finite dimensional realization as in (110)-(111), then every forward rate curve that will be produced by the model is of the form $x \longmapsto G(z, x)$ for some choice of z. Thus there exists a finite dimensional invariant submanifold $\mathcal{G}$ containing the initial forward rate curve r^o, namely $\mathcal{G} = ImG$. $\square$

Using Theorem 3.1 we immediately obtain the following geometric characterization of the existence of a finite realization.

Corollary 4.1 *The forward rate process possesses a finite dimensional realization if and only if there exists a finite dimensional manifold $\mathcal{G}$ containing r^o, such that, for each $r \in \mathcal{G}$ the following conditions hold.*

$$\mu(r) \in T_{\mathcal{G}}(r),$$
$$\sigma(r) \in T_{\mathcal{G}}(r).$$

Here $T_{\mathcal{G}}(r)$ denotes the tangent space to $\mathcal{G}$ at the point r, and the vector fields μ and σ are as above. The tangency condition for σ is as usual interpreted component wise.

4.3 The Main Result

Given the volatility vector fields $\sigma_1, \ldots, \sigma_m$, and hence also the field μ, we now are faced with the problem of determining if there exists a finite dimensional manifold $\mathcal{G}$ with the property that μ and $\sigma_1, \ldots, \sigma_m$ are tangential to $\mathcal{G}$ at each point of $\mathcal{G}$. In the case when the underlying space is finite dimensional, this is a standard problem in differential geometry, and we will now give the heuristics.

To get some intuition we start with a simpler problem and therefore consider the space $\mathcal{H}$ (or any other Hilbert space), and a smooth vector field f on the space. For each fixed point $r^o \in \mathcal{H}$ we now ask if there exists a finite dimensional manifold $\mathcal{G}$ with $r^o \in \mathcal{G}$ such that f is tangential to $\mathcal{G}$ at every point. The answer to this question is yes, and the manifold can in fact be chosen to be one-dimensional. To see this, consider the infinite dimensional ODE

$$\frac{dr_t}{dt} = f(r_t), \tag{112}$$
$$r_0 = r^o. \tag{113}$$

If r_t is the solution, at time t, of this ODE, we use the notation

$$r_t = e^{ft} r^o.$$

We have thus defined a group of operators $\{e^{ft} : t \in \mathbb{R}\}$, and we note that the set $\{e^{ft} r^o : t \in \mathbb{R}\} \subseteq \mathcal{H}$ is nothing else than the integral curve of the vector field f, passing through r^o. If we define $\mathcal{G}$ as this integral curve, then our problem is solved, since f will be tangential to $\mathcal{G}$ by construction.

Let us now take two vector fields f_1 and f_2 as given, where the reader informally can think of f_1 as σ (in the case of a scalar Wiener process) and f_2 as μ. We also fix an initial point $r^o \in \mathcal{H}$ and the question is if there exists a finite dimensional manifold $\mathcal{G}$, containing r^o, with the property that f_1 and f_2 are both tangential to $\mathcal{G}$ at each point of $\mathcal{G}$. We call such a manifold a **tangential manifold** for the vector fields. At a first glance it would seem that there always exists a tangential manifold, and that it can even be chosen to be two-dimensional. The geometric idea is that we start at r^o and let f_1 generate the integral curve $\{e^{f_1 s} r^o : s \geq 0\}$. For each point $e^{f_1 s} r^o$ on this curve we now let f_2 generate the integral curve starting at that point. This gives us the object $e^{f_2 t} e^{f_1 s} r^o$ and thus it seems that we sweep out a two dimensional surface $\mathcal{G}$ in $\mathcal{H}$. This is our obvious candidate for a tangential manifold.

In the general case this idea will, however, not work, and the basic problem is as follows. In the construction above we started with the integral curve generated by f_1 and then applied f_2, and there is of course no guarantee that we will obtain the same surface if we start with f_2 and then apply f_1. We thus have some sort of commutativity problem, and the key concept is the **Lie bracket**.

Definition 4.1 *Given smooth vector fields f and g on $\mathcal{H}$, the Lie bracket $[f, g]$ is a new vector field defined by*

$$[f, g](r) = f'(r)g(r) - g'(r)f(r) \tag{114}$$

The Lie bracket measures the lack of commutativity on the infinitesimal scale in our geometric program above, and for the procedure to work we need a condition which says that the lack of commutativity is "small". It turns out that the relevant condition is that the Lie bracket should be in the linear hull of the vector fields.

Definition 4.2 *Let $f_1, \ldots, f_n$ be smooth independent vector fields on some space X. Such a system is called a **distribution**, and the distribution is said to be **involutive** if*

$$[f_i, f_j](x) \in span\{f_1(x), \quad , f_n(x)\}, \quad \forall i, j,$$

where the span is the linear hull over the real numbers.

We now have the following basic result, which extends a classic result from finite dimensional differential geometry (see [37]).

Theorem 4.1 (Frobenius) *Let $f_1, \ldots, f_k$ and be independent smooth vector fields in $\mathcal{H}$ and consider a fixed point $r^o \in \mathcal{H}$. Then the following statements are equivalent.*

- *For each point r in a neighborhood of r^o, there exists a k-dimensional tangential manifold passing through r.*

- *The system $f_1, \ldots, f_k$ of vector fields is involutive.*

Proof. See [9], which provides a self contained proof of the Frobenius Theorem in Banach space. $\square$

Let us now go back to our interest rate model. We are thus given the vector fields μ, σ, and an initial point r^o, and the problem is whether there exists a finite dimensional tangential manifold containing r^o. Using the infinite dimensional Frobenius theorem, this situation is now easily analyzed. Suppose for simplicity that $m = 1$ i.e. that we only have one scalar driving Wiener process. Now; if $\{\mu, \sigma\}$ is involutive then there exists a two dimensional tangential manifold. If $\{\mu, \sigma\}$ is not involutive, this means that the Lie bracket $[\mu, \sigma]$ is not in the linear span of μ and σ, so then we consider the system $\{\mu, \sigma, [\mu, \sigma]\}$. If this system is involutive there exists a three dimensional tangential manifold. If it is not involutive at least one of the brackets $[\mu, [\mu, \sigma]], [\sigma, [\mu, \sigma]]$ is not in the span of $\{\mu, \sigma, [\mu, \sigma]\}$, and we then adjoin this (these) bracket(s). We continue in this way, forming brackets of brackets, and adjoining these to the linear hull of the previously obtained vector fields, until the point when the system of vector fields thus obtained actually is closed under the Lie bracket operation.

Definition 4.3 *Take the vector fields* $f_1, \ldots, f_k$ *as given. The* **Lie algebra** *generated by* $f_1, \ldots, f_k$ *is the smallest linear space (over* $\mathbb{R}$*) of vector fields which contains* $f_1, \ldots, f_k$ *and is closed under the Lie bracket. This Lie algebra is denoted by*

$$\mathcal{L} = \{f_1, \ldots, f_k\}_{LA}$$

The **dimension** *of* $\mathcal{L}$ *is defined, for each point* $r \in \mathcal{H}$ *as*

$$dim\,[\mathcal{L}(r)] = dim\,span\,\{f_1(r), \ldots, f_k(r)\}\,.$$

Putting all these results together, we can now state the main result on finite dimensional realizations. As can be seen from the arguments above, the fact that we have been studying the particular case of the forward rate equation is not at all essential: all results will continue to hold for any SDE with smooth drift and diffusion vector fields, evolving on a Hilbert space. We therefore state the main realization theorem for an arbitrary SDE in Hilbert space.

Theorem 4.2 (Main Result) *Consider the following Stratonovich SDE, evolving in a given Hilbert space* $\mathcal{H}$.

$$dr = \mu(r_t)dt + \sigma(r_t) \circ dW_t. \tag{115}$$

We assume that the drift and diffusion terms μ *and* σ *are smooth vector fields on* $\mathcal{H}$.

Then the SDE (115) generically admits a finite dimensional realization if and only if

$$dim\,\{\mu, \sigma_1, \ldots, \sigma_m\}_{LA} < \infty$$

in a neighborhood of r^o.

The result above thus provides a general solution to Problem 2 from Section 1.3. For any given specification of forward rate volatilities, the Lie algebra can in principle be computed, and the dimension can be checked. Note, however, that the theorem is a pure existence result. If, for example, the Lie algebra has dimension five, then we know that there exists a five-dimensional realization, but the theorem does not directly tell us how to construct a concrete realization. This is the subject of Section 5 below. Note also that realizations are not unique, since any diffeomorphic mapping of the factor space $\mathbb{R}^d$ onto itself will give a new equivalent realization.

When computing the Lie algebra generated by μ and σ, the following observations are often useful.

Lemma 4.1 *Let us assume tht the vector fields $f_1, \ldots, f_k$ as given. The Lie algebra $\mathcal{L} = \{f_1, \ldots, f_k\}_{LA}$ remains unchanged under the following operations.*

- *The vector field $f_i(r)$ may be replaced by $\alpha(r)f_i(r)$, where α is any smooth nonzero scalar field.*

- *The vector field $f_i(r)$ may be replaced by*

$$f_i(r) + \sum_{j \neq i} \alpha_j(r) f_j(r),$$

where α_j is any smooth scalar field.

Proof. The first point is geometrically obvious, since multiplication by a scalar field will only change the length of the vector field f_i, and not its direction, and thus not the tangential manifold. Formally it follows from the "Leibnitz rule" $[f, \alpha g] = \alpha[f, g] - (\alpha' f)g$. The second point follows from the bilinear property of the Lie bracket together with the fact that $[f, f] = 0$. □

4.4 Constructing the Invariant Manifold

As we have seen above, there exists generically an FDR for (115) if and only if there exists, for any initial point near r^o an invariant manifold containing the initial point, and this manifold will also be a tangential manifold for all vector fields in the Lie algebra $\{\mu, \sigma\}_{LA}$. In this section we provide a concrete parameterization of this invariant manifold. This result will be used in connection with the construction problem treated in Section 5.

Proposition 4.2 *Consider the SDE (115), and assume that the Lie algebra $\mathcal{L} = \{\mu, \sigma\}_{LA}$ is finite dimensional near r^o. Assume furthermore that we have chosen an involutive system of independent vector fields $f_1, \ldots, f_n$ such that $\mathcal{L} = \text{span}\{f_1, \ldots, f_n\}$. Now choose an initial point $r_0 \in \mathcal{H}$ near r^o. Denote the induced invariant (and thus tangential) manifold through r_0, by $\mathcal{G}$. Define the mapping $G : \mathbb{R}^n \to X$ by*

$$G(z_1, \ldots z_n) = e^{f_n z_n} \ldots e^{f_1 z_1} r_0.$$

Then G is a local parameterization of $\mathcal{G}$. Furthermore, the inverse of G restricted to V is a local coordinate system for $\mathcal{G}$ at r_0.

Proof. It follows directly from the definition of a tangential manifold that $G(z) \in \mathcal{G}$ for all z near 0 in $\mathbb{R}^n$. Furthermore it is easy to see that $G'(0)h = \sum_{i=1}^{n} h_i f_i(x_0)$ and , since $f_1, \ldots, f_n$ are independent, $G'(0)$ is injective. The inverse function theorem does the rest. $\square$

With this machinery we can also very easily solve a related question. Consider a fixed interest rate model, specified by the volatility σ and also a fixed family of forward rate curves parameterized by the mapping $G_0 : \mathbb{R}^k \to \mathcal{H}$. Now, if $\mathcal{G}_0 = Im[G_0]$ is invariant, then the interest rate model will, given any initial point r^o in $\mathcal{G}_0$, only produce forward rate curves belonging to $\mathcal{G}_0$ or, in the terminology of Section 3, the given model and the family $\mathcal{G}_0$ are **consistent**. If the family is not consistent, then an initial forward rate curve in $\mathcal{G}_0$ may produce future forward rate curves outside $\mathcal{G}_0$, and the question arises how to construct the smallest possible family of forward rate curves which contains the initial family $\mathcal{G}_0$, and is consistent (i.e. invariant) w.r.t the interest rate model. As a concrete example, one may want to find the minimal extension of the Nelson-Siegel family of forward rate curves (see [34], [5]) which is consistent with the Hull-White (extended Vasiček) model. In particular one would like to know under what conditions this minimal extension of $\mathcal{G}_0$ is finite dimensional.

In geometrical terms we thus want to construct the minimal manifold containing $\mathcal{G}_0$, which is tangential w.r.t. the vector fields $\mu, \sigma_1, \ldots, \sigma_m$. The solution is obvious: For very point on $\mathcal{G}_0$ we construct the minimal tangential manifold through that point, and then we define the extension $\mathcal{G}$ as the union of all these fibers. Thus we have the following result, the proof of which is obvious. Concrete applications will be given below.

Proposition 4.3 *Consider a fixed volatility mapping σ, and let $\mathcal{G}_0$ be a k-dimensional submanifold parameterized by $G_0 : \mathbb{R}^k \to \mathcal{H}$. Then $\mathcal{G}_0$ can be extended to a finite dimensional invariant submanifold $\mathcal{G}$, if and only if*

$$\dim \{\mu, \sigma_1, \ldots, \sigma_m\}_{LA} < \infty.$$

Moreover, if $\mathcal{G}_0$ is transversal to $\{\mu, \sigma\}_{LA}$ and if the Lie algebra is spanned by the independent vector fields $f_1, \ldots, f_d$, then $\dim \mathcal{G} = k + d$ and a parameterization of $\mathcal{G}$ is given by the map $G : \mathbb{R}^{k+d} \to \mathcal{H}$, defined by

$$G(z_1, \ldots, z_k, y_1, \ldots, y_d) = e^{f_d y_d} \ldots e^{f_1 y_1} G_0(z_1, \ldots, z_k). \tag{116}$$

Remark 4.2 The term "transversal" above means that no vector in the Lie algebra $\mathcal{L}(\mu, \sigma)$ is contained the tangent space of $\mathcal{G}_0$ at any point of $\mathcal{G}_0$. This prohibits an integral curve of $\mathcal{L}$ to be contained in $\mathcal{G}_0$, which otherwise would lead to an extension with lower dimension than $d + k$. In such a case the parameterization above would amount to an over parameterization in the sense that G would not be injective.

4.5 Applications

In this section we give some simple applications of the theory developed above. For more examples and results, see [9].

Constant Volatility: Existence of FDRs

We start with the simplest case, which is when the volatility $\sigma(r, x)$ does not depend on r. In other words, σ is of the form $\sigma(r, x) = \sigma(x)$, and σ is thus a constant vector field on $\mathcal{H}$. We assume for brevity of notation that we have only one driving Wiener process. Since σ is deterministic we have no Stratonovich correction term and the vector fields are given by

$$\mu(r, x) = \mathbf{F}r(x) + \sigma(x) \int_0^x \sigma^*(s)ds,$$

$$\sigma(r, x) = \sigma(x),$$

where as before $\mathbf{F} = \partial/\partial x$.

The Frechet derivatives are trivial in this case. Since $\mathbf{F}$ is linear (and bounded in our space), and σ is constant as a function of r, we obtain

$$\mu_r' = \mathbf{F},$$
$$\sigma_r' = 0.$$

Thus the Lie bracket $[\mu, \sigma]$ is given by

$$[\mu, \sigma] = \mathbf{F}\sigma,$$

and in the same way we have

$$[\mu, [\mu, \sigma]] = \mathbf{F}^2\sigma.$$

Continuing in the same manner it is easily seen that the relevant Lie algebra $\mathcal{L}$ is given by

$$\mathcal{L} = \{\mu, \sigma\}_{LA} = span\left\{\mu, \sigma, \mathbf{F}\sigma, \mathbf{F}^2\sigma, \ldots\right\} = span\left\{\mu, \mathbf{F}^n\sigma\,; n = 0, 1, 2, \ldots\right\}$$

and it is thus clear that $\mathcal{L}$ is finite dimensional (at each point r) if and only if the function space
$$span\left\{\mathbf{F}^n\sigma;\ n = 0, 1, 2, \ldots\right\}$$

is finite dimensional. This, on the other hand, occurs if and only if each component of σ solves a linear ODE with constant coefficients. This argument is easily extended to the case of a multidimensional driving Wiener process so, using Lemma 2.1, we can finally state the existence result for constant volatility models.

Proposition 4.4 *Assume that the volatility components* $\sigma_1, \ldots, \sigma_m$ *are deterministic, i.e. of the form*

$$\sigma_i(r, x) = \sigma_i(x), \quad i = 1, \ldots, m.$$

Then there exists a finite dimensional realization if and only if the function space

$$\text{span}\{\mathbf{F}^n \sigma_i; \ i = 1, \ldots, m; \ n = 0, 1, 2, \ldots\}$$

is finite dimensional. This occurs if and only if each component of σ *is a quasi-exponential function.*

Constant Volatility: Invariant Manifolds

For models with constant volatility vector fields, we now turn to the construction of invariant manifolds, and to this end we assume that the Lie algebra above is finite dimensional. Thus it is spanned by a finite number of vector fields as

$$\{\mu, \sigma\}_{LA} = \text{span}\left\{\mu, \sigma_i^{(k)}; \ i = 1, \ldots, m; \ k = 0, 1, \ldots, n_i\right\},$$

where

$$\sigma_i^{(k)}(x) = \frac{\partial^k \sigma_i}{\partial x^k}(x).$$

In order to apply Proposition 4.2 and Proposition 4.3, we have to compute the operators $\exp[\mu t]$ and $\exp\left[\sigma_i^{(k)} t\right]$, i.e. we have to solve $\mathcal{H}$-valued ODEs. We recall that

$$\mu(r) = \mathbf{F}r + D,$$

where the constant field D is given

$$D(x) = \sum_i^m \sigma_i(x) \int_0^x \sigma_i^\star(s) ds,$$

which can be written as

$$D(x) = \frac{1}{2} \frac{\partial}{\partial x} \|S(x)\|^2,$$

where $S(x) = \int_0^x \sigma(s) ds$. Thus $e^{\mu t}$ is obtained by solving

$$\frac{dr}{dt} = \mathbf{F}r + D.$$

This is a linear equation, and from Proposition 2.1 we obtain the solution

$$r_t = e^{\mathbf{F}t} r_0 + \int_0^t e^{\mathbf{F}(t-s)} D ds$$

so

$$\left(e^{\mu t} r_0\right)(x) = r_0(x+t) + \frac{1}{2}\left(\|S(x+t)\|^2 - \|S(x)\|^2\right).$$

The vector fields $\sigma_i^{(k)}$ are constant, so the corresponding ODEs are trivial. We have

$$e^{\sigma_i^{(k)} t} r_0 = r_0 + \sigma_i^{(k)} t.$$

We thus have the following results on the parameterization of invariant manifolds. For a given mapping $G : \mathbb{R}^n \to \mathcal{H}$, we write $G(z)(x)$ or $G(z, x)$ to denote the function $G(z) \in \mathcal{H}$ evaluated at $x \in \mathbb{R}_+$.

Proposition 4.5 *The invariant manifold generated by the initial forward rate curve r_0 is parameterized as*

$$G(z_0, z_i^k;\ i = 1, \ldots, m;\ k = 0, \ldots, n_i)(x)$$
$$= r_0(x + z_0) + \frac{1}{2}\left(\|S(x+t)\|^2 - \|S(x)\|^2\right) + \sum_{i=1}^{m}\sum_{k=0}^{n_i}\sigma_i^{(k)}(x)z_i^k.$$

If the k-dimensional manifold $\mathcal{G}_0$ is transversal to $\mathcal{L}\{\mu, \sigma\}$ and parameterized by $G_0(y_1, \ldots, y_k)$, then the minimal consistent (i.e. invariant) extension is parameterized as

$$G(y_1, \ldots, y_k, z_0, z_i^k;\ i = 1, \quad, m;\ k = 0, \ldots, n_i)(x)$$
$$= G_0(y_1, \ldots, y_k)(x + z_0) + \frac{1}{2}\left(\|S(x+z_0)\|^2 - \|S(x)\|^2\right) + \sum_{i=1}^{m}\sum_{k=0}^{n_i}\sigma_i^{(k)}(x)z_i^k.$$

Note that if $\mathcal{G}_0$ is invariant under shift in the x-variable (this is in fact the typical case), then a simpler parameterization of $\mathcal{G}$ is given by

$$G(y_1, \ldots, y_k, z_0, z_i^k;\ i = 1, \ldots, m;\ k = 0, \ldots, n_i)(x)$$
$$= G_0(y_1, \ldots, y_k)(x) + \frac{1}{2}\left(\|S(x+z_0)\|^2 - \|S(x)\|^2\right) + \sum_{i=1}^{m}\sum_{k=0}^{n_i}\sigma_i^{(k)}(x)z_i^k.$$

As a concrete application let us consider the simple case when $m = 1$ and

$$\sigma(x) = \sigma e^{-ax},$$

where, with a slight abuse of notation, a and σ denote positive constants. As is well known, this is the HJM formulation of the Hull-White extension of the Vasiček model [28],[36]. In this case we have

$$S(x) = \frac{\sigma}{a}\left[1 - e^{-ax}\right].$$

The relevant function space

$$\{\mathbf{F}^n \sigma; \ n \geq 0\} = \left\{ \frac{\partial^n}{\partial x^n} e^{-ax}; \ n \geq 0 \right\}$$

is obviously one-dimensional and spanned by the single function e^{-ax}, so the Lie algebra is two-dimensional.

As the given manifold $\mathcal{G}_0$ we take the Nelson-Siegel ([34]) family of forward rate curves, parameterized as

$$G_0(y_1, \ldots, y_4)(x) = y_1 + y_2 e^{-y_4 x} + y_3 x e^{-y_4 x}.$$

This family is obviously invariant under shift in x, so we have the following result.

Proposition 4.6 *For a given initial forward rate curve r_0, the invariant manifold generated by the Hull-White extended Vasiček model is parameterized by*

$$G(z_0, z_1)(x) = r_0(x+z_0) + e^{-ax} \frac{\sigma^2}{a^2} [1 - e^{-az_0}] - e^{-2ax} \frac{\sigma^2}{2a^2} [1 - e^{-2az_0}] + z_1 e^{-ax}.$$

The minimal extension of the NS family consistent with the Hull-White extended Vasiček model is parameterized by

$$G(z_0, z_1, y_1, \ldots, y_4)(x) = y_1 + y_2 e^{-y_4 x} + y_3 x e^{-y_4 x}$$
$$+ e^{-ax} \frac{\sigma^2}{a^2} [1 - e^{-az_0}] - e^{-2ax} \frac{\sigma^2}{2a^2} [1 - e^{-2az_0}] + z_1 e^{-ax}.$$

Constant Direction Volatility

We go on to study the most natural extension of the deterministic volatility case namely the case when the volatility is of the form

$$\sigma(r, x) = \varphi(r)\lambda(x). \tag{117}$$

We restrict ourselves to the case of a scalar Wiener process. In this case the individual vector field σ has the constant direction $\lambda \in \mathcal{H}$, but is of varying length, determined by φ, where φ is allowed to be any smooth functional of the entire forward rate curve. In order to avoid trivialities we make the following assumption.

Assumption 4.1 *We assume that $\varphi(r) \neq 0$ for all $r \in \mathcal{H}$.*

After a simple calculation the drift vector μ turns out to be

$$\mu(r) = \mathbf{F}r + \varphi^2(r)D - \frac{1}{2}\varphi'(r)[\lambda]\varphi(r)\lambda, \tag{118}$$

where $\varphi'(r)[\lambda]$ denotes the Frechet derivative $\varphi'(r)$ acting on the vector λ, and where the constant vector $D \in \mathcal{H}$ is given by

$$D(x) = \lambda(x) \int_0^x \lambda(s)ds.$$

We now want to know under what conditions on φ and λ we have a finite dimensional realization, i.e. when the Lie algebra generated by

$$\mu(r) = \mathbf{F}r + \varphi^2(r)D - \frac{1}{2}\varphi'(r)[\lambda]\varphi(r)\lambda,$$
$$\sigma(r) = \varphi(r)\lambda,$$

is finite dimensional. Under Assumption 4.1 we can use Lemma 4.1, to see that the Lie algebra is in fact generated by the simpler system of vector fields

$$f_0(r) = \mathbf{F}r + \Phi(r)D,$$
$$f_1(r) = \lambda,$$

where we have used the notation

$$\Phi(r) = \varphi^2(r).$$

Since the field f_1 is constant, it has zero Frechet derivative. Thus the first Lie bracket is easily computed as

$$[f_0, f_1](r) = \mathbf{F}\lambda + \Phi'(r)[\lambda]D$$

The next bracket to compute is $[[f_0, f_1], f_1]$ which is given by

$$[[f_0, f_1], f_1] = \Phi''(r)[\lambda; \lambda]D.$$

Note that $\Phi''(r)[\lambda; \lambda]$ is the second order Frechet derivative of Φ operating on the vector pair $[\lambda; \lambda]$. This pair is to be distinguished from (notice the semicolon) the Lie bracket $[\lambda, \lambda]$ (with a comma), which if course would be equal to zero. We now make a further assumption.

Assumption 4.2 *We assume that $\Phi''(r)[\lambda; \lambda] \neq 0$ for all $r \in \mathcal{H}$.*

Given this assumption we may again use Lemma 4.1 to see that the Lie algebra is generated by the following vector fields

$$f_0(r) = \mathbf{F}r,$$
$$f_1(r) = \lambda,$$
$$f_3(r) = \mathbf{F}\lambda,$$
$$f_4(r) = D.$$

Of these vector fields, all but f_0 are constant, so all brackets are easy. After elementary calculations we see that in fact

$$\{\mu, \sigma\}_{LA} = span\{\mathbf{F}r, \mathbf{F}^n\lambda,\ \mathbf{F}^n D;\ n = 0, 1, \ldots\}.$$

From this expression it follows immediately that a necessary condition for the Lie algebra to be finite dimensional is that the vector space spanned by $\{\mathbf{F}^n\lambda;\ n \geq 0\}$ is finite dimensional. This occurs if and only if λ is quasi-exponential (see Remark 2.2). If, on the other hand, λ is quasi-exponential, then we know from Lemma 2.1, that also D is quasi-exponential, since it is the integral of the QE function λ multiplied by the QE function λ. Thus the space $\{\mathbf{F}^n D;\ n = 0, 1, \ldots\}$ is also finite dimensional, and we have proved the following result.

Proposition 4.7 *Under Assumptions 4.1 and 4.2, the interest rate model with volatility given by $\sigma(r, x) = \varphi(r)\lambda(x)$ has a finite dimensional realization if and only if λ is a quasi-exponential function. The scalar field φ is allowed to be any smooth field.*

When is the Short Rate a Markov Process?

One of the classical problems concerning the HJM approach to interest rate modeling is that of determining when a given forward rate model is realized by a short rate model, i.e. when the short rate is Markovian. We now briefly indicate how the theory developed above can be used in order to analyze this question. For the full theory see [9].

Using the results above, we immediately have the following general necessary condition.

Proposition 4.8 *The forward rate model generated by σ is a generic short rate model, i.e the short rate is generically a Markov process, only if*

$$dim\{\mu, \sigma\}_{LA} \leq 2 \tag{119}$$

Proof. If the model is really a short rate model, then bond prices are given as $p_t(x) = F(t, R_t, x)$ where F solves the term structure PDE. Thus bond prices, and forward rates are generated by a two dimensional factor model with time t and the short rate R as the state variables. $\square$

Remark 4.3 The most natural case is clearly $dim\{\mu, \sigma\}_{LA} = 2$. However, it is an open problem whether there exists a non-deterministic generic short rate model with $dim\{\mu, \sigma\}_{LA} = 1$.

Note that condition (119) is only a sufficient condition for the existence of a short rate realization. It guarantees that there exists a two-dimensional realization, but the question remains whether the realization can chosen in such a way that the short rate and running time are the state variables. This question is completely resolved by the following central result.

Theorem 4.3 *Assume that the model is not deterministic, and take as given a time invariant volatility* $\sigma(r, x)$. *Then there exists a short rate realization if and only if the vector fields* $[\mu, \sigma]$ *and* σ *are parallel, i.e. if and only if there exists a scalar field* $\alpha(r)$ *such that the following relation holds (locally) for all* r.

$$[\mu, \sigma](r) = \alpha(r)\sigma(r). \tag{120}$$

Proof. See [9].   □

It turns out that the class of generic short rate models is very small indeed. We have, in fact, the following result, which was first proved in [31] (using techniques different from those above). See [9] for a proof based on Theorem 4.3.

Theorem 4.4 *Consider a HJM model with one driving Wiener process and a volatility structure of the form*

$$\sigma(r, x) = g(R, x).$$

where $R = r(0)$ *is the short rate. Then the model is a generic short rate model if and only if* g *has one of the following forms.*

- *There exists a constant* c *such that*

$$g(R, x) \equiv c.$$

- *There exist constants* u *and* c *such that.*

$$g(R, x) = ce^{-ax}.$$

- *There exist constants* a *and* b, *and a function* $\alpha(x)$, *where* α *satisfies a certain Riccati equation, such that*

$$g(R, x) = \alpha(x)\sqrt{aR + b}$$

We immediately recognize these cases as the Ho-Lee model, the Hull-White extended Vasiček model, and the Hull-White extended Cox-Ingersoll-Ross model. Thus, in this sense the only generic short rate models are the affine ones, and the moral of this, perhaps somewhat surprising, result is that most short rate models considered in the literature are not generic but "accidental". To understand the geometric picture one can think of the following program.

1. Choose an arbitrary short rate model, say of the form

$$dR_t = a(R_t)dt + b(R_t)dW_t$$

   with a fixed initial point R_0.

2. Solve the associated PDE in order to compute bond prices. This will also produce:

- An initial forward rate curve $\hat{r}^o(x)$.

- Forward rate volatilities of the form $g(R, x)$.

3. Forget about the underlying short rate model, and take the forward rate volatility structure $g(R, x)$ as given in the forward rate equation.

4. Initiate the forward rate equation with an arbitrary initial forward rate curve $r^o(x)$

The question is now whether the thus constructed forward rate model will produce a Markovian short rate process. Obviously, if you choose the initial forward rate curve r^o as $r^o = \hat{r}^o$, then you are back where you started, and everything is OK. If, however, you choose another initial forward rate curve than $\hat{r}^o$, say the observed forward rate curve of today, then it is no longer clear that the short rate will be Markovian. What the theorem above says, is that only the models listed above will produce a Markovian short rate model for all initial points in a neighborhood of $\hat{r}^o$. If you take another model (like, say, the Dothan model) then a generic choice of the initial forward rate curve will produce a short rate process which is not Markovian.

4.6 Notes

The section is based on [9] where full proofs and further results can be found, and where also the time varying case is considered. In our study of the constant direction model above, φ was allowed to be any smooth functional of the entire forward rate curve. The simpler special case when φ is a point evaluation of the short rate, i.e. of the form $\varphi(r) = h(r(0))$ has been studied in [1], [29] and [35]. All these cases fall within our present framework and, the results are included as special cases of the general theory above. A different case, treated in [14], occurs when σ is a finite point evaluation, i.e. when $\sigma(t, r) = h(t, r(x_1), \dots r(x_k))$ for fixed benchmark maturities $x_1, \dots, x_k$. In [14] it is studied when the corresponding finite set of benchmark forward rates is Markovian.

A classic paper on Markovian short rates is [13], where a deterministic volatility of the form $\sigma(t, x)$ is considered. Theorem 4.4 was first stated and proved in [31]. See [18] for an example with a driving Levy process.

The geometric ideas presented above and in [9] are intimately connected to controllability problems in systems theory, where they have been used extensively (see [30]). They have also been used in filtering theory, where the problem is to find a finite dimensional realization of the unnormalized conditional density process, the evolution of which is given by the Zakai equation. See [12] for an overview of these areas.

5 Constructing Realizations

The purpose of this section is to present a systematic procedure for the construction of finite dimensional realizations for *any* model possessing a finite dimensional realization.

5.1 The Construction Algorithm

The method basically works as follows: From Theorem 4.2 we know that there exists an FDR if and only if the Lie algebra $\{\mu, \sigma\}_{LA}$ is finite dimensional. Given a set of generators for this Lie algebra we now show how to construct an FDR by essentially solving a finite number of ordinary differential equations in Hilbert space. The method will work for any Hilbert space SDE of the form (115) with smooth drift and diffusion vector fields, and in particular it can be applied to the forward rate equation.

Let us assume that the Lie algebra $\{\mu, \sigma\}_{LA}$, is finite dimensional near the point r^o. Then a finite dimensional realization can be constructed in the following way:

- Choose an involutive system of independent vector fields $f_1, \ldots, f_d$ which span $\{\mu, \sigma\}_{LA}$. Lemma 4.1 is often useful for simplifying the vector fields.

- Compute the invariant manifold $G(z_1, \ldots, z_d)$ using Proposition 4.2.

- Since $\mathcal{G}$ is invariant under r, we now know that $r_t = G(Z_t)$ for some state process Z. We thus make the following *Ansatz* for the dynamics of the state space variables Z

$$dZ_t = a(Z_t)dt + b(Z_t) \circ dW_t.$$

- From the Stratonovich version of the Itô formula it then follows that

$$G_\star a = \mu, \qquad G_\star b = \sigma. \tag{121}$$

- Use the equations in (121) to solve for the vector fields a and b.

Before going on to concrete applications, let us make some remarks.

Remark 5.1

- *We know that there will always exist solutions, a and b, to (121).*

- *It may be that the equations in (121) do not have unique solutions, but for us it is enough to find one solution, and any solution will do.*

- *Although we have to solve for the Stratonovich dynamics of the state variables, it turns out that the Itô-dynamics are typically much nicer looking (see below). This is not surprising since this is also true for the forward rate dynamics themselves.*

Again we emphasize that this method can be applied quite mechanically, the only choice to be made is that of vector fields which span the Lie algebra $\{\mu, \sigma\}_{LA}$. Generally we will want to choose these vector fields as simple as possible and to do this we use Lemma 4.1. The reason why we want simple vector fields is that this simplifies the computation of the parameterization of the forward rate curves in the next step (recall that this requires solving $\mathcal{H}$-valued ODEs with right hand sides equal to the generating vector fields).

In the next few sections we will apply this scheme repeatedly to various volatilities σ and derive finite dimensional realizations.

5.2 Deterministic Volatility

Assume that
$$\sigma(r, x) = \sigma(x), \tag{122}$$
where each component of the vector σ is of the following form
$$\sigma_i(x) = \sigma_i \lambda_i(x), \qquad i = 1, \ldots, m \tag{123}$$

Here, with a slight abuse of notation, σ_i on the right hand side denotes a constant, and λ_i is a constant vector field. We know from Proposition 4.4 that the forward rate equation generated by this volatility structure has a finite dimensional realization if and only if
$$dim(span\,\{\sigma, \mathbf{F}\sigma, \mathbf{F}^2\sigma, \ldots\}) < \infty,$$
where $\mathbf{F}$ denotes the operator $\frac{\partial}{\partial x}$. We therefore assume that λ_i solves the ODE
$$\mathbf{F}^{n_i+1}\lambda_i(x) = \sum_{k=0}^{n_i} c_k^i \mathbf{F}^k \lambda_i(x), \tag{124}$$

where the c_k^i:s are constants. Since the Lie algebra spanned by μ and σ for this case is given by
$$\{\mu, \sigma\}_{LA} = span\{\mu, \sigma, \mathbf{F}\sigma, \mathbf{F}^2\sigma, \ldots\},$$
we can choose the following generator system for the Lie algebra
$$\{\mu, \sigma\}_{LA} = span\{\mu, \mathbf{F}^k \lambda_i;\ i = 1, \ldots, m;\ k = 0, 1, \ldots, n_i\}.$$

The next step in constructing a finite dimensional realization is to compute the invariant manifold $G(z_0, z_k^i;\ i = 1, \ldots, m;\ k = 0, 1, \ldots, n_i)$. This means computing the operators $\exp\{\mu t\}$ and $\exp\{\mathbf{F}^k \lambda_i\}$, $i = 1, \ldots, m$, $k = 0, \ldots, n_i$. This has been done in Proposition 4.5 and the invariant manifold generated by the initial forward rate curve r_0 is parameterized as
$$G(z_0, z_k^i;\ i = 1, \ldots, m;\ k = 0, 1, \ldots, n_i)(x)$$
$$= r_0(x + z_0) + \frac{1}{2}(\|S(x + z_0)\|^2 - \|S(x)\|^2) + \sum_{i=1}^{m}\sum_{k=0}^{n_i} \mathbf{F}^k \lambda_i(x) z_k^i, \tag{125}$$

where

$$S(x) = \int_0^x \sigma(u)du.$$

We now proceed to the last step of the procedure, which is finding the dynamics of the state space variables. This means solving the equations (121). We therefore need the Frechet derivative G' of G. Simple calculations give

$$G'(z_0, z_k^i; i = 1, \ldots, m; k = 0, 1, \ldots, n_i) \begin{pmatrix} h_0 \\ h_0^1 \\ h_1^1 \\ \vdots \\ h_{n_m}^m \end{pmatrix}(x)$$

$$= \frac{\partial}{\partial x} r_0(x + z_0)h_0 + D(x + z_0)h_0 + \sum_{i=1}^{m} \sum_{k=0}^{n_i} \mathbf{F}^k \lambda_i(x) h_k^i,$$

where D is the constant field given by

$$D(x) = \sum_{i=1}^{m} \sigma_i^2 \lambda_i(x) \int_0^x \lambda_i(u)du.$$

Since for this model the Frechet derivative with respect to r of each component of the volatility is zero, i.e. $\sigma_i'(r, x) = 0$, we obtain the following expression for μ.

$$\mu(r) = \mathbf{F}r + D.$$

If we use that $r = G(z)$ we can obtain an expression for $\mathbf{F}r$, and the equation $G_* a = \mu$ then reads

$$\frac{\partial}{\partial x} r_0(x + z_0)a_0 + D(x + z_0)a_0 + \sum_{j=1}^{m} \sum_{k=0}^{n_j} \mathbf{F}^k \lambda_j(x) a_{jk}$$

$$= \frac{\partial}{\partial x} r_0(x + z_0) + D(x + z_0) + \sum_{j=1}^{m} \sum_{k=0}^{n_j} \mathbf{F}^{k+1} \lambda_j(x) z_k^j.$$

Since this equality is to hold for all x, and a is not allowed to depend on x it is possible to identify what a must look like. If we recall that λ_i solves the ODE defined in (124) we obtain

$$a_0 = 1,$$

$$a_{j0} = c_0^j z_{n_j}^j, \qquad j = 1, \ldots, m,$$

$$a_{jk} = c_k^j z_{n_j}^j + z_{k-1}^j, \qquad j = 1, \ldots, m; \ k = 1, \ldots, n_j.$$

From $G_* b^i(z)(x) = \sigma_i(x)$ we obtain the equation

$$\frac{\partial}{\partial x} r_0(x + z_0) b_0^i + D(x + z_0) b_0^i + \sum_{j=1}^{m} \sum_{k=0}^{n_j} \mathbf{F}^k \lambda_j(x) b_{jk}^i$$

$$= \sigma_i \lambda_i(x),$$

where σ_i denotes a constant. Therefore we have that

$$b_{jk}^i = \sigma_i, \quad j = i, \ k = 0,$$

$$b_{jk}^i = 0, \quad \text{all other } j \text{ and } k.$$

From this we see that to each Wiener process there corresponds one state variable which is driven by this, and only this, Wiener process. The dynamics for these state variables are given by

$$dZ_0^j = c_0^j Z_{n_j}^j dt + \sigma_j \circ dW_t^j, \quad j = 1, \dots, m.$$

Since σ_j is a constant, the Itô-dynamics will look the same, and we have thus proved the following proposition.

Proposition 5.1 *Given the initial forward rate curve r_0 the forward rate system generated by the volatilities described in equations (122) through (124) has a finite dimensional realization given by*

$$r_t = G(Z_t),$$

where G was defined in (125) and the dynamics of the state space variables Z are given by

$$\begin{cases} dZ_0 = dt, \\ dZ_0^j = c_0^j Z_{n_j}^j dt + \sigma_j dW_t^j, \ j = 1, \dots, m, \\ dZ_k^j = (c_k^j Z_{n_j}^j + Z_{k-1}^j) dt, \ j = 1, \dots, m; \ k = 1, \dots, n_j. \end{cases}$$

Remark 5.2 Note that the first state space variable represents running time. This will be the case for all realizations derived below.

Ho-Lee

As a special case of the deterministic volatilities studied in the previous section consider a volatility given by

$$\sigma(x) = \sigma, \tag{126}$$

where σ is a scalar constant, that is we have only one driving Wiener process. In the formalism of the previous paragraph we have $\lambda(x) \equiv 1$, which satisfies the trivial ODE $\mathbf{F}\lambda(x) = 0$. A direct application of Proposition 5.1 gives the following result.

Proposition 5.2 *Given the initial forward rate curve r_0 the forward rate system generated by the volatility of equation (126) has a finite dimensional realization given by*

$$r_t = G(Z_t),$$

where G is given by

$$G(z_0, z_1)(x) = r_0(x + z_0) + \sigma^2 \left(x z_0 + \frac{1}{2} z_0^2 \right) + z_1,$$

and the dynamics of the state space variables Z are given by

$$\begin{cases} dZ_0(t) = dt, \\ dZ_1(t) = \sigma dW_t. \end{cases}$$

Hull-White

Another special case of deterministic volatilities is

$$\sigma(x) = \sigma e^{-cx}, \tag{127}$$

where σ and c are scalar constants, so again there is only one driving Wiener process. This time we have $\lambda(x) = e^{-cx}$, which satisfies the ordinary differential equation $\mathbf{F}\lambda(x) = -c\lambda(x)$. Applying Proposition 5.1 once more we obtain the following.

Proposition 5.3 *Given the initial forward rate curve r_0 the forward rate system generated by the volatility of equation (127) has a finite dimensional realization given by*

$$r_t = G(Z_t),$$

where G is given by

$$G(z_0, z_1)(x) = r_0(x + z_0) + \frac{\sigma^2}{c^2} \left(e^{-cx}(1 - e^{-cz_0}) + \frac{e^{-2cx}}{2}(e^{-2cz_0} - 1) \right) + z_1,$$

and the dynamics of the state space variables Z are given by

$$\begin{cases} dZ_0(t) = dt, \\ dZ_1(t) = -cZ_1(t)dt + \sigma dW_t. \end{cases}$$

5.3 Deterministic Direction Volatility

Consider a volatility structure of the form

$$\sigma(r, x) = \varphi(r)\lambda(x). \tag{128}$$

Here φ is a smooth functional of r, and λ is a constant vector field. Note that we are now dealing with the case with only one driving Wiener process. Depending on whether φ satisfies a certain non-degeneracy condition or not we get two cases. We next study these two cases separately.

The Generic Case

In the generic case φ satisfies the following assumption.

Assumption 5.1 *We assume that*

- $\varphi(r) \neq 0$ *for all* $r \in \mathcal{H}$ *and for all* $i = 1, \dots, m$.
- $\Phi''(r)[\lambda; \lambda] \neq 0$ *for all* $r \in \mathcal{H}$, *where* $\Phi(r) = \varphi^2(r)$ *and* $\Phi''(r)[\lambda; \lambda]$ *denotes the second order Frechet derivative of* Φ *operating on* $[\lambda; \lambda]$.

Given these assumptions, Proposition 6.1 in [9] states that the system of forward rates generated by the volatility (128) possesses a finite dimensional realization if and only if λ is a quasi-exponential function, i.e. of the form $\lambda(x) = ce^{Ax}b$, where c is a row vector, A is a square matrix and b is a column vector. We will therefore assume that λ is of the form

$$\lambda(x) = p(x)e^{\alpha x}, \tag{129}$$

where p is a polynomial of degree n and α is a scalar constant.

It is also shown in [9] that, given Assumption 5.1, the Lie algebra generated by μ and σ is given by

$$\{\mu, \sigma\}_{LA} = span\{\mathbf{F}r, \mathbf{F}^i\lambda, \mathbf{F}^iD; \ i = 0, 1, \dots\},$$

where

$$D(x) = \lambda(x) \int_0^x \lambda(u)du. \tag{130}$$

We may now note that λ, regardless of what p looks like, satisfies the following ODE of order $n+1$

$$(\mathbf{F} - \alpha)^{n+1}\lambda(x) = 0.$$

This can also be written in the following way

$$\mathbf{F}^{n+1}\lambda(x) = -\sum_{i=0}^n \binom{n+1}{i} (-\alpha)^{n+1-i}\mathbf{F}^i\lambda(x). \tag{131}$$

Partial integration reveals that D can be written as $D(x) = u(x)e^{2\alpha x} + \gamma\lambda(x)$, where u is a polynomial of degree $q = 2n$ and γ is a constant. Using Lemma 4.1 we see that we can use $\widetilde{D}$ instead of D to generate the Lie algebra, where $\widetilde{D}$ is given by

$$\widetilde{D}(x) = D(x) - \left[\sum_{i=0}^n \left(\frac{-1}{\alpha}\right)^{i+1} \mathbf{F}^i p(0)\right] \cdot \lambda(x).$$

Here the sum on the right hand side equals γ. Therefore $\widetilde{D}(x) = u(x)e^{2\alpha x}$ and thus $\widetilde{D}$ satisfies the following ODE of order $q+1$

$$(\mathbf{F} - 2\alpha)^{q+1}\widetilde{D}(x) = 0,$$

which we can also write as

$$\mathbf{F}^{q+1}\widetilde{D}(x) = -\sum_{j=0}^{q} \binom{q+1}{j} (-2\alpha)^{q+1-j}\mathbf{F}^j\widetilde{D}(x). \tag{132}$$

After these considerations we choose the following generator system for the Lie algebra

$$\{\mu, \sigma\}_{LA} = span\{\mathbf{F}r, \mathbf{F}^i\lambda, \mathbf{F}^j\widetilde{D}; \ i = 0, 1, \ldots, n; \ j = 0, 1, \ldots, q\},$$

We now turn to the task of finding a parameterization of the invariant manifold $G(z_0, z_i^1, z_j^2; \ i = 0, 1, \ldots, n; \ j = 0, 1, \ldots, q)$, which amounts to computing the operators $\exp\{\mathbf{F}rt\} \exp\{\mathbf{F}^i\lambda t\}$, $i = 0, 1, \ldots, n$ and $\exp\{\mathbf{F}^j\widetilde{D}t\}$, $j = 0, 1, \ldots, q$. The operator $\exp\{\mathbf{F}rt\}$ is obtained as the solution to

$$\frac{dy_t}{dt} = \mathbf{F}r.$$

This is a linear equation and the solution is

$$y_t = e^{\mathbf{F}t}y_0,$$

which means that

$$(e^{\mathbf{F}t}r_0)(x) = r_0(x + t).$$

Since the rest of the generating fields are constant, the corresponding ODEs are trivial, and we have

$$(e^{\mathbf{F}^i\lambda t}r_0)(x) = r_0(x) + \mathbf{F}^i\lambda t,$$

and

$$(e^{\mathbf{F}^j\widetilde{D}t}r_0)(x) = r_0(x) + \mathbf{F}^j\widetilde{D}t,$$

respectively. The invariant manifold generated by the initial forward rate curve r_0 is thus parameterized as

$$G(z_0, z_i^1, z_j^2; \ i = 0, 1, \ldots, n; \ j = 0, 1, \ldots, q)(x)$$

$$= r(x + z_0) + \sum_{i=0}^{n} \mathbf{F}^i\lambda(x)z_i^1 + \sum_{j=0}^{q} \mathbf{F}^j\widetilde{D}(x)z_j^2. \tag{133}$$

To obtain the state space dynamics we solve the equations (121). The Frechet derivative G' of G is given by

$$G'(z_0, z_i^1, z_j^2; \ i = 0, 1, \ldots, n; \ j = 0, 1, \ldots, q) \begin{pmatrix} h_0 \\ h_0^1 \\ h_1^1 \\ \vdots \\ h_q^2 \end{pmatrix} (x)$$

$$= \frac{\partial}{\partial x}r_0(x + z_0)h_0 + \sum_{i=0}^{n} \mathbf{F}^i\lambda(x)h_i^1 + \sum_{j=0}^{q} \mathbf{F}^j\widetilde{D}(x)h_j^2.$$

We have the following expression for μ

$$\mu(r) = \mathbf{F}r + \varphi^2(r)D - \frac{1}{2}\varphi'(r)[\lambda]\varphi(r)\lambda,$$

where D was defined in (130). Using that $r = G(z)$, the equation $G_\star a = \mu$ reads

$$\frac{\partial}{\partial x}r_0(x+z_0)a_0 + \sum_{i=0}^{n}\mathbf{F}^i\lambda(x)a_i^1 + \sum_{j=0}^{q}\mathbf{F}^j\widetilde{D}(x)a_j^2$$

$$= \frac{\partial}{\partial x}r_0(x+z_0) + \sum_{i=0}^{n}\mathbf{F}^{i+1}\lambda(x)z_i^1 + \sum_{j=0}^{q}\mathbf{F}^{j+1}\widetilde{D}(x)z_j^2$$

$$+ \varphi^2(G(z))D(x) - \frac{1}{2}\varphi'(G(z))[\lambda]\varphi(G(z))\lambda(x).$$

This equality has to hold for all x, and a is not allowed to depend on x. This allows us to identify what a must look like. Recall that λ solves the ODE defined in (131), and that $\widetilde{D}$ solves the ODE in (132). Furthermore, recall that $D(x) = \widetilde{D}(x) + \gamma\lambda(x)$, and let

$$c_i = -\binom{n+1}{i}(-\alpha)^{n+1-i} \quad \text{and} \quad d_j = -\binom{q+1}{j}(-2\alpha)^{q+1-j}. \quad (134)$$

We then obtain

$$a_0 = 1,$$

$$a_0^1 = c_0 z_n^1 + \gamma\varphi^2(G(z)) - \frac{1}{2}\varphi'(G(z))[\lambda]\varphi(G(z)),$$

$$a_i^1 = c_i z_n^1 + z_{i-1}^1, \qquad\qquad i = 1,\ldots,n,$$

$$a_0^2 = d_0 z_q^2 + \varphi^2(G(z)),$$

$$a_j^2 = d_j z_q^2 + z_{j-1}^2, \qquad\qquad j = 1,\ldots,q.$$

From $G_\star b = \sigma$ we obtain the equation

$$\frac{\partial}{\partial x}r_0(x+z_0)b_0 + \sum_{i=0}^{n}\mathbf{F}^i\lambda(x)b_i^1 + \sum_{j=0}^{q}\mathbf{F}^j\widetilde{D}(x)b_j^2$$

$$= \varphi(G(z))\lambda(x),$$

where we have used that $r = G(z)$. This gives us

$$b_0 = 0,$$

$$b_j^i = \varphi(G(z)), \quad i = 1,\, j = 0,$$

$$b_j^i = 0, \qquad\qquad \text{all other } i,\, j.$$

Just as for the case with deterministic volatilities we see that the Wiener process only drives one of the state variables. On Stratonovich form the dynamics of Z_0^1 are

$$dZ_0^1 = \left(c_0 Z_n^1 + \gamma\varphi^2(G(Z)) - \frac{1}{2}\varphi'(G(Z))[\lambda]\varphi(G(Z)) \right) dt + \varphi(G(Z)) \circ dW_t.$$

Changing to Itô-dynamics for Z_0^1 we have the following proposition.

Proposition 5.4 *Given the initial forward rate curve r_0 the forward rate system generated by the volatility defined by the equations (128) and (129) has a finite dimensional realization given by*

$$r_t = G(Z_t),$$

where G was defined in (133) and the dynamics of the state space variables Z are given by

$$\begin{cases} dZ_0 = dt, \\[2mm] dZ_0^1 = [c_0 Z_n^1 + \gamma\varphi^2(G(Z))]dt + \varphi(G(Z))dW_t, \\[2mm] dZ_i^1 = (c_i Z_n^1 + Z_{i-1}^1)dt, \qquad i = 1, \ldots, n, \\[2mm] dZ_0^2 = [d_0 Z_q^2 + \varphi^2(G(Z))]dt, \\[2mm] dZ_j^2 = (d_j Z_q^2 + Z_{j-1}^2)dt, \qquad j = 1, \ldots, q. \end{cases}$$

Here c_i and d_j are given by (134)

6 The Filipović and Teichmann Extension

While in one sense the general FDR problem is more or less completely solved using the Lie algebra methodology of [9] described above, we still have a major technical problem to tackle. This has to do with the fact that in the approach above, the framework was that of strong solutions of infinite dimensional SDEs in Hilbert space and this forced us to construct the particular Hilbert space $\mathcal{H}$ of real analytic functions as the space of forward rate curves. While serving reasonably well, it was even at an early stage clear that this particular space was very small, and in particular it was pointed out by Filipović and Teichmann that the space does not include the forward rate curves generated by the Cox-Ingersoll-Ross model (see [15]). It was therefore necessary to extend the theory to a larger space but such an extension is far from to trivial to carry out, the problem being that on a larger Hilbert space you will loose the smoothness of the differential operator $\partial/\partial x$ appearing in the drift term of the forward rate equation. This problem was overcome with great elegance by Filipović and Teichmann who, partly building on the geometric and analytic results from [22], in [23] managed to extend the Lie algebraic FDR theory to a much larger space of forward rate curves than the space $\mathcal{H}$ considered in [9]. In doing so, Filipović and Teichmann first extend the space of [9] to a much larger Hilbert space. On the new space, however, the operator $\partial/\partial x$ becomes unbounded so they then change the topology on the space, thus making it into a Frechet space where the operator in fact is bounded. This approach, however, leads to new problems, since on a Frechet space there is no easy way of introducing differential calculus–in fact there is even no obvious way of defining the concept of smoothness which is necessary in order to have a Frobenius theorem. In order to overcome this problem, Filipović and Teichmann used the framework of so called "convenient spaces" developed some ten years ago (see [23] for references) in order to carry out analysis on the enlarged space. The main result of all this is that the Lie algebra conditions obtained by Björk and Svensson are shown to still hold in this more general setting. At this point it is worth mentioning that the technical price one has to pay for going into the deep parts of the theory of convenient analysis is quite high. It is therefore fortunate that the Lie algebraic machinery of [23] can be used without going into these (sometimes very hard) technical details. In fact, one of the main result of [23] can be formulated in the following pedestrian terms for the working mathematician: "When you are searching for FDRs for equations of HJM type, you can compute the relevant Lie algebra without worrying about the space, since Filipović and Teichmann will always provide you with a convenient space to work in". In [23] and in the follow up papers [24] and [25] the extended Lie algebra theory in [23] is used in to analyze a number of concrete problems concerning the forward rate equation: In particular, Filipović and Teichmann prove the remarkable result that any forward rate model admitting an FDR must necessarily have an affine term structure.

7 Stochastic Volatility Models

We now extend the theory developed above to include stochastic volatility models. More precisely we will study HJM models of the forward rates in which the volatility, apart from being dependent on the present forward rate curve, also is allowed to be modulated by a k dimensional hidden Markov process y. The model is defined as follows.

Definition 7.1 *The Itô formulation of the* **stochastic volatility model** *(henceforth SVM) is defined as the process pair* (r, y), *where the* $\mathbb{Q}$-*dynamics of* r *and* y *are defined by the following system of SDEs.*

$$dr_t(x) = \left\{ \frac{\partial}{\partial x} r_t(x) + \mathbf{H}\sigma(r_t, y_t, x) \right\} dt + \sigma(r_t, y_t, x) dW_t \tag{135}$$

$$dy_t = a^0(y_t) dt + b(y_t) dW_t, \tag{136}$$

where $\mathbf{H}$ is defined by

$$\mathbf{H}\sigma(r, y, x) = \sigma(r, y, x) \int_0^x \sigma^*(r, y, s) ds, \tag{137}$$

and * denotes transpose.

In this specification we consider the following objects as given a priori:

- The volatility structure σ for the forward rates, i.e. a deterministic mapping

$$\sigma : \mathcal{H} \times \mathbb{R}^k \times \mathbb{R}_+ \to \mathbb{R}^m.$$

- The drift vector field a^0 for y, i.e. a deterministic mapping

$$a^0 : \mathbb{R}^k \to \mathbb{R}^k.$$

(The superscript on a^0 will be explained below)

- The volatility vector field b for y, i.e. a deterministic mapping

$$b : \mathbb{R}^k \to M(k, m).$$

where $M(k, m)$ denotes the set of $k \times m$ matrices.

We view σ as a row vector

$$\sigma(r, y, x) = [\sigma_1(r, y, x), \ldots, \sigma_m(r, y, x)],$$

the drift a^0 is viewed as a column vector and the volatility b as a matrix:

$$a^0(y) = \begin{bmatrix} a_1^0(y) \\ \vdots \\ a_k^0(y) \end{bmatrix}, \quad b(y) = \begin{bmatrix} b_{11}(y) & b_{12}(y) & \cdots & b_{1m}(y) \\ b_{21}(y) & b_{22}(y) & \cdots & b_{2m}(y) \\ \vdots & \vdots & & \vdots \\ b_{k1}(y) & b_{k2}(y) & \cdots & b_{km}(y) \end{bmatrix}.$$

We note in particular that the forward rate volatility σ is allowed to be an arbitrary functional of the entire forward rate curve r, as well as a function of the k-dimensional variable y. We may also view each component of σ as a mapping from $\mathcal{H} \times \mathbb{R}^k$ to a space of functions (parameterized by x), and we will in fact assume that each σ_i, viewed in this way, is a smooth mapping with values in $\mathcal{H}$, i.e.

$$\sigma_i : \mathcal{H} \times \mathbb{R}^k \to \mathcal{H}.$$

We make the following regularity assumptions.

Assumption 7.1 *From now on we assume that:*

- *The mappings $\sigma_i : \mathcal{H} \times \mathbb{R}^k \to \mathcal{H}$ are smooth for $i = 1, \ldots, m$.*
- *The mapping $\mathbf{H}\sigma : \mathcal{H} \times \mathbb{R}^k \to \mathcal{H}$, defined by (137) is smooth.*
- *The mappings a^0 and b are smooth on $\mathbb{R}^k$.*

In the forward rate dynamics (135) we recognize the drift term in the r-dynamics above as the HJM drift condition, transferred into the Musiela parameterization. Note the particular structure of the equations (135)-(136): The y-process is feeding the drift and diffusion terms of the r-dynamics, but the r-process does not appear in the y-dynamics. Thus the y process is a Markov process in its own right, but this is not the case for the r-process. The extended process $\hat{r} = (r, y)$ is, however, Markovian.

In many applications it is natural to study, not only the full SVM above but also a restricted model, where we forget about the dynamics of y and consider y as a constant parameter. In this way we obtain a *parameterized* model, and the formal definition is as follows.

Definition 7.2 *Consider the SVM defined by (135)-(136) above. For any **fixed** value of $y \in \mathbb{R}^k$, the induced **parameterized forward rate model** is defined by the dynamics*

$$dr_t^y(x) = \left\{ \frac{\partial}{\partial x} r_t^y(x) + \mathbf{H}\sigma(r_t^y, y, x) \right\} dt + \sigma(r_t^y, y, x) dW_t. \tag{138}$$

Note that in the parameterized model, the forward rate process r^y itself is Markovian, whereas this is not the case in the full stochastic volatility model. For ease of reading we will sometimes drop the superscript y.

7.1 Problem Formulation

The basic problem to be discussed is under what conditions the, inherently infinite dimensional, SVM defined above by (135)-(136), with given initial conditions $r_0 = r^0, y_0 = y^0$, admits a generic finite dimensional Markovian realization in the sense of Section 4. More precisely we thus want to investigate under what conditions the extended process $\hat{r}_t = (r_t, y_t)$ possesses a local representation of the form

$$\hat{r}_t = \hat{G}(Z_t), \quad \mathbb{Q} - a.s. \tag{139}$$

where, for some d, Z satisfies a d-dimensional SDE of the form

$$\begin{cases} dZ_t = A_0(Z_t)dt + B(Z_t)dW_t, \\ Z_0 = z_0. \end{cases} \tag{140}$$

and where $\hat{G}$ is a smooth map $G : \mathbb{R}^d \to \mathcal{H} \times \mathbb{R}^k$. The drift and diffusion terms A_0 and B are assumed to be smooth and of suitable dimensions.

In a realization of this kind, the objects $\hat{G}$, A_0, B and z_0 will typically depend upon the choice of starting point (r^0, y^0). We recall that the term "generic" above means that we demand that there exists a realization, not only for the given initial point (r^0, y^0), but in fact for all initial points (r_0, y_0) in a neighborhood of (r^0, y^0). When we speak of realizations in the sequel we always intend this to mean generic realizations.

Note that the state process Z above is driven by the same Wiener process as the $\hat{r}$ system, and that the realization above is assumed to hold almost surely and trajectory wise.

We may now formulate some natural problems: **Main problems:**

- Find necessary and sufficient conditions for the existence of an FDR for a given stochastic volatility model.

- Assuming the existence of an FDR has been guaranteed, how do you construct it?

- How is the existence of an FDR for the full stochastic volatility model related to the existence of an FDR for the induced parameterized model? More precisely: is the existence of an FDR for the parameterized model necessary and/or sufficient for the existence of an FDR for the full model?

7.2 Test Examples: I.

To illustrate technique, we now present four simple recurrent test examples. In all cases we assume a scalar driving Wiener process W^r for the forward rates, a scalar y process and a scalar driving Wiener process W^y for the y process. Furthermore we assume that W^r and W^y are independent. To motivate our choice of examples we recall (see [2]) the following well known (non stochastic) HJM volatilities for the forward rates.

I. Hull-White extended Vasiček:

$$\sigma(r, x) = \sigma e^{-ax}. \tag{141}$$

Here a and the right hand side occurrence of σ are real constants. This HJM model has a short rate realization of the form.

$$dR_t = \{\Phi(t) - aR_t\}\, dt + \sigma dW_t, \tag{142}$$

where the deterministic function Φ depends on the initial term structure (see [2]). The parameters σ and a are the same as in (141).

II. Hull-White extended Cox-Ingersoll-Ross:

$$\sigma(r, x) = \sigma\sqrt{r(0)} \cdot \lambda(x, \sigma, a), \tag{143}$$

Here a and the right hand side occurrence of σ are real constants, whereas λ is given by

$$\lambda(x, \sigma, a) = -\frac{\partial}{\partial x}\left(\frac{2(e^{\gamma x} - 1)}{(\gamma + a)(e^{\gamma x} - 1) + 2\gamma}\right), \tag{144}$$

where

$$\gamma = \sqrt{a^2 + 2\sigma^2}.$$

Also this HJM model admits a short rate realization, namely

$$dR_t = \{\Phi(t) - aR_t\}\, dt + \sigma\sqrt{R_t}dW_t \tag{145}$$

The role of Φ is as in the extended Vasiček model above.

It is now natural to ask if we can extend these models by allowing one or several parameters to be stochastic, and still retain the existence of a finite dimensional realization.

We consider the following extensions of the above volatility structures. In all cases we assume that the scalar y process has dynamics of the form

$$dy_t = a_0(y_t)dt + b(y_t)dW_t^y,$$

with $b(y) \neq 0$ for all y.

1. HW with stochastic a:
$$\sigma(r, y, x) = \sigma e^{-yx} \tag{146}$$

2. HW with stochastic σ:
$$\sigma(r, y, x) = ye^{-ax} \tag{147}$$

3. CIR with stochastic σ:
$$\sigma(r, y, x) = y\sqrt{r(0)} \cdot \lambda(x, y, a) \tag{148}$$

4. CIR with stochastic a:
$$\sigma(r, y, x) = \sigma\sqrt{r(0)} \cdot \lambda(x, \sigma, y) \tag{149}$$

For all these models, the induced parameterized model admits, by construction, an FDR. It is now reasonable to ask if this also holds for the corresponding stochastic volatility models.

7.3 Finite Realizations for General Stochastic Volatility Models

In order to solve the FDR problem for stochastic volatility models we will of course use the Lie algebra theory for the existence of FDRs in Hilbert space, developed in Section 4.

Lie Algebra Conditions for the Existence of an FDR

Our problem is to study the existence of an FDR for a stochastic volatility model of the form

$$dr_t = \mu_0(r_t, y_t)dt + \sigma(r_t, y_t)dW_t \qquad (150)$$

$$dy_t = a^0(y_t)dt + b(y_t)dW_t. \qquad (151)$$

In the particular case of a forward rate model, the drift term is given by

$$\mu_0(r, y, x) = \frac{\partial}{\partial x}r(x) + \mathbf{H}\sigma(r, y, x) \qquad (152)$$

but none of the results in this section does in fact depend upon this particular structure of μ_0. Therefore we will, for the rest of the section, consider a general abstract stochastic volatility model of the form (150)-(151).

To apply our earlier Lie algebra results to the present situation we proceed in the following way.

- Define the Hilbert space $\hat{\mathcal{H}}$ by $\hat{\mathcal{H}} = \mathcal{H} \times \mathbb{R}^k$.
- Define the $\hat{\mathcal{H}}$-valued process $\hat{r}$ by

$$\hat{r}_t = \begin{bmatrix} r_t \\ y_t \end{bmatrix} \qquad (153)$$

- Write the dynamics of $\hat{r}$ on Stratonovich form instead of the original Itô form.
- Use the abstract Lie algebraic result from Theorem 4.2 on the process $\hat{r}$.

We will thus view $\hat{r}$ as an infinite dimensional "column vector" process, and we will henceforth always write it on block vector form as above.

The Stratonovich dynamics of $\hat{r}$ are routinely derived as

$$dr_t = \mu(r_t, y_t)dt + \sigma(r_t, y_t) \circ dW_t \qquad (154)$$

$$dy_t = a(y_t)dt + b(y_t) \circ dW_t, \qquad (155)$$

where

$$\mu(r, y) = \mu_0(r, y) - \frac{1}{2}\sigma_r(r, y)\sigma(r, y) - \frac{1}{2}\sigma_y(r, y)b(y) \qquad (156)$$

$$a(y) = a^0(y) - \frac{1}{2}b_y(y)b(y). \qquad (157)$$

Here σ_r denotes the partial Frechet derivative of σ w.r.t. the vector variable r and similarly for the other terms.

Written as a single equation on $\hat{\mathcal{H}}$ we thus have

$$d\hat{r}_t = \hat{\mu}(\hat{r})dt + \hat{\sigma}(\hat{r}) \circ dW_t, \tag{158}$$

where $\hat{\mu}$ and $\hat{\sigma}$ are given by

$$\hat{\mu}(r,y) = \begin{bmatrix} \mu(r,y) \\ a(y) \end{bmatrix} \tag{159}$$

$$\hat{\sigma}(r,y) = \begin{bmatrix} \hat{\sigma}_1(r,y), \ldots, \hat{\sigma}_m(r,y) \end{bmatrix} \tag{160}$$

Here the vector fields $\hat{\sigma}_1, \ldots, \hat{\sigma}_m$ are defined by

$$\hat{\sigma}_i(r,y) = \begin{bmatrix} \sigma_i(r,y) \\ b_i(y) \end{bmatrix} \tag{161}$$

where b_i is the $i : th$ column of the b matrix.

We make the following standing regularity assumption which is assumed to hold throughout the entire chapter.

Assumption 7.2 *We assume that the dimension (evaluated pointwise) of the Lie algebra*

$$\{\hat{\mu}, \hat{\sigma}_1, \ldots, \hat{\sigma}_m\}_{LA} < \infty, \tag{162}$$

is constant in a neighborhood of $\hat{r}_0 \in \hat{\mathcal{H}}$

Our first general result now follows immediately from Theorem 4.2. Note that when we below speak about the dimension of a Lie algebra, this is always to be interpreted in terms of pointwise evaluation.

Theorem 7.1 *Under Assumption 7.2, the stochastic volatility model (150)-(151) will have a generic FDR at the point $\hat{r}_0$ if and only if*

$$dim\,\{\hat{\mu}, \hat{\sigma}_1, \ldots, \hat{\sigma}_m\}_{LA} < \infty, \tag{163}$$

in a neighborhood of $\hat{r}_0 \in \hat{\mathcal{H}}$.

For simplicity of notation we will often use the shorthand notation $\{\hat{\mu}, \hat{\sigma}\}_{LA}$ for the Lie algebra $\{\hat{\mu}, \hat{\sigma}_1, \ldots, \hat{\sigma}_m\}_{LA}$.

Geometric Intuition

At this level of generality it is hard to obtain more concrete results. As an example: there seems to be no simple result connecting the existence of an FDR for the full model with existence of an FDR for the parameterized model. The geometric intuition behind this is roughly as follows.

- From Proposition 4.1 we know that existence of an FDR for $\hat{r}$ is equivalent to the existence of a finite dimensional invariant manifold in $\hat{\mathcal{H}}$ passing through $\hat{r}_0$.

- If the parameterized model admits a generic FDR then, for every fixed y near y^0, there exists an invariant manifold $\mathcal{G}$ in $\mathcal{H}$ through r_0. Thus one would perhaps guess that the manifold $\mathcal{G} \times \mathbb{R}^k$ would be invariant for $\hat{r}$, thus implying the existence of an FDR for $\hat{r}$.

- However, the manifold $\mathcal{G}$ above will generically depend on y. Writing it as $\mathcal{G}^y$, what may (and generically will) happen is that, as $\hat{r}_t$ moves around in $\hat{\mathcal{H}}$, y_t will move in $\mathbb{R}^k$ and the family $\{\mathcal{G}^{y_t}; \ t \geq 0\}$ may sweep out an infinite dimensional manifold in $\hat{\mathcal{H}}$. Thus the existence of an FDR for the parameterized model is not sufficient for the existence of an FDR for the full model.

- Conversely, the existence of an FDR for the parameterized model does not even seem to be necessary for the existence of an FDR for the full model. Suppose for example that, for each y, there does not exist an invariant manifold for the parameterized model. This means that the parameterized model does not possess an FDR. Despite this it could well happen that the process $\hat{r}$ **does** live on a finite dimensional invariant manifold (and thus possesses an FDR). The reason for this is that there could be a subtle interplay between the dynamics of r and y, and in particular one might intuitively expect this interplay to be possible if there is strong correlation between the Wiener process components driving r and y.

- From the argument above we are led to guess that the simplest structural situation occurs when r and y are driven by independent Wiener processes. Since in this case, the evolution of y is independent of the present state of r, we may even guess (bravely) that any FDR properties of the full model will be "uniform" w.r.t. y in the sense that the results will not depend much on the particular dynamics of y.

As we shall see below, the intuition outlined above is basically substantiated.

7.4 General Orthogonal Noise Models

Based on the informal arguments in the previous section we now go on to study the case when r and y are driven by independent Wiener processes. We will refer to this type of model as an "orthogonal noise model". We consider the case of a general SDE in Hilbert space.

Model Specification and Preliminary Results

Assumption 7.3 *For the rest of the section we assume that we can write the Wiener process W on block vector form as*

$$W_t = \begin{bmatrix} W_t^r \\ W_t^y \end{bmatrix}$$

where W^r and W^y are vector Wiener processes of dimensions m_r and m_y respectively. Furthermore we assume that the (r, y) dynamics are of the particular form

$$dr_t = \mu_0(r_t, y_t)dt + \sigma(r_t, y_t)dW_t^r \tag{164}$$

$$dy_t = a^0(y_t)dt + b(y_t)dW_t^y, \tag{165}$$

where the coefficients satisfy suitable smoothness conditions (see Section 7.3).

Under this assumption r and y are driven by orthogonal noise terms, and this leads to an important simplification of the geometric structure of the model.

Lemma 7.1 The Stratonovich formulation of (164)-(165) is given by

$$dr_t = \mu(r_t, y_t)dt + \sigma(r_t, y_t) \circ dW_t^r \tag{166}$$

$$dy_t = a(y_t)dt + b(y_t) \circ dW_t^y, \tag{167}$$

where

$$\mu(r, y) = \mu_0(r, y) - \frac{1}{2}\sigma_r(r, y)\sigma(r, y) \tag{168}$$

$$a(y) = a^0(y) - \frac{1}{2}b_y(y)b(y). \tag{169}$$

Proof. In order to find the Stratonovich form of the r dynamics we need to compute

$$d\langle \sigma, W^r \rangle_t = d\sigma(r_t, y_t).$$

The infinite dimensional Itô formula gives us

$$d\sigma(r_t, y_t) = (dt\text{-terms}) + \sigma_r(r_t, y_t)\sigma(r_t, y_t)dW_t^r + \sigma_y(r_t, y_t)b(y_t)dW_t^y$$

We thus obtain

$$d\langle \sigma, W^r \rangle_t = \sigma_r(r_t, y_t)\sigma(r_t, y_t)d\langle W^r, W^r \rangle_t + \sigma_y(r_t, y_t)b(y_t)d\langle W^y, W^r \rangle_t$$

Since W^r and W^y are independent this simplifies to

$$d\langle \sigma, W^r \rangle_t = \sigma_r(r, y)\sigma(r, y)dt. \quad \square$$

In order to see more clearly the geometric structure of the orthogonal noise model we write it on block operator form as

$$d\begin{bmatrix} r_t \\ y_t \end{bmatrix} = \begin{bmatrix} \mu(r_t, y_t) \\ a(y_t) \end{bmatrix} dt + \begin{bmatrix} \sigma(r_t, y_t) \\ 0 \end{bmatrix} \circ dW_t^r + \begin{bmatrix} 0 \\ b(y_t) \end{bmatrix} \circ dW_t^y \tag{170}$$

We thus have the following immediate and preliminary result.

Proposition 7.1 *The orthogonal noise model (164)-(165) admits an FDR if and only if the Lie algebra generated by the vector fields*

$$\begin{bmatrix} \mu(r,y) \\ a(y) \end{bmatrix}, \ \begin{bmatrix} \sigma_1(r,y) \\ 0 \end{bmatrix}, \ \ldots, \ \begin{bmatrix} \sigma_{m_r}(r,y) \\ 0 \end{bmatrix}, \ \begin{bmatrix} 0 \\ b_1(y) \end{bmatrix}, \ \ldots, \ \begin{bmatrix} 0 \\ b_{m_y}(y) \end{bmatrix}$$

is finite dimensional at $\hat{r}_0$.

More compactly we will often write the generators of the Lie algebra above as $\hat{\mu}$, $\hat{\sigma}$, and $\hat{b}$ where,

$$\hat{\mu}(r,y) = \begin{bmatrix} \mu(r,y) \\ a(y) \end{bmatrix}, \quad \hat{\sigma}(r,y) = \begin{bmatrix} \sigma(r,y) \\ 0 \end{bmatrix}, \quad \hat{b}(y) = \begin{bmatrix} 0 \\ b(y) \end{bmatrix} \tag{171}$$

A very useful property of the orthogonal noise model is the simple structure of the Stratonovich formulation of the parameterized model. The proof is trivial.

Lemma 7.2 *For the orthogonal noise model (164)-(165), the Itô formulation of the parameterized model is defined by*

$$dr_t = \mu_0(r_t, y)dt + \sigma(r_t, y)dW_t^r, \tag{172}$$

and the Stratonovich formulation of the parameterized model is given by

$$dr_t = \mu(r_t, y)dt + \sigma(r_t, y) \circ dW_t^r, \tag{173}$$

with μ defined by (168).

The point of this Lemma is that it shows that, for orthogonal noise models, the operations "restrict to the parameterized model" and "compute the Stratonovich dynamics" commute, i.e. the Stratonovich formulation of the parameterized model is identical to the parameterized version of the Stratonovich formulation of the original model.

In order to obtain easily verifiable necessary and sufficient conditions for the existence of an FDR we will in the next sections introduce some further structural assumptions. In doing this we will have to deal with Lie brackets in several spaces, so we have to clarify some notation.

Definition 7.3 *From now on, the following notation is in force:*

- *For any vector smooth fields $\hat{f}(r,y)$ and $\hat{g}(r,y)$ on $\hat{\mathcal{H}}$, the expression $\left[\hat{f}, \hat{g}\right]$ denotes the Lie bracket in $\hat{\mathcal{H}}$.*

- *For any smooth mapping $f(r, y)$ where $f : \hat{\mathcal{H}} \to \mathcal{H}$ and for any fixed $y \in \mathbb{R}$, the* **parameterized vector field** $f^y : \mathcal{H} \to \mathcal{H}$ *is defined by* $f^y(r) = f(r, y)$

- *For any smooth mappings $f, g : \hat{\mathcal{H}} \to \mathcal{H}$, the expression $[f^y, g^y]$ denotes the Lie bracket on $\mathcal{H}$ between f^y and g^y. This Lie bracket will sometimes also be denoted by $[f(\cdot, y), g(\cdot, y)]_{\mathcal{H}}$.*

- *For vector fields $c(y)$ and $d(y)$ on $\mathbb{R}^k$, the notation $[c, d]$ denotes the Lie bracket on $\mathbb{R}^k$.*

Necessary Conditions

It turns out that, in order to obtain easy necessary condition, a crucial role is played by the geometric relation between the drift vector field $a(y)$ and the Lie algebra on $\mathbb{R}^k$ generated by the diffusion vector fields $b_1(y), \ldots, b_{m_y}$.

Our first result relates the stochastic volatility model to the corresponding parameterized model.

Proposition 7.2 *Consider the model (164)-(165). Assume that*

$$a \in \left\{ b_1, \ldots, b_{m_v} \right\}_{LA} \tag{174}$$

in a neighborhood of y^0. Under this assumption, a necessary condition for the existence of an FDR for the stochastic volatility model is that the corresponding parameterized model

$$dr_t = \mu(r_t, y)dt + \sigma(r_t, y)dW_t^r \tag{175}$$

admits a generic FDR at y_0.

Proof. We assume that the full stochastic volatility model admits an FDR, and we also assume that (174) is satisfied. We now have to show that, under these assumptions, the parameterized model admits and FDR, i.e. that the Lie algebra (on $\mathcal{H}$) of the parameterized model is finite dimensional near r_0, for every fixed y near y_0. From Lemma 7.2 we know that the Stratonovich formulation of the parameterized model is given by

$$dr_t = \mu(r_t, y)dt + \sigma(r_t, y) \circ dW_t^r, \tag{176}$$

which we write as

$$dr_t = \mu^y(r_t)dt + \sigma^y(r_t) \circ dW_t^r. \tag{177}$$

Our task is now to show that

$$\{\mu^y, \sigma^y\}_{LA}$$

is finite dimensional near r_0 for all y near y_0.

Since we assumed that the full model possessed an FDR we know that the Lie algebra

$$\left\{\hat{\mu},\hat{\sigma},\hat{b}\right\}_{LA} = \left\{\begin{bmatrix} \mu(r,y) \\ a(y) \end{bmatrix}, \begin{bmatrix} \sigma(r,y) \\ 0 \end{bmatrix}, \begin{bmatrix} 0 \\ b(y) \end{bmatrix}\right\}_{LA}$$

is finite dimensional near $\hat{r}_0$. We now have the trivial inclusion

$$\left\{\hat{b}\right\}_{LA} \subseteq \left\{\hat{\mu},\hat{\sigma},\hat{b}\right\}_{LA},$$

and we go on to compute $\left\{\hat{b}\right\}_{LA} = \left\{\hat{b}_1,\ldots,\hat{b}_{m_y}\right\}_{LA}$. For any i and j, let us thus compute the Lie bracket $\left[\hat{b}_i,\hat{b}_j\right]$. We easily obtain the block matrix form for the Frechet derivative of $\hat{b}_i$ on $\hat{\mathcal{H}}$ as

$$\hat{b}_i'(y) = \begin{bmatrix} 0 & 0 \\ 0 & b_i'(y) \end{bmatrix}$$

where b_i' denotes the Frechet derivative on $\mathbb{R}^k$ of the vector field b_i. Performing the same calculation for $\hat{b}_j$ we obtain

$$\left[\hat{b}_i,\hat{b}_j\right]_{\hat{\mathcal{H}}} = \begin{bmatrix} 0 & 0 \\ 0 & b_i' \end{bmatrix}\begin{bmatrix} 0 \\ b_j \end{bmatrix} - \begin{bmatrix} 0 & 0 \\ 0 & b_j' \end{bmatrix}\begin{bmatrix} 0 \\ b_i \end{bmatrix} = \begin{bmatrix} 0 \\ [b_i,b_j]_{\mathbb{R}^k} \end{bmatrix}.$$

Continuing in this way by taking repeated brackets, wee see that if $\hat{\beta}$ denotes a generic element of $\left\{\hat{b}\right\}_{LA}$ then it has the form

$$\hat{\beta} = \begin{bmatrix} 0 \\ \beta \end{bmatrix}$$

where β denotes a generic element of $\{b\}_{LA}$. We can formally write this as

$$\left\{\hat{b}\right\}_{LA} = \left\{\hat{b}_1,\ldots,\hat{b}_{m_y}\right\}_{LA} = \begin{bmatrix} 0 \\ \{b_1,\ldots,b_{m_y}\}_{LA} \end{bmatrix} = \begin{bmatrix} 0 \\ \{b\}_{LA} \end{bmatrix}$$

We assumed that $a \in \{b\}_{LA}$, so there exists vector fields $c_1(y),\ldots,c_n(y)$ in $\{b\}_{LA}$ and scalar fields $\alpha_1(y),\ldots,\alpha_n(y)$ on $\mathbb{R}^k$ such that

$$a(y) = \sum_1^n \alpha_i(y)c_i(y)$$

for all y near y_0. Since $\left\{\hat{b}\right\}_{LA} \subseteq \left\{\hat{\mu},\hat{\sigma},\hat{b}\right\}_{LA}$ we see from the above that the vector fields $\hat{c}_1,\ldots,\hat{c}_n$ where

$$\hat{c}_i = \begin{bmatrix} 0 \\ c_i(y) \end{bmatrix}$$

all lie in $\left\{\hat{\mu},\hat{\sigma},\hat{b}\right\}_{LA}$. From [9] we know that we are allowed to perform Gaussian elimination. More precisely, we may replace $\hat{\mu}$ by $\hat{\mu} - \sum_1^n \alpha_i\hat{c}_i$, and we obtain

$$\hat{\mu} - \sum_{1}^{n} \alpha_i \hat{c}_i = \begin{bmatrix} \mu \\ a \end{bmatrix} - \sum_{1}^{n} \alpha_i \begin{bmatrix} 0 \\ c_i \end{bmatrix} = \begin{bmatrix} \mu \\ 0 \end{bmatrix}.$$

From this we see that the Lie algebra $\left\{ \hat{\mu}, \hat{\sigma}, \hat{b} \right\}_{LA}$ for the full model is in fact generated by the much simpler system $\hat{m}$, $\hat{\sigma}$ and $\hat{b}$ where $\hat{m}$ is defined by

$$\hat{m} = \begin{bmatrix} \mu \\ 0 \end{bmatrix}.$$

Since we assumed that $\left\{ \hat{\mu}, \hat{\sigma}, \hat{b} \right\}_{LA}$ was finite dimensional, then also the smaller Lie algebra

$$\{\hat{m}, \hat{\sigma}\}_{LA} = \left\{ \begin{bmatrix} \mu \\ 0 \end{bmatrix}, \begin{bmatrix} \sigma \\ 0 \end{bmatrix} \right\}_{LA}$$

is necessarily also finite dimensional. In computing this latter Lie algebra we may now argue as for $\{b\}_{LA}$ above. Let us, for example, compute the Lie bracket $[\hat{m}, \hat{\sigma}_i]$. We easily obtain

$$[\hat{m}, \hat{\sigma}_i] = \begin{bmatrix} \mu_r & \mu_y \\ 0 & 0 \end{bmatrix} \begin{bmatrix} \sigma_i \\ 0 \end{bmatrix} - \begin{bmatrix} \sigma_{ir} & \sigma_{iy} \\ 0 & 0 \end{bmatrix} \begin{bmatrix} \mu \\ 0 \end{bmatrix} = \begin{bmatrix} \mu_r \sigma_i - \sigma_{ir} \mu \\ 0 \end{bmatrix}$$

where subindex r and y denotes the partial Frechet derivative w.r.t r and y. Now we observe that $\mu_r(r, y)\sigma_i(r, y) - \sigma_{ir}(r, y)\mu(r, y) = [\mu^y, \sigma_i^y](r)$ so we have

$$[\hat{m}, \hat{\sigma}_i](r, y) = \begin{bmatrix} [\mu^y, \sigma_i^y] \\ 0 \end{bmatrix}(r),$$

and continuing in this way we obtain

$$\{\hat{m}, \hat{\sigma}\}_{LA}(r, y) = \left\{ \begin{bmatrix} \mu \\ 0 \end{bmatrix}, \begin{bmatrix} \sigma \\ 0 \end{bmatrix} \right\}_{LA}(r, y) = \begin{bmatrix} \{\mu^y, \sigma^y\}_{LA} \\ 0 \end{bmatrix}(r)$$

Since $\{\hat{m}, \hat{\sigma}\}_{LA}$ is finite dimensional for all (r, y) near (r_0, y_0) we thus see that $\{\mu^y, \sigma^y\}_{LA}$ has to be finite dimensional near r_0 for all y near y_0. This however is equivalent to the existence of an FDR for the parameterized model. $\square$

We have the following obvious corollary, which seems to be enough for many concrete applications.

Corollary 7.1 *Assume that the Lie algebra generated by b in $\mathbb{R}^k$ is full, i.e. that*

$$\{b_1, \ldots, b_{m_y}\}_{LA} = \mathbb{R}^k. \tag{178}$$

Then, regardless of the form of a, the existence of an FDR for the parameterized model is necessary for the existence of an FDR for the full model. In particular, the assumption above is valid, and thus the conclusion holds, for the following special cases.

- $m_y = k$ and the $k \times k$ diffusion matrix $b(y)$ is invertible near y_0.

- y is scalar and driven by a scalar Wiener process (i.e. $k = m_y = 1$), and the scalar field $b(y)$ is nonzero near y_0.

We now go on to obtain more precise (but still easily verifiable) necessary conditions, and the simplest case is when the diffusion matrix b is square and invertible. Since the multidimensional case is a bit messy we start with the scalar case, and we will in fact use the scalar result in the proof of the multidimensional case.

Proposition 7.3 *Assume that y and W^y are scalar and that the (scalar) diffusion term $b(y)$ is nonzero near y_0. Then the following conditions are necessary for the existence of an FDR for the full model.*

- *For every fixed r and y near (r_0, y_0) the partial derivatives of μ and $\sigma_i(r, y)$ $i = 1, \ldots, m_r$ w.r.t y span a finite dimensional space in $\mathcal{H}$, i.e. there exists a finite number N such that for every (r, y)*

$$dim\ span \left\{ \frac{\partial^n \mu}{\partial y^n}(r, y); \quad n = 0, 1, 2, \ldots \right\} \leq N \tag{179}$$

and

$$dim\ span \left\{ \frac{\partial^n \sigma_i}{\partial y^n}(r, y); \quad n = 0, 1, 2, \right\} \leq N \tag{180}$$

for every $i = 1, \ldots m_r$.

- *The drift term μ, and each volatility component σ_i have the form*

$$\mu(r, y, x) = \sum_{j=1}^{n_0} c_{0j}(r, y)\lambda_{0j}(r, x). \tag{181}$$

and

$$\sigma_i(r, y, x) = \sum_{j=1}^{n_i} c_{ij}(r, y)\lambda_{ij}(r, x). \tag{182}$$

Proof. In order to obtain necessary conditions we assume that the full model admits an FDR, and for simplicity of notation we assume that $m_r = 1$ (this will not affect the proof). The Lie algebra for the full model is then finite dimensional and it is generated by

$$\hat{\mu} = \begin{bmatrix} \mu(r, y) \\ a(y) \end{bmatrix}, \quad \hat{\sigma} = \begin{bmatrix} \sigma \\ 0 \end{bmatrix}, \quad \hat{b} = \begin{bmatrix} 0 \\ b(y) \end{bmatrix}.$$

Since b is scalar and nonzero we can use Gaussian elimination and locally replace $\hat{b}$ by

$$\frac{1}{b(y)}\hat{b}(y) = \begin{bmatrix} 0 \\ 1 \end{bmatrix},$$

and, with further elimination, we see that the full Lie algebra is in fact generated by

$$\hat{\mu} = \begin{bmatrix} \mu(r,y) \\ 0 \end{bmatrix}, \quad \hat{\sigma} = \begin{bmatrix} \sigma \\ 0 \end{bmatrix}, \quad \hat{1} = \begin{bmatrix} 0 \\ 1 \end{bmatrix}.$$

We start by proving (180), the proof for (179) being identical. Since the full algebra is finite dimensional, also the smaller Lie algebra generated by $\hat{\sigma}$ and $\hat{1}$ has to be finite dimensional. In particular the space spanned in $\mathcal{H}$ by the vector fields

$$\hat{\sigma}, \quad [\hat{\sigma}, \hat{1}], \quad [[\hat{\sigma}, \hat{1}], \hat{1}], \quad [[[\hat{\sigma}, \hat{1}], \hat{1}], \hat{1}], \dots$$

obtained by starting with $\hat{\sigma}$ and then taking repeated brackets with $\hat{1}$, has to be finite dimensional at every point (r, y) near $\hat{r}_0$. We can write these vectors more compactly as

$$ad_{\hat{1}}^0(\sigma), \quad ad_{\hat{1}}^1(\sigma), \quad ad_{\hat{1}}^2(\sigma), \dots$$

where for any vector field $\hat{f}$ the operators $ad_{\hat{f}}^n : \mathcal{H} \to \mathcal{H}$ are defined recursively by

$$ad_{\hat{f}}^0(\hat{g}) = \hat{g},$$

$$ad_{\hat{f}}^1(\hat{g}) = \left[\hat{g}, \hat{f}\right],$$

$$ad_{\hat{f}}^{n+1}(\hat{g}) = \left[ad_{\hat{f}}^n(\hat{g}), \hat{f}\right].$$

We easily obtain the Frechet derivatives of $\hat{\sigma}$ and $\hat{1}$ as

$$\hat{\sigma}' = \begin{bmatrix} \partial_r \sigma & \partial_y \sigma \\ 0 & 0 \end{bmatrix}, \quad \hat{1}' = \begin{bmatrix} 0 & 0 \\ 0 & 0 \end{bmatrix},$$

where ∂_r and ∂_r denotes the corresponding partial Frechet derivatives. Thus we have

$$ad_{\hat{1}}^1(\hat{\sigma}) = [\hat{\sigma}, \hat{1}] = \begin{bmatrix} \partial_r \sigma & \partial_y \sigma \\ 0 & 0 \end{bmatrix} \begin{bmatrix} 0 \\ 1 \end{bmatrix} - \begin{bmatrix} 0 & 0 \\ 0 & 0 \end{bmatrix} \begin{bmatrix} \sigma \\ 0 \end{bmatrix} = \begin{bmatrix} \partial_y \sigma \\ 0 \end{bmatrix}$$

Similarly we have

$$\{ad_{\hat{1}}^1(\hat{\sigma})\}' = \begin{bmatrix} \partial_r \partial_y \sigma & \partial_y^2 \sigma \\ 0 & 0 \end{bmatrix}$$

and thus

$$ad_{\hat{1}}^2(\hat{\sigma}) = [ad_{\hat{1}^1}(\hat{\sigma}), \hat{1}] = \begin{bmatrix} \partial_r \partial_y \sigma & \partial_y^2 \sigma \\ 0 & 0 \end{bmatrix} \begin{bmatrix} 0 \\ 1 \end{bmatrix} - \begin{bmatrix} 0 & 0 \\ 0 & 0 \end{bmatrix} \begin{bmatrix} \partial_y \sigma \\ 0 \end{bmatrix} = \begin{bmatrix} \partial_y^2 \sigma \\ 0 \end{bmatrix}$$

Continuing this way we see by induction that

$$ad_{\hat{1}}^n(\hat{\sigma}) = \begin{bmatrix} \partial_y^n \sigma \\ 0 \end{bmatrix}.$$

Since, by the argument above, $\left\{ ad_{\hat{1}}^n(\hat{\sigma})(r,y); \quad n \geq 0 \right\}$ span a finite dimensional subspace of $\hat{\mathcal{H}}$ for all (r,y) near $\hat{r}_0$, we thus see that

$$\left\{ \partial_y^n \sigma(r,y), \quad n \geq 0 \right\}$$

must span a finite dimensional subspace in $\mathcal{H}$ for all (r,y) near $\hat{r}_0$. We have thus proved (180) for the case when W^r is scalar. The general case is proved by applying the above argument for each component of σ.

We now go on to prove the necessary condition (182) and we will in fact show that (182) follows from (180). Again we carry out a separate argument for each component σ_i, so without loss of generality we may assume that σ only has a single component (i.e that $m_r = 1$). Now, if (180) holds and we denote the dimension of the spanned subspace by $n+1$, there exists scalar fields $a_j(r,y); \quad j = 0, \ldots n$, such that we have the following $\mathcal{H}$-valued vector identity holding locally at $\hat{r}_0$

$$\partial_y^{n+1} \sigma(r,y) = \sum_{j=0}^n a_j(r,y) \partial_y^j \sigma(r,y) \tag{183}$$

We now fix an arbitrary r, and for this fixed r we define the $\mathcal{H}$-vector functions $Z_0(y), Z_1(y), \ldots Z_n(y)$ by

$$Z_0(y) = \sigma(r,y),$$
$$Z_1(y) = \partial_y \sigma(r,y),$$

$$\vdots \quad \vdots$$

$$Z_n(y) = \partial_y^n \sigma(r,y),$$

and the $\mathcal{H}^{n+1}$-valued block vector function $Z(y)$ by

$$Z(y) = \begin{bmatrix} Z_0(y) \\ Z_1(y) \\ \vdots \\ Z_n(y) \end{bmatrix}$$

The point of this is that we can now write equation (183) as the linear ODE

$$\frac{dZ(y)}{dy} = (A(y) \otimes I) Z(y) \tag{184}$$

where $\otimes$ denotes the Kronecker product, and the $(n+1) \times (n+1)$ matrix function A is defined as the companion matrix

$$A(y) = \begin{bmatrix} 0 & 1 & 0 & \ldots & 0 \\ 0 & 0 & 1 & \ldots & 0 \\ \vdots & \vdots & & & 1 \\ a_0(y) & a_1(y) & a_2(y) & \ldots & a_n(y) \end{bmatrix}.$$

As one would perhaps guess, the solution of (184) can be shown to have the representation

$$Z(y) = [\Phi(y, y_0) \otimes I] Z(y_0),$$ (185)

where Φ is the transition matrix induced by A. In particular we thus obtain

$$Z_0(y) = \sum_{j=0}^{n} c_j(y) Z_j(y_0)$$

where $c_j(y) = \Phi(y, y_0)_{1,j}$. Recalling that there is a suppressed r and that $Z_j(y) = \partial_y^j \sigma(r, 0)$ we obtain

$$\sigma(r, y) = \sum_{j=0}^{n} c_j(r, y) \partial_y^j \sigma(r, 0),$$ (186)

which proves (182). The proof for (181) is identical.   □

In order to state the corresponding multidimensional result we need to introduce some notation.

Definition 7.4 *A **multi index** $\alpha \in \mathbb{Z}_+^k$ is any k-vector with nonnegative integer elements. For a multi index $\alpha = (\alpha_1, \ldots, \alpha_k)$ the differential operator ∂_y^α is defined by*

$$\partial_y^\alpha = \frac{\partial^{\alpha_1}}{\partial y_1^{\alpha_1}} \frac{\partial^{\alpha_2}}{\partial y_2^{\alpha_2}} \cdots \frac{\partial^{\alpha_k}}{\partial y_k^{\alpha_k}}$$

We can now state multidimensional version of the theorem above. The crucial assumption needed is that the Lie algebra generated by the diffusion matrix $b(y)$ spans the entire space $\mathbb{R}^k$. For the proof see [8].

Proposition 7.4 *Assume that the condition*

$$\{b_1, \ldots b_{m_y}\}_{LA} = \mathbb{R}^k,$$ (187)

is satisfied near y_0.

Then the following conditions are necessary for the existence of an FDR for the stochastic volatility model.

- *For every fixed r and y near (r_0, y_0) the partial derivatives of $\mu(r, y)$ and $\sigma_i(r, y)$ w.r.t y span a uniformly finite dimensional space in $\mathcal{H}$, i.e. there exists a number N such that for every (r, y)*

$$dim \; span \left\{ \partial_y^\alpha \mu(r, y); \quad \alpha \in \mathbb{Z}_+^k \right\} \leq N$$ (188)

and

$$dim \; span \left\{ \partial_y^\alpha \sigma_i(r, y); \quad \alpha \in \mathbb{Z}_+^k \right\} \leq N$$ (189)

for every $i = 1, \ldots m_r$.

- *The drift μ and every volatility component σ_i have the form*

$$\mu(r, y, x) = \sum_{j=1}^{n_i} c_{ij}(r, y)\lambda_{ij}(r, x). \tag{190}$$

$$\sigma_i(r, y, x) = \sum_{j=1}^{n_i} c_{ij}(r, y)\lambda_{ij}(r, x). \tag{191}$$

Test Examples: II.

We illustrate the necessary conditions obtained so far by studying the test examples (146)-(149) of Section 7.2. By the assumptions of Section 7.2, all three examples are within the class of orthogonal noise models. We may thus directly apply Proposition 7.2, or (since we have a scalar model) Corollary 7.1 and check whether the corresponding parameterized models possess finite dimensional realizations. In all these cases, however, this test is trivially satisfied since the volatility structures were constructed directly from HJM models possessing short rate realizations. Thus all the models pass this necessary conditions.

We now go on to the necessary conditions of Proposition (7.3). From (182) and ocular inspection of the examples above we immediately have the following result.

Proposition 7.5 *Assuming a scalar y-process with non zero diffusion term, the stochastic volatilities in (146), (148) and (149) do not admit an FDR.*

Thus (146), (148) and (149) are out of the race. In particular we note that there is no stochastic volatility extension of the CIR forward rate volatility for which there exists a finite dimensional realization. In fact, it is easy to see that we in fact have the following stronger result where we allow both the parameters a and σ to depend upon the process y.

Proposition 7.6 *Consider any stochastic volatility extension of the CIR model of the form*

$$\sigma(r, y, x) = \sigma(y)\sqrt{r(0)} \cdot \lambda(x, \sigma(y), a(y)) \tag{192}$$

where the functions $\sigma(y)$ and $a(y)$ are assumed to be non-constant and where the y process is assumed to have non zero diffusion term. Then the stochastic volatility model does not possess an FDR.

It remains to study the volatility structure (146) in more detail, and this will be done below.

Necessary and Sufficient Conditions

In this section we provide necessary and sufficient conditions for the existence of an FDR in the case of an orthogonal noise model, thus improving upon the general results of Theorem 7.1.

We need the following definition.

Definition 7.5 *Define, for each y, the parameterized Lie algebra $\mathcal{L}^y$ on $\mathcal{H}$ by*

$$\mathcal{L}^y = \left\{ \partial_y^\alpha \mu^y, \partial_y^\alpha \sigma_1^y, \ldots, \partial_y^\alpha \sigma_{m_r}^y; \quad \alpha \in \mathbb{Z}_+^k \right\}_{LA}$$

In this expression $\partial_y^\alpha \mu^y$ is, for each fixed y, considered as a (parameterized) vector field on $\mathcal{H}$, and correspondingly for the σ components.

In order to obtain reasonably concrete results we need to assume that the Lie algebra generated by the b matrix is full dimensional, leaving the general case as an open problem.

Proposition 7.7 *Assume that*

$$dim \left\{ b_1, \ldots, b_{m_y} \right\}_{LA} = k. \tag{193}$$

Under this assumption, a necessary and sufficient condition for the existence of an FDR for the stochastic volatility model is that, for each y, we have

$$dim \, \mathcal{L}^y < \infty \tag{194}$$

near r^0.

Proof. From proposition 7.1 we know that there exists an FDR if and only if the Lie algebra $\mathcal{L}$ on $\hat{\mathcal{H}}$ generated by

$$\begin{bmatrix} \mu(r,y) \\ a(y) \end{bmatrix}, \begin{bmatrix} \sigma_1(r,y) \\ 0 \end{bmatrix}, \ldots, \begin{bmatrix} \sigma_{m_r}(r,y) \\ 0 \end{bmatrix}, \begin{bmatrix} 0 \\ b_1(y) \end{bmatrix}, \ldots, \begin{bmatrix} 0 \\ b_{m_y}(y) \end{bmatrix}$$

is finite dimensional. Under the assumption (193), and using Gaussian elimination, we see that $\mathcal{L}$ is generated by

$$\begin{bmatrix} \mu(r,y) \\ 0 \end{bmatrix}, \begin{bmatrix} \sigma_1(r,y) \\ 0 \end{bmatrix}, \ldots, \begin{bmatrix} \sigma_{m_r}(r,y) \\ 0 \end{bmatrix}, \begin{bmatrix} 0 \\ I_k \end{bmatrix},$$

where I_k is the identity matrix on $\mathbb{R}^k$. Using the fact that repeated bracketing of a vector field of the form

$$\begin{bmatrix} f(r,y) \\ 0 \end{bmatrix}$$

with different columns in

$$\begin{bmatrix} 0 \\ I_k \end{bmatrix}$$

will produce a vector field of the form

$$\begin{bmatrix} \partial_y^\alpha f(r,y) \\ 0 \end{bmatrix}$$

it now follows that $\mathcal{L}$ is in fact generated by

$$\begin{bmatrix} \partial_y^\alpha \mu(r,y) \\ 0 \end{bmatrix}, \begin{bmatrix} \partial_y^\alpha \sigma(r,y) \\ 0 \end{bmatrix}, \dots, \begin{bmatrix} \partial_y^\alpha \sigma_{m_r}(r,y) \\ 0 \end{bmatrix}, \begin{bmatrix} 0 \\ I_k \end{bmatrix}; \quad \alpha \in \mathbb{Z}_+^k$$

From this it is clear that $\mathcal{L}$ is generated by

$$\begin{bmatrix} \mathcal{L}^y \\ 0 \end{bmatrix}, \begin{bmatrix} 0 \\ I_k \end{bmatrix}; \quad y \in \mathbb{R}^k,$$

and the proof is finished if we can show that for each multi index α we have

$$\partial_y^\alpha \mathcal{L}^y \subseteq \mathcal{L}^y. \tag{195}$$

It follows by induction that in order to prove (195) we may WLOG assume that $k = 1$ (i.e. y is scalar) and that it is in fact enough to prove that

$$\partial_y \mathcal{L}^y \subset \mathcal{L}^y. \tag{196}$$

Now, it is easily seen that

$$\mathcal{L}^y = \bigcup_{k=0}^{\infty} L_k^y,$$

where

$$L_0^y = span\left\{ \partial_y^n \mu^y, \partial_y^n \sigma_1^y, \dots, \partial_y^n \sigma_{m_r}^y; \quad n \geq 0 \right\}$$
$$L_{k+1}^y = span\left\{ L_k^y, [L_k^y, L_k^y] \right\}, \quad k = 0, 1, \dots$$

so it is enough to prove that each L_k^y is invariant under ∂_y and we prove this by induction. The case $k = 0$ is clear, so assume that

$$\partial_y L_n^y \subseteq L_n^y$$

for all $n \leq k$. Now fix an arbitrary $f \in L_{k+1}^y$. We start by considering two cases: the case when $f \in L_k^y$ and the case when $f = [g, h]$ with $g, h \in L_k^y$. If $f \in L_k^y$ then $\partial_y \in L_k^y$ by the induction assumption, so $\partial_y \in L_{k+1}^y$. If $f = [g, h]$ with $g, h \in L_k^y$ then an easy calculation shows that

$$\partial_y f = [\partial_y g, h] + [g, \partial_y h]$$

which is in $[L_k^y, L_k^y]$ by the induction assumption. Thus also in this case we have $\partial_y f \in L_{k+1}^y$. A generic $f \in L_{k+1}^y$ is, by definition, a linear combination of terms of the above type so we are finished. $\square$

A Simple Sufficient Condition

The object of this section is to show that, under some rather restrictive but nontrivial assumptions, it is possible to derive an extremely simple sufficient condition for the existence of an FDR for the full stochastic volatility model in terms of the FDR for the parameterized model. Furthermore; under these assumptions the realization for the full model can be constructed directly, and in a trivial manner, from the realization for the parameterized model.

Assumption 7.4

1. *The Ito formulation of the r-dynamics of the stochastic volatility model is of the form*

$$dr_t = \mu_0(r_t, y_t)dt + \sigma_t(r_t, y_t)dW_t. \tag{197}$$

2. *We assume that y is independent of W. Apart from this assumption, the process y is allowed to be an arbitrary semimartingale with values in $\mathbb{R}^k$.*

3. *For any fixed y, the parameterized r-model is assumed to possess an FDR of the form*

$$r_t^y = G(Z_t^y), \tag{198}$$

$$dZ_t^y = A(Z_t^y, y)dt + B(Z_t^y, y) \circ dW_t, \tag{199}$$

where Z^y is $\mathbb{R}^d$ valued and G is a smooth mapping $G : \mathbb{R}^d \to \mathcal{H}$.

The important part of this assumption is that, for the parameterized model, the parameter y only appears in the Z^y dynamics, but not the output mapping G. We will discuss the geometric significance of this below, but first we state the result.

Proposition 7.8 *Under Assumption 7.4, the stochastic volatility model possesses an FDR, and a concrete realization is in fact given by*

$$r_t = G(Z_t), \tag{200}$$

$$dZ_t = A(Z_t, y_t)dt + B(Z_t, y_t) \circ dW_t, \tag{201}$$

With G, A and B as in (198)-(199).

Proof. From the independence between y and W it follows that the Stratonovich formulation of the r-dynamics is given by

$$dr_t = \mu(r_t, y_t)dt + \sigma(r_t, y_t) \circ dW_t, \tag{202}$$

where

$$\mu(r, y) = \mu_0(r, y) - \frac{1}{2}\sigma_r(r, y)\sigma(r, y).$$

Now let us consider (200)-(201) as an Ansatz. The r-dynamics induced by (200)-(201) are given by

$$dr_t = G'(Z_t)A(Z_t, y_t)dt + G'(Z_t)B(Z_t, y_t) \circ dW_t, \tag{203}$$

so it follows that (200)-(201) is a realization of (202) if and only if

$$\mu(r, y) = G_\star A(r, y), \tag{204}$$
$$\sigma(r, y) = G_\star B(r, y). \tag{205}$$

We thus have to prove that (204)-(205) hold, and to this end we use the fact that, by assumption, (198)-(199) is a realization for the parameterized model. The Stratonovich formulation for the parameterized model is easily seen to be given by

$$dr_t^y = \mu(r_t^y, y)dt + \sigma(r_t^y, y) \circ dW_t, \tag{206}$$

and the important point here is that this is precisely the parameterized version of the Stratonovich formulation of the original r-dynamics. The r^y-dynamics induced by (198)-(199) are given by

$$dr_t^y = G'(Z_t^y)A(Z_t^y, y)dt + G'(Z_t^y)B(Z_t^y, y) \circ dW_t, \tag{207}$$

and since this was assumed to be a realization of (206) we thus have

$$\mu(r, y) = G_\star A(r, y),$$
$$\sigma(r, y) = G_\star B(r, y),$$

which was to be proved.   □

Remark 7.1 If the Stratonovich differential in (199) is replaced by an Itô differential i.e. by

$$dZ_t^y = A(Z_t^y, y)dt + B(Z_t^y, y)dW_t,$$

then the conclusion of Proposition 7.8 still holds if the Stratonovich differential in (201) is replaced by an Itô differential, i,.e. by

$$dZ_t = A(Z_t, y_t)dt + B(Z_t, y_t)dW_t.$$

This is useful if the realization of the parameterized model is originally given in Itô form.

This, very strong but also very restrictive, result has a clear and simple geometric interpretation. First, we know from general (orthogonal noise) theory that a necessary condition for an FDR is that the parameterized model possesses an FDR. In general, the realization for the parameterized model will of course be of the form

$$r_t^y = G(Z_t^y, y), \tag{208}$$
$$dZ_t^y = A(Z_t^y, y)dt + B(Z_t^y, y) \circ dW_t, \tag{209}$$

where the output function G as well as the drift term A and diffusion term B depend upon y, but in Proposition 7.8 we have assumed that G does not in fact depend on y. To understand the geometric meaning of this assumption we recall from Proposition 4.1 that the parameterized model, for a fixed y, admits an FDR if and only if there exists an invariant manifold $\mathcal{G}^y$ passing through r^0, and in the generic case this invariant manifold will of course depend upon y. The relation between $\mathcal{G}^y$ and the realization (208)-(209) is that

$$\mathcal{G}^y = Im\ G^y,$$

where the mapping $G^y : \mathbb{R}^d \to \mathcal{H}$ is defined by $G^y(z) = G(z, y)$. Thus; assuming that G does not depend upon the parameter y is equivalent to assuming that the invariant manifold for the parameterized model passing through r^0 does not depend upon y. In that case, denoting the invariant manifold by $\mathcal{G}$ it is of course geometrically obvious that $\mathcal{G} \times \mathbb{R}^k$ will be a finite dimensional invariant manifold for the process (r_t, y_t) thus guaranteeing the existence of an FDR for the full model.

Furthermore, it follows from Proposition 4.2 that the invariant manifold $\mathcal{G}^y$ is determined uniquely by the parameterized Lie algebra

$$\mathcal{L}^y = \left\{ \mu^y, \sigma_1^y, \ldots, \sigma_{m_r}^y \right\}_{LA}, \tag{210}$$

so if $\mathcal{L}^y$ does not depend upon y then neither will $G(z, y)$. We thus have the following result.

Proposition 7.9 *Assume that*

- *The process y is an $\mathbb{R}^k$-valued semimartingale which is independent of W.*

- *The parameterized model admits an FDR for every fixed y.*

- *Lie algebra $\mathcal{L}^y$ defined in (210) does not depend upon the parameter y.*

Then the full model will possess an FDR.

We finish this discussion by noticing that for the general Lie algebraic machinery to work it is essential that all processes are Wiener driven. The geometric reason for this is that the Wiener process acts locally in space (the infinitesimal generator is a partial differential operator) and this allows us to analyze the realization problems using differential geometry (i.e. local analysis). It is therefore noteworthy that in the simple situation discussed above in this section, we did not have to assume that y is driven by a Wiener process – it can also have jumps.

An Example

As an application of the results in Section 7.4, we consider the following volatility structure for a standard forward rate model driven by a scalar Wiener process W^r,

$$\sigma(r, x) = \varphi(r)e^{-\alpha x}. \tag{211}$$

Here φ is assumed to be an arbitrarily chosen smooth scalar field, and α is a positive constant. This is an extension of the model investigated in [35], where an FDR was constructed for the case when φ was assumed to be of the particular form $\varphi(r) = g(r(0))$, for some smooth function $g : \mathbb{R} \to \mathbb{R}$. As was shown in Section 5.3, the model admits an FDR of the following form.

Define the mapping $G : \mathbb{R}_+ \times \mathbb{R}^2 \to \mathcal{H}$ by

$$G(t, z_1, z_2)(x) = r^0(x + t) + z_1 e^{-\alpha x} + z_2 e^{-2\alpha x} \tag{212}$$

The realization is then given by

$$r_t(x) = G(t, Z_1(t), Z_2(t))(x), \tag{213}$$

$$dZ_1(t) = \left\{ \frac{1}{\alpha} \varphi^2 [G_t] - \alpha Z_1(t) \right\} dt + \varphi [G_t] dW_t^r, \tag{214}$$

$$dZ_2(t) = - \left\{ 2\alpha Z_1(t) + \frac{1}{\alpha} \varphi^2 [G_t] \right\} dt. \tag{215}$$

where we have used the shorthand notation

$$G_t = G(t, Z_1(t), Z_2(t)).$$

The important point to notice is that the mapping G in (212) does not involve φ. We may now extend the model above to a stochastic volatility model with an arbitrary scalar y-process (assumed to be independent of W^r), by defining the volatility structure as

$$\sigma(r, y, x) = \varphi(r, y)e^{-\alpha x}. \tag{216}$$

where φ is an arbitrarily chosen scalar field.

By construction, the parameterized model admits an FDR of the form (213)-(215) where G is exactly as above, and where $\varphi [G_t]$ is replaced by $\varphi [G_t, y]$. The point is again that G does not involve y, so it now follows immediately from Proposition 7.8 that a realization for the stochastic volatility model is given by

$$r_t(x) = G(t, Z_1(t), Z_2(t))(x),$$

$$dZ_1(t) = \left\{ \frac{1}{\alpha} \varphi^2 [G_t, y_t] - \alpha Z_1(t) \right\} dt$$
$$\qquad + \varphi [G_t, y_t] dW_t^r,$$

$$dZ_2(t) = - \left\{ 2\alpha Z_1(t) + \frac{1}{\alpha} \varphi^2 [G_t, y_t] \right\} dt.$$

Remark 7.2 In this example we have used the Itô dynamics instead of the Stratonovich dynamics. The reason is that the Itô dynamics of the realization are simpler than the Stratonovich dynamics.

7.5 Forward Rate Stochastic Volatility Models

We now go on to apply the general results above to the more concrete case of forward rate models. we recall that the Ito formulation of the stochastic volatility forward rate model is given by

$$dr_t(x) = \left\{ \frac{\partial}{\partial x} r_t(x) + \mathbf{H}\sigma(r_t, y_t, x) \right\} dt + \sigma(r_t, y_t, x) dW_t \qquad (217)$$

$$dy_t = a^0(y_t)dt + b(y_t)dW_t, \qquad (218)$$

where $\mathbf{H}$ is defined in (137). On Stratonovich form the model has the form

$$dr_t = \mu(r_t, y_t)dt + \sigma(r_t, y_t) \circ dW_t \qquad (219)$$
$$dy_t = a(y_t)dt + b(y_t) \circ dW_t, \qquad (220)$$

where

$$\mu(r, y) = \mathbf{F}r + \mathbf{H}\sigma(r, y) - \frac{1}{2}\sigma_r(r, y)\sigma(r, y) - \frac{1}{2}\sigma_y(r, y)b(y) \qquad (221)$$

$$a(y) = a^0(y) - \frac{1}{2}b_y(y)b(y). \qquad (222)$$

As usual $\mathbf{F}$ denotes the operator $\partial/\partial x$, σ_r denotes the partial Frechet derivative of σ w.r.t. the vector variable r and similarly for σ_y.

Necessary Conditions for Orthogonal Noise Models

In the orthogonal noise case the model has the following Stratonovich form

$$dr_t = \mu(r_t, y_t)dt + \sigma(r_t, y_t) \circ dW_t^r \qquad (223)$$
$$dy_t = a(y_t)dt + b(y_t) \circ dW_t^y, \qquad (224)$$

where

$$\mu(r, y) = \mathbf{F}r + \mathbf{H}\sigma(r, y) - \frac{1}{2}\sigma_r(r, y)\sigma(r, y) \qquad (225)$$

$$a(y) = a^0(y) - \frac{1}{2}b_y(y)b(y). \qquad (226)$$

We now have the following surprisingly restrictive result.

Proposition 7.10 *Assume the following:*

- *The model is an orthogonal noise model.*

- *The condition*

$$\{b_1, \ldots b_{m_y}\}_{LA} = \mathbb{R}^k, \tag{227}$$

is satisfied near y^0.

Then, a necessary condition for the existence of an FDR is that the volatility structure has the form

$$\sigma_i(r, y, x) = \sum_{j=1}^{N} \varphi_{ij}(r, y)\lambda_j(x), \quad i = 1, \ldots, m_r, \tag{228}$$

where $\lambda_1, \ldots, \lambda_N$ are constant vector fields, and φ_{ij} are smooth scalar fields.

Proof. Since we have assumed orthogonal noise, Proposition 7.2 implies that a necessary condition for the existence of an FDR is that the parameterized model admits an FDR. Furthermore; applying Theorem 4.13 of [23] to the parameterized model it follows that the volatility must be of the form

$$\sigma_i(r, y, x) = \sum_{j=1}^{N} \varphi_{ij}(r, y)\lambda_j(y, x). \tag{229}$$

Given this expression, an application of Proposition 7.4 finishes the proof. $\quad\square$

Given a volatility structure of the form (228) we now go on to find sufficient conditions for the existence of an FDR.

Sufficient Conditions for the General Noise Models

We now consider a multidimensional forward rate model of the form

$$dr_t = \mu(r_t, y_t)dt + \sigma(r_t, y_t) \circ dW_t \tag{230}$$

$$dy_t = a(y_t)dt + b(y_t) \circ dW_t. \tag{231}$$

where W is assumed to be m-dimensional, and y is as usual k-dimensional. We will assume that the volatility structure is of the form (228), but we stress the fact that we do not restrict ourselves to the orthogonal noise model.

We recall from Section 2.2 that a real valued function $f : \mathbb{R} \rightarrow \mathbb{R}$ is said to be *quasi exponential* if it is the solution of a linear ODE with constant coefficients, alternatively that it can be written as

$$f(x) = \sum_{i} e^{\gamma_i x} + \sum_{j} e^{\alpha_j x} \left[p_j(x) \cos(\omega_j x) + q_j(x) \sin(\omega_j x) \right], \tag{232}$$

where $\gamma_i, \alpha_j, \omega_j$ are real numbers, whereas p_j and q_j are real polynomials.

The main result is as follows.

Proposition 7.11 *Consider the model (230)-(231) and assume that the components of σ are of the form*

$$\sigma_i(r, y, x) = \sum_{j=1}^{N} \varphi_{ij}(r, y)\lambda_j(x), \quad i = 1, \ldots, m. \tag{233}$$

Under this assumption a sufficient condition for the existence of an FDR is that $\lambda_1(x), \ldots, \lambda_m(x)$ are quasi exponential. The scalar fields $\varphi_{ij}(x)$ are allowed to be arbitrary.

Proof. In order to avoid to much and messy notation, we give the proof only for the simplified case when

$$\sigma_i(r, y, x) = \varphi_i(r, y)\lambda_i(x).$$

The arguments in the general case are almost identical. Under the given assumption the Stratonovich drift term of r is given by

$$\mu = \mathbf{F}r + \sum_{i=1}^{m} \Phi_i D_i - \frac{1}{2}\sum_{i=1}^{m} \varphi_{ir}[\lambda_i]\varphi_i\lambda_i - \frac{1}{2}\sum_{i=1}^{m} \varphi_{iy}[b_i]\lambda_i \tag{234}$$

where b_i denotes the i.th column of the matrix b. The Lie algebra $\mathcal{L}$ under study is the one generated by the vector fields

$$\begin{bmatrix} \mu \\ a \end{bmatrix}, \begin{bmatrix} \varphi_1\lambda_1 \\ b_1 \end{bmatrix}, \ldots, \begin{bmatrix} \varphi_m\lambda_m \\ b_m \end{bmatrix}.$$

Obviously, $\mathcal{L}$ is included in the larger algebra $\mathcal{L}_1$, generated by

$$\begin{bmatrix} \mu \\ 0 \end{bmatrix}, \begin{bmatrix} \varphi_1\lambda_1 \\ 0 \end{bmatrix}, \ldots, \begin{bmatrix} \varphi_m\lambda_m \\ 0 \end{bmatrix}, \begin{bmatrix} 0 \\ a \end{bmatrix}, \begin{bmatrix} 0 \\ b_1 \end{bmatrix}, \ldots, \begin{bmatrix} 0 \\ b_m \end{bmatrix}.$$

Using the structure of μ we can reduce this generator system to

$$\begin{bmatrix} \mathbf{F}r + \sum_{i=1}^{m} \Phi_i D_i \\ 0 \end{bmatrix}, \begin{bmatrix} \lambda_1 \\ 0 \end{bmatrix}, \ldots, \begin{bmatrix} \lambda_m \\ 0 \end{bmatrix}, \begin{bmatrix} 0 \\ a \end{bmatrix}, \begin{bmatrix} 0 \\ b \end{bmatrix}.$$

From this we see that $\mathcal{L}_1$ is included in the algebra $\mathcal{L}_2$, generated by

$$\begin{bmatrix} \mathbf{F}r \\ 0 \end{bmatrix}, \begin{bmatrix} D_1 \\ 0 \end{bmatrix}, \ldots, \begin{bmatrix} D_m \\ 0 \end{bmatrix}, \begin{bmatrix} \lambda_1 \\ 0 \end{bmatrix}, \ldots, \begin{bmatrix} \lambda_m \\ 0 \end{bmatrix}, \begin{bmatrix} 0 \\ a \end{bmatrix}, \begin{bmatrix} 0 \\ b \end{bmatrix}.$$

As in Section 4.5 it now follows that $\mathcal{L}_2$ is finite dimensional if and only if $\lambda_1, \ldots, \lambda_m$ are quasi exponential. □

The Scalar Case

We finish by a reasonably complete investigation of the most important special case, which occurs when y is scalar, r and y are driven by scalar Wiener processes, and the volatility has the form

$$\sigma(r, y, x) = \varphi(r, y)\lambda(x). \tag{235}$$

Such a model will have the form

$$dr_t(x) = \{\mathbf{F}r_t(x) + \Phi(r, y)D(x)\}\, dt + \varphi(r, y)\lambda(x)dW_t^r$$

$$dy_t = a^0(y_t)dt + b(y_t)dW_t^y.$$

where

$$\Phi(r, y) = \varphi^2(r, y),$$

$$D(x) = \lambda(x)\int_0^x \lambda(s)ds.$$

In order to allow for a correlation, ρ, between W^r and W^y we write them as

$$W_t^r = \rho W_t^1 + \sqrt{1 - \rho^2}W_t^2,$$
$$W_t^y = W_t^1$$

where W^1 and W^2 are independent Wiener processes. We then have the dynamics

$$dr_t = \{\mathbf{F}r_t + \Phi D\}\, dt + \varphi\lambda\rho W_t^1 + \varphi\lambda\sqrt{1 - \rho^2}W_t^2$$

$$dy_t = a^0 dt + bdW_t^1.$$

We can now prove the following main result for the scalar case.

Proposition 7.12 *Assume that $\varphi_y(r, y) \neq 0$, and that $b(y) \neq 0$ i.e. that the model is non trivial. Then the following hold.*

- *In the non-perfectly correlated case $|\rho| < 1$, a necessary and sufficient condition for the existence of an FDR is that the vector field λ is quasi exponential. The scalar field $\varphi(r, y)$ is allowed to be arbitrary.*

- *In the perfectly correlated case $|\rho| = 1$, the condition above is sufficient.*

Proof. The Stratonovich dynamics of the model are given by

$$dr_t = \left\{\mathbf{F}r_t + \Phi D - \frac{1}{2}\varphi_r[\lambda]\varphi\lambda - \frac{1}{2}\varphi_y b\lambda\right\} dt + \varphi\lambda \circ W_t^1 + \sqrt{1 - \rho^2}\varphi\lambda \circ W_t^2$$

$$dy_t = adt + b \circ dW_t^1.$$

Thus the relevant Lie algebra $\mathcal{L}$ on $\hat{\mathcal{H}}$ is generated by the vector fields

$$\begin{bmatrix} \mathbf{F}r + \Phi D - \frac{1}{2}\varphi_r[\lambda]\varphi\lambda - \frac{1}{2}\varphi_y b\lambda \\ a \end{bmatrix}, \begin{bmatrix} \rho\varphi\lambda \\ b \end{bmatrix}, \begin{bmatrix} \sqrt{1-\rho^2}\varphi\lambda \\ 0 \end{bmatrix},$$

We start with the non-perfectly correlated case, so we assume that $|\rho| < 1$. Then, by Gaussian elimination, the system of generators can immediately be reduced to

$$\begin{bmatrix} \mathbf{F}r + \Phi D \\ 0 \end{bmatrix}, \begin{bmatrix} 0 \\ 1 \end{bmatrix}, \begin{bmatrix} \lambda \\ 0 \end{bmatrix}$$

The Lie bracket between the first two vector fields gives us

$$\begin{bmatrix} \Phi_y D \\ 0 \end{bmatrix},$$

so after reducing this field we have the generators

$$\begin{bmatrix} \mathbf{F}r + \Phi D \\ 0 \end{bmatrix}, \begin{bmatrix} D \\ 0 \end{bmatrix}, \begin{bmatrix} \lambda \\ 0 \end{bmatrix}, \begin{bmatrix} 0 \\ 1 \end{bmatrix},$$

which finally reduce to

$$\begin{bmatrix} \mathbf{F}r \\ 0 \end{bmatrix}, \begin{bmatrix} D \\ 0 \end{bmatrix}, \begin{bmatrix} \lambda \\ 0 \end{bmatrix}, \begin{bmatrix} 0 \\ 1 \end{bmatrix}.$$

From this it follows immediately that the Lie algebra is finite dimensional if an only if the linear span of

$$\{\mathbf{F}^n\lambda, \quad \mathbf{F}^n D; \quad n \geq 0\}$$

is a finite dimensional subspace in $\mathcal{H}$. It is however easily seen that this happens if and only if λ is quasi exponential.

In the perfectly correlated case $|\rho| = 1$ we can WLOG assume that $\rho = 1$ and we are left with the following generators for the Lie algebra $\mathcal{L}$.

$$\begin{bmatrix} \mathbf{F}r + \Phi D - \frac{1}{2}\varphi_r[\lambda]\varphi\lambda - \frac{1}{2}\varphi_y b\lambda \\ a \end{bmatrix}, \begin{bmatrix} \varphi\lambda \\ b \end{bmatrix},$$

There seems tho be no easy way of reducing this set of generators, but it is obvious that $\mathcal{L}$ is included in the Lie algebra $\mathcal{L}_{ext}$ generated by the fields

$$\begin{bmatrix} \mathbf{F}r + \Phi D - \frac{1}{2}\varphi_r[\lambda]\varphi\lambda - \frac{1}{2}\varphi_y b\lambda \\ a \end{bmatrix}, \begin{bmatrix} \varphi\lambda \\ 0 \end{bmatrix}, \begin{bmatrix} 0 \\ b \end{bmatrix}$$

Thus a sufficient condition for an FDR is that the larger Lie algebra $\mathcal{L}_{ext}$ is finite dimensional. It is however easily seen that $\mathcal{L}_{ext}$ is identical with the algebra discussed in the non-perfectly correlated case above, so we are finished. $\square$

Test Examples: III.

We can now continue our study of the test examples of Section 7.2. In fact, only one example is left in the race, namely

2. HW with stochastic σ:

$$\sigma(r, y, x) = ye^{-ax}. \tag{236}$$

We now have the following result, which is immediately obtained from Proposition 7.12.

Proposition 7.13 *The stochastic volatility version of the Hull-White extended Vasiček volatility structure with stochastic σ, as in (236) admits an FDR.*

Construction of Realizations

In the previous sections we have provided existence results for FDRs, but so far we have not actually constructed any concrete realizations. However, the construction technique outlined in Section 5 can immediately be adapted to the stochastic volatility framework, and we only give an illustrative example. The example is the Hull-White extended Vasiček model with stochastic σ as in (236) above. We already know that the forward rate model of the form (230)-(231), with volatilities given by (236), has a finite dimensional realization. Not surprisingly, y can be chosen as one of the state variables, and a concrete realization can be shown to be given by

$$\hat{r}_t = \widehat{G}(Z_t, y_t). \tag{237}$$

Here $\widehat{G}$ is defined by

$$\widehat{G}(z_0, z_1, z_2, y) = \begin{bmatrix} G(z_0, z_1, z_2, y) \\ y_0 + y \end{bmatrix}, \tag{238}$$

where G is given by

$$G(z_0, z_1, z_2, y)(x) = r_0(x + z_0) + e^{-\alpha x}z_1 - \frac{e^{-2\alpha x}}{\alpha}z_2. \tag{239}$$

The dynamics of the state space variables are given by

$$\begin{cases} dZ_0 = dt, \\[2mm] dZ_1 = \left[-\alpha Z_1 + \frac{1}{\alpha}(y_0 + y)^2)\right] dt + (y_0 + y)dW_t, \\[2mm] dZ_2 = \left[-2\alpha Z_2 + (y_0 + y)^2\right] dt \\[2mm] dy = a_0(y)dt + b(y)dW_t. \end{cases} \tag{240}$$

Here $a_0(y) = a(y) + \frac{1}{2}b_y(y)b(y)$.

References

1. BHAR, R., AND CHIARELLA, C. Transformation of Heath-Jarrow-Morton models to Markovian systems. *The European Journal of Finance* **3** (1997), 1–26.
2. BJÖRK, T. *Arbitrage Theory in Continuous Time.* Oxford University Press, 1998.
3. BJÖRK, T. A geometric view of interest rate theory. In *Option Pricing, Interest Rates and Risk Management.*, E. Jouini, J. Cvitanic, and M. Musiela, Eds. Cambridge University Press, 2001.
4. BJÖRK, T. Interest rate theory. In *Financial Mathematics, Springer Lecture Notes in Mathematics*, Vol **1656**, W. Runggaldier, Ed. Springer Verlag, 2001.
5. BJÖRK, T., AND CRISTENSEN, B. Interest rate dynamics and consistent forward rate curves. *Mathematical Finance* **9**, (1999), 323–348.
6. BJÖRK, T., AND GOMBANI, A. Minimal realizations of interest rate models. *Finance and Stochastics* **3**, 4 (1999), 413–432.
7. BJÖRK, T., AND LANDÉN, C. On the construction of finite dimensional realizations for nonlinear forward rate models. *Finance and Stochastics* **6**, 3 (2002), 303–331.
8. BJÖRK, T., LANDÉN, C., AND SVENSSON, L. Finite dimensional markovian realizations for stochastic volatility forward rate models. In: *Proceedings of the Royal Society,*. Vol. **460**, No. 2041, 53–84, 2004.
9. BJÖRK, T., AND SVENSSON, L. On the existence of finite dimensional realizations for nonlinear forward rate models. *Mathematical Finance* **11**, 2 (2001), 205–243.
10. BRACE, A., AND MUSIELA, M. A multifactor Gauss Markov implementation of Heath, Jarrow, and Morton. *Mathematical Finance* **4** (1994), 259–283.
11. BROCKET, P. *Finite Dimensional Linear Systems.* Wiley, 1970.
12. BROCKET, P. Nonlinear systems and nonlinear estimation theory. In *Stochastic systems: The mathematics of filtering and identification and applications*, M. Hazewinkel and J. Willems, Eds. Reidel, 1981.
13. CARVERHILL, A. When is the spot rate Markovian? *Mathematical Finance* **4** (1994), 305–312.
14. CHIARELLA, C., AND KWON, O. K. Forward rate dependent Markovian transformations of the Heath-Jarrow-Morton term structure model. *Finance and Stochastics* **5** (2001), 237–257.
15. COX, J., INGERSOLL, J., AND ROSS, S. A theory of the term structure of interest rates. *Econometrica* **53** (1985), 385–408.
16. DA PRATO, G., AND ZABZCYK, J. *Stochastic Equations in Infinite Dimensions.* Cambridge University Press, 1992.
17. DUFFIE, D., AND KAN, R. A yield factor model of interest rates. *Mathematical Finance* **6**, 4 (1996), 379–406.
18. EBERLEIN, E., AND RAIBLE, S. Term structure models driven by general Levy processes. *Mathematical Finance* **9**, 1 (1999), 31–53.
19. FILIPOVIĆ, D. A note on the nelson-siegel family. *Mathematical Finance* **9**, 4 (1999), 349–359.
20. FILIPOVIĆ, D. Exponential-polynomial families and the term structure of interest rates. *Bernoulli* **6** (2000), 1–27.
21. FILIPOVIĆ, D. Invariant manifolds for weak solutions of stochastic equations. *Probability Theory and Related Fields* **118** (2000), 323–341.
22. FILIPOVIĆ, D. *Consistency Problems for Heath-Jarrow-Morton Interest Rate Models.* Springer Lecture Notes in Mathematics, Vol. **1760**. Springer Verlag., 2001.

23. FILIPOVIĆ, D., AND TEICHMANN, J.   Existence of finite dimensional realizations for stochastic equations, 2001. Forthcoming in *J. Funct. Anal.*

24. FILIPOVIĆ, D., AND TEICHMANN, J.   On finite dimensional term structure models, 2002. Working paper.

25. FILIPOVIĆ, D., AND TEICHMANN, J.   On the geometry of the term structure of interest rates, 2003. In: *Proceedings of the Royal Society*, Vol. **460**, No. 2041, 129–168, 2004.

26. HEATH, D., JARROW, R., AND MORTON, A.   Bond pricing and the term structure of interest rates: a new methodology for contingent claims valuation. *Econometrica* **60** (1992), 77–105.

27. HO, T., AND LEE, S.   Term structure movements and pricing interest rate contingent claims. *Journal of Finance* **41** (1986), 1011–1029.

28. HULL, J., AND WHITE, A.   Pricing interest-rate-derivative securities. *Review of Financial Studies* **3** (1990), 573–592.

29. INUI, K., AND KIJIMA, M.   A Markovian framework in multi-factor Heath-Jarrow-Morton models. *Journal of Financial and Quantitative Analysis* **33** (1998), 423–440.

30. ISIDORI, A. *Nonlinear Control Systems.* Springer Verlag, 1989.

31. JEFFREY, A.   Single factor Heath-Jarrow-Morton term structure models based on Markov spot interest rate dynamics. *Journal of Financial and Quantitative Analysis* **30** (1995), 619–642.

32. MUSIELA, M. Stochastic PDE:s and term structure models. Preprint, 1993.

33. MUSIELA, M., AND RUTKOWSKI, M.   *Martingale Methods in Financial Modelling.* Springer-Verlag, 1997.

34. NELSON, C., AND SIEGEL, A.   Parsimonious modelling of yield curves. *Journal of Business* **60** (1987), 473–489.

35. RITCHKEN, P., AND SANKARASUBRAMANIAN, L.   Volatility structures of forward rates and the dynamics of the term structure. *Mathematical Finance* **5**, 1 (1995), 55–72.

36. VASIČEK, O.   An equilibrium characterization of the term stucture. *Journal of Financial Economics* **5** (1977), 177–188.

37. WARNER, F.   *Foundations of differentiable manifolds and Lie groups.* Scott, Foresman, Hill, 1979.

38. ZABCZYK, J.   Stochastic invariance and conistency of financial models. Preprint. Scuola Normale Superiore, Pisa, 2001.

Heterogeneous Beliefs, Speculation and Trading in Financial Markets

José Scheinkman and Wei Xiong

Bendheim Center for Finance
Princeton University
Princeton, NJ 08540
email: joses@princeton.edu
email: wxiong@princeton.edu

Summary. We survey recent developments in finance that analyze how heterogeneous beliefs among investors generate speculation and trading. We describe the joint effects of heterogeneous beliefs and short-sales constraints on asset prices, using both static and dynamic models, discuss the no-trade theorem in the rational expectations framework, and present investor overconfidence as a potential source of heterogeneous beliefs. We review recent results of Scheinkman and Xiong (2003) modeling the resale option that is embedded in share prices in the presence of short-sale constraints and heterogeneous beliefs, highlighting the implied correlation between stock prices and trading volume. Finally, we discuss the survival of investors with incorrect beliefs.

Key words: Heterogeneous beliefs, speculation, trading, overconfidence, resale option, bubble, optimal stopping, survival.
MSC 2000 subject classification. 91B24, 91B28, 91B44.

Acknowledgements: Research partially supported by the NSF through grant 0001647.

1 Introduction

Standard asset pricing theories have difficulty explaining episodes of asset price bubbles such as the one that seems to have occurred in the market for US internet stocks during the period of 1998-2000. In addition to asset prices that are difficult to justify by fundamentals such as expected future dividends, one typically observes inordinate increases in trading volume. For instance, Ofek and Richardson (2003) document that during the internet bubble of the late 90's, Internet stocks represented six percent of the market capitalization but accounted for 20% of the publicly traded volume of the U.S. stock market. Cochrane (2002) provides additional support for the correlation between bubbles and trading volume for US stocks during the late 90's. This evidence indicates that a satisfactory theory of bubbles should be able to explain simultaneously the level of prices and trading volume.

Several papers have been written over the last couple of decades that emphasize the role of heterogeneous beliefs in generating higher levels of asset prices and trading volume. In this chapter we present a selective survey of this literature.

We start by expositing a simple point made by Miller (1977), who argued that, if agents have heterogeneous beliefs about an asset's fundamentals and short sales are not allowed, equilibrium prices would, if opinions diverge enough, reflect the opinion of the more optimistic investor.

The Miller model is static and cannot be used to discuss the dynamics of trading. Harrison and Kreps (1978) exploit the *dynamic* consequences of heterogenous beliefs. Since an investor knows that, in the future, there may be other investors that value the asset more than he does, the investor is willing to pay more for an asset than he would pay if he was forced to hold the asset forever. The difference between the investor's willingness to pay, and his own discounts expected dividends reflects a speculative motive, the willingness to pay more than the intrinsic value of an asset because the ownership of the asset gives the owner the right to sell it in the future. To make this right valuable, short sales must be costly - in the Harrison and Kreps model it is simply assumed that short sales are not possible.

In the Harrison-Kreps model there is a single unit of an asset and several classes of risk neutral traders that disagree about the probability distribution of future dividends. The reservation price of a buyer is the supremum, over all stopping times, of the discounted cumulative dividends until the stopping time plus the discounted (ex-dividend) price that the owner can obtain by selling the asset at the stopping time. At each time, the agent in the group with highest reservation price would buy the asset. The equilibrium price process satisfies a simple recursive relationship. Furthermore, if there is a positive probability that at some future time a group that is currently not holding the asset would have a higher reservation price than the current owner, then the current price has to strictly exceed the maximum of the discounted expected future dividends among all groups - that is the current price exceeds the maximum fundamental valuation of any agent in the economy. This difference between the current price and the maximal valuation can be identified as a bubble. Section 3 contains a summary of the Harrison-Kreps theory.

Harrison and Kreps do not discuss the source of heterogeneous beliefs. Although private information seems to be a natural source of disagreement and of trading, a series of results, known as "no-trade theorems" appear in the economics literature showing that if all agents are rational and share identical prior beliefs, heterogeneity of information cannot generate trading or cause a price bubble. These results are discussed in Section 4. Several possibilities exist to avoid these no-trade results, including the presence of noise-traders, who trade for liquidity reasons, or heterogeneous priors. Another option is to assume that agents have behavioral biases that preclude "full rationality." Several behavioral biases are suggested by the psychology literature. See Hirshleier (2001) and Barberis and Thaler (2003) for detailed reviews of these biases.

Overconfidence, the tendency to overestimate the precision of own's opinion is a well documented behavioral bias. Scheinkman and Xiong (2003) use overconfidence as a convenient way to generate a parameterized model of heterogeneous beliefs. They adopt a continuous time framework describing a market for a single risky-asset with a limited supply, and many risk-neutral agents who can borrow and lend at a fixed rate of interest r. The current dividend of the asset is a noisy observation of a fundamental variable that will determine future dividends. More precisely:

$$dD_t = f_t dt + \sigma_D dZ_t^D,$$

where Z^D is a standard Brownian motion and f is not observable. However, it satisfies:

$$df_t = -\lambda(f_t - \bar{f})dt + \sigma_f dZ_t^f,$$

In addition to the dividends, there are two other sets of information available at each instant. These signals again satisfy the linear SDEs:

$$ds_t^A = f_t dt + \sigma_s dZ_t^A$$
$$ds_t^B = f_t dt + \sigma_s dZ_t^B.$$

The information is available to all agents. However, agents are divided in two groups A and B, and group A (B) has more confidence in signal s^A (resp. s^B.) As a consequence, when forecasting future dividends, each group of agents place different weights in the three sets of information, resulting in different forecasts. In the parametric structure that Scheinkman and Xiong (2003) consider, linear filtering applies and the conditional beliefs are normally distributed with a common variance and means f_t^A and f_t^B. Although agents in the model know exactly the amount by which their forecast of the fundamental variables exceed that of agents in the other group, behavioral limitations lead them to continue to disagree. As information flows, the mean forecasts by agents of the two groups fluctuate, and the group of agents that is at one instant relatively more optimistic, may become in a future date less optimistic than the agents in the other group. These changes in relative opinion generate trades.

Each agent in the model understands that the agents in the other group are placing different weights on the different sources of information. When deciding the value of the asset, agents consider their own view of the fundamentals as well as the fact that the owner of the asset has an option to sell the asset in the future to the agents in the other group. This option can be exercised at any time by the current owner, and the new owner gets in turn another option to sell the asset in the future. In the parametric example discussed by Scheinkman and Xiong (2003) it is natural to look for an equilibrium where the value of this option for the current owner, is a function of differences in opinions, that is if he belongs to group A (B,) this value equals $q = q(f_t^B - f_t^A)$ (resp. $q = q(f_t^B - f_t^A)$.) This option is "American," and hence the value of the option is the value function of an optimal stopping problem. Since the buyer's willingness to pay is a function of the value of the option that he acquires, the payoff from stopping is, in turn, related to the value of the option. This gives rise to a fixed point problem that the option value must satisfy. Scheinkman and Xiong (2003) show that the function q must, in the absence of trading costs, satisfy:

$$q(x) = \sup_{\tau \geq 0} \mathbb{E}^o \left[\left(\frac{x_\tau}{r + \lambda} + q(x_\tau) \right) e^{-r\tau} \right],$$

where $\mathbb{E}^o$ represents the expected value using the beliefs of the current owner. Scheinkman and Xiong (2003) write down an "explicit" solution to this equation that involves Kummer functions.

In equilibrium an asset owner will sell the asset to agents in the other group, whenever his view of the fundamental is surpassed by the view of agents in the other group by a critical amount. When there are no trading costs, this critical amount is zero - it is optimal to sell the asset immediately after the valuation of the fundamentals of the asset owner is "crossed" by the valuation of agents in the other group. The agents' beliefs satisfy simple stochastic differential equations and it is a consequence of properties of Brownian motion, that once the beliefs of agents cross, they will cross infinitely many times in any finite period of time right afterwards. This results in a trading frenzy. Although agents' profit from exercising the resale option is infinitesimal, the net value of the option is large because of the high frequency of trades. When trading costs are positive the duration between trades increases but in a continuous manner. In this way the model predicts large trading volume in markets with small transaction costs.

When a trade occurs the buyer has the highest fundamental valuation among all agents, and because of the re-sale option the price he pays exceeds his fundamental valuation. Agents pay prices that exceed their own valuation of future dividends, because they believe that in the future they will find a buyer willing to pay even more. This difference between the transaction price and the highest fundamental valuation can be reasonably called a bubble. Sections 5 and 6 contain an exposition of the model in Scheinkman and Xiong (2004).

The bubble in the Scheinkman and Xiong model, based on the expectations of traders to take advantage of future differences in opinions is quite different from the more traditional "rational bubbles" that are discussed, for instance, in Blanchard and Watson (1982) or Santos and Woodford (1997). In the typical rational bubble model agents have identical rational expectations, but prices include an extra bubble component that is always expected to grow at a rate equal to the risk free rate. Models of rational bubbles are incapable of explaining the increase in trading volume that is typically observed in the historic bubble episodes, and have non-stationarity properties that are at odds with empirical observations. Nonetheless, because these models constituted the first attempt to develop equilibrium models of price bubbles we briefly discuss them in Section 7.1.

While rational bubble models study asset prices while not considering questions on trading volume, a complementary literature uses heterogeneous beliefs to study trading, without dealing with the impact of heterogeneous beliefs on asset prices. Section 7.2 discusses these models using a paper by Harris and Raviv (1993) as an example.

We also discuss two other related questions that have played an important role in the economics literature on asset pricing. The first one considers what economists have dubbed the equity premium puzzle. This puzzle is the observation that stock returns over the last 50 years have been too high to be justified by standard asset pricing models with reasonable risk aversion parameters for investors. Section 7.3 briefly describes this puzzle and reviews several models that use heterogeneous beliefs to explain it.

The second question that dates back at least to Friedman (1953), is whether traders with irrational beliefs will lose money trading with rational traders and eventually disappear from the market. In Section 8, we discuss a model by Kogan, Ross, Wang and Westerfield (2004) who analyze this issue using a continuous-time equilibrium setup. In their model, in addition to a risk-free bond, there is a stock that pays a dividend D_T at the consumption date T. This terminal dividend satisfies:

$$dD_t = D_t(\mu dt + \sigma dZ_t).$$

There are two groups of traders of equal size. One group the "rational" traders, knows the true probability P determining the distribution of the final dividend. The second group believes incorrectly on a probability Q that is absolutely continuous with respect to P. As a result, the two groups disagree on the drift of the process determining D_T. Irrational traders believe that

$$dD_t = D_t[(\mu + \sigma^2\eta)dt + \sigma dZ_t^Q],$$

where Z_t^Q is a Brownian motion under Q. Irrational traders are optimistic (pessimistic) if $\eta > 0$ (*resp.* $\eta < 0$.) All agents maximize a constant relative risk aversion utility function that depends on final consumption, that is they choose a trading strategy to maximize:

$$\mathbb{E}_0^{\Pi}\left[\frac{1}{1-\gamma}C_T^{1-\gamma}\right]$$

where C_T is the consumption at T, and the measure $\Pi = P$ for rational agents and equals Q for irrational agents. Kogan *et al.* use the fact that, since there is only one source of uncertainty and two traded assets, it is natural to expect that the markets are complete and as a consequence that the equilibrium is Pareto efficient. They find the set of all Pareto optimal allocations of final consumption across types, by maximizing the weighted sum of utilities and derive the corresponding support prices. Using these support prices and the budget constraint Kogan *et al.* characterize the particular Pareto efficient allocation that corresponds to a competitive equilibrium, and show that the corresponding support prices are such that markets are complete. They then combine the expression for the equilibrium allocation with the strong law of large numbers to show that, if traders are sufficiently risk averse and irrational traders are moderately optimistic, irrational traders may survive in the long run. The reason is that they may earn higher returns albeit by bearing excessive risk.

The papers we discuss in this chapter were chosen mainly on the basis of our views concerning their contribution to the understanding of speculation and trading. Some such as Miller (1977) involve very little mathematics. Nonetheless we believe that the area is ripe for a more rigorous mathematical treatment. In fact, in section 9 we discuss some open problems related to the speculative behavior of investors and the resale option component in asset prices. Typically these problems involve optimal stopping problems with nonlinear filtering and multiple state variables.

2 A Static Model with Heterogeneous Beliefs and Short-Sales Constraints

Miller (1977) argued that short-sales constraints can cause stocks to be overpriced when investors have heterogeneous beliefs about stock fundamentals. In the presence of short-sale constraints, stock prices reflect the views of the more optimistic participants. If some of the pessimistic investors that would like to short are not allowed to do it, prices will in general be higher than the price that would prevail in the absence of short-sale constraints.

We illustrate Miller's argument using a version of Lintner (1969) model, where we add short-sales constraints. Related models can also be found in Jarrow (1980), Varian (1989), Chen, Hong and Stein (2002), and Gallmeyer and Hollifield (2004). The model has one period with two dates: $t = 0, 1$. There is one risky asset which will be liquidated at $t = 1$. The final liquidation value is

$$\tilde{f} = \mu + \epsilon,$$

where μ is a constant unknown to investors and ϵ is normally distributed with mean zero. Investors have diverse opinions concerning the distribution of liquidation values. Investor i believes that $\tilde{f}$ has a normal distribution with mean μ_i and variance σ^2. Since all investors share the same views concerning the variance, we index investors by their mean beliefs μ_i, and assume that μ_i is uniformly distributed around μ in an interval $[\mu - \kappa, \mu + \kappa]$. The parameter κ measures the heterogeneity of beliefs. In addition, we assume that all investors can borrow or lend at a risk-free interest rate of zero, short-sales of the risky asset are prohibited, and the total supply of the asset is Q.

At $t = 0$, each investor chooses his asset demand to maximize his expected utility at $t = 1$:

$$\max_{x_i} \mathbb{E}\left[-e^{-\gamma(W_0 + x_i(\tilde{f} - p_0))}\right],$$

where γ is the investor's risk aversion, W_0 is the initial wealth, p_0 is the market price of the asset, and x_i is the investor's asset demand, subject to $x_i \geq 0$. It is immediate that

$$x_i = \max\left\{\frac{\mu_i - p_0}{\gamma\sigma^2}, 0\right\}.$$

The investor's demand, in the absence of the short-sale constraint would be $(\mu_i - p_0)/(\gamma\sigma^2)$. When these constraints are present, investors with mean beliefs μ_i below the market price stay out of the market. The market clearing condition, $\int_i x_i = Q$ thus implies that

$$\int_{\max\{p_0,\mu-\kappa\}}^{\mu+\kappa} \frac{\mu_i - p_0}{\gamma\sigma^2} \frac{d\mu_i}{2\kappa} = Q,$$

and the equilibrium price

$$p_0 = \begin{cases} \mu - \gamma\sigma^2 Q & \text{if } \kappa < \gamma\sigma^2 Q \\ \mu + \kappa - 2\sqrt{\kappa\gamma\sigma^2 Q} & \text{if } \kappa \geq \gamma\sigma^2 Q \end{cases}$$

In the absence of short-sales constraints the equilibrium price would be $\mu - \gamma\sigma^2 Q$. Thus, the short-sales constraints cause the asset price to become higher when the heterogeneity of investors' beliefs κ is greater than $\gamma\sigma^2 Q$. If heterogeneity of beliefs is small enough than the no-short-sale constraint is not binding for any investor, and the equilibrium price is not affected by the presence of the constraint.

This simple model shows that short-sales constraints combined with heterogeneous beliefs can cause asset prices to become higher than they would be in the absence of the short-sales constraints. When beliefs are sufficiently heterogeneous, short-sale constraints insure that asset prices reflect the opinion of the more optimistic investors. However, because of its static nature, the model has no prediction concerning the dynamics of trading. In the following sections, we discuss the effects of heterogeneous beliefs and short-sales constraints on asset prices and share turnovers in dynamic models.

3 A Dynamic Model in Discrete Time with Short-Sales Constraints

Harrison and Kreps (1978) say that investors exhibit *speculative behavior* if the right to resell an asset makes them willing to pay more for it than they would pay if obliged to hold it forever. This definition is particularly compelling when agents are risk-neutral since in this case no risk-sharing benefits arise from trading. Harrison and Kreps constructed a model where, because risk-neutral agents have heterogeneous expectations and face short-sales constraints, speculative behavior arises.

In the model, there exists one unit of an asset that pays a non-negative random dividend d_t at each time t. All agents are risk-neutral, and discount future revenues at a constant rate $\gamma < 1$ or equivalently can borrow and lend at a rate $r = \frac{1-\gamma}{\gamma}$. Agents are divided into groups that differ on their views on the distribution of the stochastic process d_t. Harrison and Kreps allow for an arbitrary number of groups of agents, but the analysis they provide is well illustrated by treating the case of two groups.

Thus, we consider two groups $\{A,B\}$, each with an infinite number of agents. For simplicity we assume that each group $C \in \{A, B\}$ views d_t as a stochastic process defined on a probability space $\{\Omega, \mathcal{F}, \mathbb{P}^C, \}$ and that $\mathbb{P}^A \sim \mathbb{P}^B$. We write $\mathbb{E}^C$ for the expected value with respect to the probability distribution shared by all agents in group $C \in \{A, B\}$.

Write $\mathcal{F}_t, t \geq 0$ for the σ-algebra generated by the realizations of $d^t \equiv (d_1, \dots, d_t)$. A price process is an $\mathcal{F}_t$ adapted non-negative process.

The owner of an asset at t must decide on a strategy to sell all or part of his holdings in the future. Since agents are risk neutral it suffices to consider the strategy of selling one unit of the asset at a (possibly infinite) stopping time. For this reason we define a feasible selling strategy from time t as a (possibly infinite) integer valued stopping time $T > t$.

Because each group has an infinite number of agents and there is a single unit of the asset, competition among buyers will lead to a price that equals the reservation price of the buyers. Since agents are risk-neutral and can choose any feasible selling strategy, the value of the asset at t for an agent in group $C \in \{A, B\}$ is given by

$$\sup_{T>t} \mathbb{E}^C \left[\sum_{k=t+1}^{T} \gamma^{k-t} d_k + \gamma^{T-t} p_T | \mathcal{F}_t \right],$$

where $\sum_{k=t+1}^{T} \gamma^{k-t} d_k$ represents the value of the discounted dividend stream received up to the moment of sale at T, and $\gamma^{T-t} p_T$ represents the discounted value from selling the asset at the prevailing market price at T. The buyers will belong to the group that places the highest valuation on the asset. Hence an equilibrium price process has to satisfy:

$$p_t = \max_{C \in \{A,B\}} \sup_{T>t} \mathbb{E}^C \left[\sum_{k=t+1}^{T} \gamma^{k-t} d_k + \gamma^{T-t} p_T | \mathcal{F}_t \right]. \tag{1}$$

Since $T = \infty$ is a feasible strategy, it follows that

$$p_t \geq \max_{C \in \{A,B\}} \mathbb{E}^C \left[\sum_{k=t+1}^{\infty} \gamma^{k-t} d_k | \mathcal{F}_t \right]. \tag{2}$$

Since the right hand side of equation (2) represents the maximal value that any agent is willing to pay for the asset if resale is impossible, speculative behavior is equivalent to a strict inequality in equation (2). Suppose $F \in \mathcal{F}_t$ is such that A realizes the maximum in the right-hand side of (1) for each $\omega \in F$. Suppose further that for some $t' > t$ and event $F' \in \mathcal{F}_{t'}$, $F' \subset F$ with $\mathbb{P}^A(F') > 0$ (and hence $\mathbb{P}^B(F') > 0$)

$$\mathbb{E}^B \left[\sum_{k=t'+1}^{\infty} \gamma^{k-t} d_k | \mathcal{F}_{t'} \right] (\omega) > \mathbb{E}^A \left[\sum_{k=t'+1}^{\infty} \gamma^{k-t} d_k | \mathcal{F}_{t'} \right] (\omega),$$

for each $\omega \in F'$. Then a strict inequality must hold in equation (2), for $\omega \in F$.

In fact,

$$\mathbb{E}^A \left[\sum_{k=t+1}^{\infty} \gamma^{k-t} d_k | \mathcal{F}_t \right]$$

$$= \mathbb{E}^A \left[\sum_{k=t+1}^{t'} \gamma^{k-t} d_k | \mathcal{F}_t \right] + \mathbb{E}^A \left[\mathbb{E}^A \left[\sum_{k=t'+1}^{\infty} \gamma^{k-t} d_k | \mathcal{F}_{t'} \right] | \mathcal{F}_t \right]$$

$$< \mathbb{E}^A \left[\sum_{k=t+1}^{t'} \gamma^{k-t} d_k | \mathcal{F}_t \right] + \mathbb{E}^A \left[\max_{C \in \{A,B\}} \mathbb{E}^C \left[\sum_{k=t'+1}^{\infty} \gamma^{k-t} d_k | \mathcal{F}_{t'} \right] | \mathcal{F}_t \right]$$

$$\leq \mathbb{E}^A \left[\sum_{k=t+1}^{t'} \gamma^{k-t} d_k | \mathcal{F}_t \right] + \mathbb{E}^A \left[\gamma^{t'-t} p_{t'} | \mathcal{F}_t \right] \leq p_t.$$

Speculative behavior arises, because the owner of the asset retains in addition to the flow of future dividends an option to resell the asset to other investors. This option will become *in-the-money* when there are investors that have a relatively more optimistic view of future dividends than the current owner.

The following proposition allows one to characterize all pricing processes that satisfy equation (1) by a two period condition.

Proposition 1. *A price process satisfies equation (1) if and only if, for each t,*

$$p_t = \max_{C \in \{A,B\}} \mathbb{E}^C \left[\gamma d_{t+1} + \gamma p_{t+1} | \mathcal{F}_t \right]. \tag{3}$$

Proof: Suppose (3) holds. Then for each C,

$$p_t \geq \mathbb{E}^C \left[\gamma d_{t+1} + \gamma p_{t+1} | \mathcal{F}_t \right].$$

Hence the process $y_t = \sum_{s=1}^{t} \gamma^s d_s + \gamma^t p_t$ is a non-negative supermartingale and hence $\lim_{t \to \infty} y_t$ exists. Doob's optional stopping theorem implies that $p_t \geq \mathbb{E}^C \left[\sum_{k=t+1}^{T} \gamma^{k-t} d_k + \gamma^{T-t} p_T | \mathcal{F}_t \right]$, what implies that (1) must hold.

Conversely, suppose (1) holds, but that:

$$p_t > \max_{C \in \{A,B\}} \mathbb{E}^C \left[\gamma d_{t+1} + \gamma p_{t+1} | \mathcal{F}_t \right].$$

The law of iterated expectations and equation (1) applied at $t+1$ implies that:

$$p_t > \max_{C \in \{A,B\}} \sup_T \mathbb{E}^C \left[\sum_{k=t+1}^{T} \gamma^{k-t} d_k + \gamma^{T-t} p_t | \mathcal{F}_t \right],$$

a contradiction. □

Suppose d_t is a time-homogeneous Markov process, that is

$$\mathbb{P}^C[d_{t+s}|\mathcal{F}_t] = \mathbb{P}^C[d_{t+s}|d_t] = \mathbb{P}^C[d_{s+1}|d_1],$$

for each $C \in \{A, B\}$. Then it is natural to search for equilibrium prices that are of the form $p_t = p(d_t)$. If we write:

$$\mathcal{T}p(d) = \max_{C \in \{A,B\}} \mathbb{E}^C\left[\gamma d_{t+1} + \gamma p(d_{t+1})|d_t = d\right]. \tag{4}$$

Then equation (3) can be rewritten as:

$$\mathcal{T}p = p. \tag{5}$$

The operator $\mathcal{T}$ has a monotonicity property. If $p \geq q$ then $\mathcal{T}(p) \geq \mathcal{T}(q)$. The existence and uniqueness of (continuous) solutions to equation (5) are guaranteed if, for instance, the process d stays in a bounded set.

Harrison and Kreps do not explicitly address the source of heterogenous beliefs among investors. In what follows we will examine specific mechanisms to generate beliefs' heterogeneity. In Section 4 we describe some results concerning the difficulty of generating heterogeneous beliefs from the presence of private information. We then discuss a model that utilizes *overconfidence*, a behavioral limitation that is suggested by psychological studies, to parameterize heterogeneous beliefs. The model will be then used to link Harrison and Kreps' speculative behavior to a resale option value and to explain some empirical regularities concerning trading volume and prices during asset "bubbles."

4 No-Trade Theorem under Rational Expectations

A possible source of heterogeneous beliefs is private information. The presence of private information suggests that investors could use their information to trade and realize a profit. However, Tirole (1982) and Milgrom and Stokey (1982) prove that this cannot happen when all agents are rational and share identical prior beliefs, the conditions that are imposed in the standard rational expectations models. Thus, private information cannot be the source of speculative trading. Results of this kind are called "no-trade theorems".

We use a static setup from Tirole (1982) to illustrate the main ideas. Consider a market with I risk neutral traders: $i = 1, ..., I$. The traders exchange claims for an asset with random payoff $\tilde{p} \in E$, which will be realized after the trading. The set $E \subset \mathbb{R}$ is the set of all possible payoffs. Claims are traded at an equilkibrium market price p. If a trader buys x units of the asset and $\tilde{p}$ obtains, the realized profit is

$$G = (\tilde{p} - p)x.$$

Each trader i receives a private signal s^i belonging to a possible set of signals S^i. Let $s = (s^1, \dots, s^I) \in S = \Pi_{i=1}^I S^i$, and $\Omega = E \times S$. Assume that all traders have a common prior ν on Ω, and that each trader can only take a bounded position.

Trader $i = 1, \dots, I$ chooses an amount x^i to maximize his conditional expected value of G. In a *Rational Expectations Equilibrium*, (REE) each trader uses all information at his disposal, including the observed market price p. In spite of its name, a REE does not involve only an equilibrium price, but a *forecast function* that maps each vector of all signals $s \in S$ into a price that establishes equilibrium in the market if the state s obtains. This forecast function is typically not one-to-one. By observing the price p as well as his own signal s^i, trader i is not able to identify the full vector of signals s. However, a forecast function $\Phi : S \to \mathbb{R}$, an observed p, and a signal s^i induce a conditional distribution on $E \times S$, Γ_{Φ,p,s^i}^i.

Definition 1: A rational expectations equilibrium (REE) is a forecast function $\Phi : S \to R$, and a set of trades $x^i(\Phi, p, s^i)$ for each trader i such that

1. $x^i(\Phi, p, s^i)$ maximizes $\mathbb{E}_\Phi(G^i|p, s^i) \equiv \int G d\Gamma_{\Phi,p,s^i}^i$

2. The market clears for each $s \in S$: $\sum_i x^i(\Phi, p, s^i) = 0$.

The next proposition is a no-trade theorem.

Proposition 2. *In a REE, $\mathbb{E}_\Phi(G^i|p, s^i) = 0$. As a consequence given an REE, there exists another REE with the same forecast function and $x^i = 0$.*

Proof: Since $x^i = 0$ is always a possible choice,

$$\mathbb{E}_\Phi(G^i|p, s^i) \geq 0. \tag{6}$$

The Law of Iterated expectations thus implies that

$$\mathbb{E}_\Phi(G^i|p) \geq 0. \tag{7}$$

The market clearing condition implies that for each realization of $\tilde{p}$ aggregate gains are null. Hence

$$\sum_i \mathbb{E}_\Phi(G^i|p) = 0,$$

and equality must hold in equations (6) and (7). $\square$

Proposition 2 rules out the possibility that investors that share the same prior can expect to profit from speculating against each other based on differences in information. As a consequence, they cannot do any better than by choosing not to trade. Although Proposition 2 only deals with the risk neutral case, it is intuitive that risk aversion would further reduce the net gain among investors from trading.

Tirole (1982) also analyzes a dynamic model with rational expectations and demonstrates that the no-trade theorem holds in dynamic setup. He further shows that the resale options suggested by Harrison and Kreps cannot arise in asset prices in such an environment even if short-sales constraints are imposed. Diamond and Verrecchia (1987) also study the effects of short-sales constraints on asset prices in a rational expectations model with asymmetric information. They show that short-sales constraints reduce the adjustment speed of prices to private information, especially to bad news, since agents with negative information are prohibited from shorting the asset. However Diamond and Verrecchia also confirm that short-sales constraints do not lead to an upward bias in prices since agents, when forming their own beliefs, could rationally take into account the fact that negative information may be not reflected in trading prices.

There are at least two ways to weaken the assumptions in Proposition 2 and avoid the no-trade result. First one might consider some agents who trade for non-speculative reasons such as diversification or liquidity. The presence of such traders would make the trading among speculators a positive-sum game. This is the approach that has been adopted in several models of market microstructure such as Grossman and Stiglitz (1980), Kyle (1985), and Wang (1993).

Another possibility is to relax the assumption that agents share the same prior beliefs. This approach is pursued by Morris (1996), Biais and Bossaerts (1998), and Brav and Heaton (2002). Finally one may assume that agents display behavioral biases. In this chapter, we discuss in detail overconfidence as a way to parameterize the dynamics of heterogeneous beliefs among agents. However many other behavioral biases may generate heterogeneity of beliefs. For instance, heterogeneous beliefs can arise if agents gain utility from adopting certain beliefs as discussed in Brunnermeier and Parker (2003).

5 Overconfidence as Source of Heterogeneous Beliefs

Overconfidence, the tendency of people to overestimate the precision of their knowledge, provides a convenient way to generate heterogeneous beliefs. Psychology studies suggest that people are overconfident. Alpert and Raiffa (1982), and Brenner et al. (1996) and other calibration studies find that people overestimate the precision of their knowledge. Camerer (1995) argues that even experts can display overconfidence. Hirshleifer (2001) and Barberis and Thaler (2003) contain extensive reviews of the literature.

In finance, researchers have developed theoretical models to analyze the implications of overconfidence on financial markets. Kyle and Wang (1997) show that overconfidence can be used as a commitment device over competitors to improve one's welfare. Daniel, Hirshleifer and Subrahmanyam (1998) use overconfidence to explain the predictable returns of financial assets. Odean (1998) demonstrates that overconfidence can cause excessive trading. Bernardo and Welch (2001) discuss the benefits of overconfidence to entrepreneurs through the reduced tendency to herd. In all these studies, overconfidence is modelled as overestimation of the precision of one's information.

In this section we exposit the model in Scheinkman and Xiong (2003), that exploits the consequences of this overestimation in a dynamic model of pricing and trading. Since overconfident investors believe more strongly in their own assessments of an asset's value than in the assessment of others, heterogeneous beliefs arise.

Consider a single risky asset with a dividend process that is the sum of two components. The first component is a fundamental variable that determines future dividends. The second is "noise". The cumulative dividend process D satisfies:

$$dD_t = f_t dt + \sigma_D dZ_t^D, \tag{8}$$

where Z^D is a standard Brownian motion and σ_D is the volatility parameter. The stochastic process of fundamentals f is not observable. However, it satisfies:

$$df_t = -\lambda(f_t - \bar{f})dt + \sigma_f dZ_t^f, \tag{9}$$

where $\lambda \geq 0$ is the mean reversion parameter, $\bar{f}$ is the long-run mean of f, $\sigma_f > 0$ is a volatility parameter and Z^f is a standard Brownian motion, uncorrelated to Z^D. The presence of dividend noise makes it impossible to infer f perfectly from observations of the cumulative dividend process.

There are two sets of risk neutral agents, who use the observations of D and any other signals that are correlated with f to infer current f and to value the asset. In addition to the cumulative dividend process, all agents observe a vector of signals s^A and s^B that satisfy:

$$ds_t^A = f_t dt + \sigma_s dZ_t^A \tag{10}$$
$$ds_t^B = f_t dt + \sigma_s dZ_t^B, \tag{11}$$

where Z^A and Z^B are standard Brownian motions, and $\sigma_s > 0$ is the common volatility of the signals. We assume that all four processes Z^D, Z^f, Z^A and Z^B are mutually independent.

Agents in group A (B) think of s^A (s^B) as their own signal although they can also observe s^B (s^A). Heterogeneous beliefs arise because each agent believes that the informativeness of his own signal is larger than its true informativeness. Agents of group A (B) believe that innovations dZ^A (dZ^B) in the signal s^A (s^B) are correlated with the innovations dZ^f in the fundamental process, with ϕ ($0 < \phi < 1$) as the correlation parameter. Specifically, agents in group A believe that the process s^A satisfies

$$ds_t^A = f_t dt + \sigma_s \phi dZ_t^f + \sigma_s \sqrt{1 - \phi^2} dZ_t^A.$$

Although agents in group A perceive the correct unconditional volatility of the signal s^A, the correlation that they attribute to innovations causes them to over-react to signal s^A. Similarly, agents in group B believe the process s^B satisfies

$$ds_t^B = f_t dt + \sigma_s \phi dZ_t^f + \sigma_s \sqrt{1 - \phi^2} dZ_t^B.$$

On the other hand, agents in group A (B) believe (correctly) that innovations to s^B (s^A) are uncorrelated with innovations to Z^B (Z^A.) We assume that the joint dynamics of the processes D, f, s^A and s^B in the mind of agents of each group is public information.

Since all variables are Gaussian, the filtering problem of the agents is standard. With Gaussian initial conditions, the conditional beliefs of agents in group $C \in \{A, B\}$ is Gaussian. Standard arguments, *e.g.* section VI.9 in Rogers and Williams (1987) and Theorem 12.7 in Liptser and Shiryayev (1977), can be used to compute the variance of the stationary solution and the evolution of the conditional mean of beliefs. The variance of this stationary solution is the same for both groups of agents and equals

$$\gamma \equiv \frac{\sqrt{(\lambda + \phi\sigma_f/\sigma_s)^2 + (1 - \phi^2)(2\sigma_f^2/\sigma_s^2 + \sigma_f^2/\sigma_D^2)} - (\lambda + \phi\sigma_f/\sigma_s)}{\frac{1}{\sigma_D^2} + \frac{2}{\sigma_s^2}}.$$

One can directly verify that the stationary variance γ decreases with ϕ. When $\phi > 0$, agents have an exaggerated view of the precision of their estimates of f. A larger ϕ leads to more overstatement of this precision. For this reason we refer to ϕ as the "overconfidence" parameter.

The conditional mean of the beliefs of agents in group A satisfies:

$$d\hat{f}_t^A = -\lambda(\hat{f}_t^A - \bar{f})dt + \frac{\phi\sigma_s\sigma_f + \gamma}{\sigma_s^2}(ds_t^A - \hat{f}_t^A dt)$$

$$+ \frac{\gamma}{\sigma_s^2}(ds_t^B - \hat{f}_t^A dt) + \frac{\gamma}{\sigma_D^2}(dD_t - \hat{f}_t^A dt). \tag{12}$$

Since f mean-reverts, the conditional beliefs also mean-revert. The other three terms represent the effects of "surprises." These surprises can be represented as standard mutually independent Brownian motions for agents in group A:

$$dW_t^{A,A} = \frac{1}{\sigma_s}(ds_t^A - \hat{f}_t^A dt), \tag{13}$$

$$dW_t^{A,B} = \frac{1}{\sigma_s}(ds_t^B - \hat{f}_t^A dt), \tag{14}$$

$$dW^{A,D} = \frac{1}{\sigma_D}(dD_t - \hat{f}_t^A dt). \tag{15}$$

Note that these processes are only Wiener processes in the mind of group A agents. Due to overconfidence ($\phi > 0$), agents in group A over-react to surprises in s^A.

Similarly, the conditional mean of the beliefs of agents in group B satisfies:

$$d\hat{f}_t^B = -\lambda(\hat{f}_t^B - \bar{f})dt + \frac{\gamma}{\sigma_s^2}(ds_t^A - \hat{f}_t^B dt)$$

$$+ \frac{\phi\sigma_s\sigma_f + \gamma}{\sigma_s^2}(ds_t^B - \hat{f}_t^B dt) + \frac{\gamma}{\sigma_D^2}(dD_t - \hat{f}_t^B dt), \tag{16}$$

and the surprise terms can be represented as mutually independent Wiener processes: $dW^{B,A} = \frac{1}{\sigma_s}(ds_t^A - \hat{f}_t^B dt)$, $dW^{B,B} = \frac{1}{\sigma_s}(ds_t^B - \hat{f}_t^B dt)$, and $dW^{B,D} = \frac{1}{\sigma_D}(dD_t - \hat{f}_t^B dt)$. These processes form a standard 3-d Brownian only for agents in group B.

Since, in the stationary solution the beliefs of all agents have constant variance, one may refer to the conditional mean of the beliefs as the agents *beliefs*. Let g_A and g_B denote the differences in beliefs:

$$g^A = \hat{f}^B - \hat{f}^A, \quad g^B = \hat{f}^A - \hat{f}^B.$$

The next proposition describes the evolution of these differences in beliefs:

Proposition 3.

$$dg_t^A = -\rho g_t^A dt + \sigma_g dW_t^{A,g}, \tag{17}$$

where

$$\rho = \sqrt{\left(\lambda + \phi\frac{\sigma_f}{\sigma_s}\right)^2 + (1 - \phi^2)\sigma_f^2 \left(\frac{2}{\sigma_s^2} + \frac{1}{\sigma_D^2}\right)}, \tag{18}$$

$$\sigma_g = \sqrt{2}\phi\sigma_f,$$

and $W^{A,g}$ is a standard Wiener process for agents in group A.

Proof: from equations (12) and (16):

$$dg_t^A = d\hat{f}_t^B - d\hat{f}_t^A = -\left[\lambda + \frac{2\gamma + \phi\sigma_s\sigma_f}{\sigma_s^2} + \frac{\gamma}{\sigma_D^2}\right] g_t^A dt + \frac{\phi\sigma_f}{\sigma_s}(ds_t^B - ds_t^A).$$

Using the formula for γ, we may write the mean-reversion parameter as in equation (18). Using equations (13) and (14),

$$dg_t^A = -\rho g_t^A dt + \frac{\phi\sigma_f}{\sigma_s} \left(\sigma_s dW^{A,B} - \sigma_s dW^{A,A}\right).$$

The result follows by writing

$$W^{A,g} = \frac{1}{\sqrt{2}} \left(W^{A,B} - W^{A,A}\right).$$

It is easy to verify that innovations to $W^{A,g}$ are orthogonal to innovations to $\hat{f}^A$ in the mind of agents in group A. $\square$

Proposition 3 implies that the difference in beliefs g^A follows a simple mean reverting diffusion process in the mind of group A agents. In particular, the volatility of the difference in beliefs is zero in the absence of overconfidence. A larger ϕ leads to greater volatility. In addition, $-\rho/(2\sigma_g^2)$ measures the pull towards the origin. A simple calculation shows that this mean-reversion decreases with ϕ. A higher ϕ causes an increase in fluctuations of opinions and a slower mean-reversion.

In an analogous fashion, for agents in group B, g^B satisfies:

$$dg_t^B = -\rho g_t^B dt + \sigma_g dW_t^{B,g}, \tag{19}$$

where $W^{B,g}$ is a standard Wiener process.

Notice that although we started with a Markovian structure on dividends and signals, the beliefs depend on the history of dividends and signals - only the vector involving dividends, all signals and beliefs is Markovian. This is a consequence of the inference problem faced by investors. In contrast, in this model, the difference in beliefs is a Markov diffusion, what greatly facilitates the analysis that follows.

6 Trading and Equilibrium Price in Continuous Time

In the previous section we specified a particular model of heterogeneous beliefs, generated by overconfidence. Equations (17) and (19) state that, in each group's mind, the difference of opinions follows a mean-reverting diffusion process. The coefficients of this process are linked to the parameters describing the original uncertainty and the degree of overconfidence. In this section we derive implications of this particular model of heterogeneity for the equilibrium prices and trading behavior. We also summarize some results from Scheinkman and Xiong (2003) concerning the effect of the parameters that determine the original uncertainty and overconfidence on the prices and trading volume that obtain in equilibrium.

As in the Harrison and Kreps model described in Section 3, assume that each group of investors is large and there is no short selling of the risky asset. To value future cash flows, assume that every agent can borrow and lend at the same rate of interest r. These assumptions facilitate the calculation of equilibrium prices.

At each t, agents in group $C = \{A, B\}$ are willing to pay p_t^C for a unit of the asset. As in the Harrison and Kreps model described in Section 3, the presence of the short-sale constraint, a finite supply of the asset, and an infinite number of prospective buyers, guarantee that any successful bidder will pay his reservation price. The amount that an agent is willing to pay reflects the agent's fundamental valuation and the fact that he may be able to sell his holdings at a later date at the demand price of agents in the other group for a profit. If $o \in \{A, B\}$ denotes the group of the current owner, $\bar{o}$ the other group, and $\mathbb{E}_t^o$ the expectation of members of group o, conditional on the information they have at t, then:

$$p_t^o = \sup_{\tau \geq 0} \mathbb{E}_t^o \left[\int_t^{t+\tau} e^{-r(s-t)} dD_s + e^{-r\tau} (p_{t+\tau}^{\bar{o}} - c) \right], \tag{20}$$

where τ is a stopping time, c is a transaction cost charged to the seller, and $p_{t+\tau}^{\bar{o}}$ is the reservation value of the buyer at the time of transaction $t + \tau$.

Using the equations for the evolution of dividends and for the conditional mean of beliefs (equations (8), (12) and (16) above), one obtains:

$$\int_t^{t+\tau} e^{-r(s-t)} dD_s = \int_t^{t+\tau} e^{-r(s-t)} [\bar{f} + e^{-\lambda(s-t)} (\hat{f}_s^o - \bar{f})] ds + M_{t+\tau},$$

where $\mathbb{E}_t^o M_{t+\tau} = 0$. Hence, we may rewrite equation (20) as:

$$p_t^o = \max_{\tau \geq 0} \mathbb{E}_t^o \left\{ \int_t^{t+\tau} e^{-r(s-t)} [\bar{f} + e^{-\lambda(s-t)} (\hat{f}_s^o - \bar{f})] ds + e^{-r\tau} (p_{t+\tau}^{\bar{o}} - c) \right\}. \quad (21)$$

Scheinkman and Xiong (2003) start by postulating a particular form for the equilibrium price function, equation (22) below. Proceeding in a heuristic fashion, they derive properties that our candidate equilibrium price function should satisfy. They then construct a function that satisfies these properties, and verify that they have produced an equilibrium.

Since all the relevant stochastic processes are Markovian and time-homogeneous, and traders are risk-neutral, it is natural to look for an equilibrium in which the demand price of the current owner satisfies

$$p_t^o = p^o(\hat{f}_t^o, g_t^o) = \frac{\bar{f}}{r} + \frac{\hat{f}_t^o - \bar{f}}{r + \lambda} + q(g_t^o). \quad (22)$$

with $q > 0$ and $q' > 0$. This equation states that prices are the sum of two components. The first part, $\frac{\bar{f}}{r} + \frac{\hat{f}_t^o - \bar{f}}{r+\lambda}$, is the expected present value of future dividends from the viewpoint of the current owner. The second is the value of the resale option, $q(g_t^o)$, which depends on the current difference between the beliefs of the other group's agents and the beliefs of the current owner. We call the first quantity the owner's fundamental valuation and the second the value of the resale option. Using (22) in equation (21) and collecting terms:

$$p_t^o = p^o(\hat{f}_t^o, g_t^o) = \frac{\bar{f}}{r} + \frac{\hat{f}_t^o - \bar{f}}{r + \lambda} + \sup_{\tau \geq 0} \mathbb{E}_t^o \left[\left(\frac{g_{t+\tau}^o}{r + \lambda} + q(g_{t+\tau}^{\bar{o}}) - c \right) e^{-r\tau} \right].$$

Equivalently, the resale option value satisfies

$$q(g_t^o) = \sup_{\tau \geq 0} \mathbb{E}_t^o \left[\left(\frac{g_{t+\tau}^o}{r + \lambda} + q(g_{t+\tau}^{\bar{o}}) - c \right) e^{-r\tau} \right]. \quad (23)$$

Hence to show that an equilibrium of the form (22) exists, it is necessary and sufficient to construct an option value function q that satisfies equation (23). This equation is similar to a Bellman equation. The current asset owner chooses an optimal stopping time to exercise his re-sale option. Upon the exercise of the option, the owner gets the "strike price" $\frac{g_{t+\tau}^o}{r+\lambda} + q(g_{t+\tau}^{\bar{o}})$, the amount of excess optimism that the buyer has about the asset's fundamental value and the value of the resale option to the buyer, minus the cost c. In contrast to the optimal exercise problem of American options, the "strike price" in this problem depends on the re-sale option value function itself.

The region where the value of the option equals that of an immediate sale is the stopping region. The complement is the continuation region. In the mind of the risk neutral asset holder, the discounted value of the option $e^{-rt}q(g_t^o)$ should be a martingale in the continuation region, and a supermartingale in the stopping region. Using Ito's lemma and the evolution equation for g^o, these conditions can be stated as:

$$q(x) \geq \frac{x}{r+\lambda} + q(-x) - c \qquad (24)$$

$$\frac{1}{2}\sigma_g^2 q'' - \rho x q' - rq \leq 0, \text{ with equality if (24) holds strictly.} \qquad (25)$$

In addition, the function q should be continuously differentiable (smooth pasting). As usual, one first shows that there exists a smooth function q that satisfies equations (24) and (25) and then uses these properties and a growth condition on q to show that in fact the function q solves (23).

To construct the function q, guess that the continuation region will be an interval $(-\infty, k^*)$, with $k^* \geq 0$. k^* is the minimum amount of difference in opinions that generates a trade. The second order ordinary differential equation that q must satisfy, albeit only in the continuation region, is:

$$\frac{1}{2}\sigma_g^2 u'' - \rho x u' - ru = 0 \qquad (26)$$

The following proposition describes all solutions of equation (26).

Proposition 4. *Let*

$$h(x) = \begin{cases} U\left(\frac{r}{2\rho}, \frac{1}{2}, \frac{\rho}{\sigma_g^2}x^2\right) & \text{if } x \leq 0 \\ \frac{2\pi}{\Gamma(\frac{1}{2}+\frac{r}{2\rho})\Gamma(\frac{1}{2})}M\left(\frac{r}{2\rho}, \frac{1}{2}, \frac{\rho}{\sigma_g^2}x^2\right) - U\left(\frac{r}{2\rho}, \frac{1}{2}, \frac{\rho}{\sigma_g^2}x^2\right) & \text{if } x > 0 \end{cases} \qquad (27)$$

where $\Gamma(\cdot)$ is the Gamma function, and $M : R^3 \to R$ and $U : R^3 \to R$ are two Kummer functions described in chapter 13 of Abramowitz and Stegum (1964). $h(x)$ is positive and increasing in $(-\infty, 0)$. In addition h solves equation (26) with

$$h(0) = \frac{\pi}{\Gamma\left(\frac{1}{2}+\frac{r}{2\rho}\right)\Gamma\left(\frac{1}{2}\right)}.$$

Any solution $u(x)$ to equation (26) that is strictly positive and increasing in $(-\infty, 0)$ must satisfy: $u(x) = \beta_1 h(x)$ for some $\beta_1 > 0$.

Proof: Let $v(y)$ be a solution to the differential equation

$$yv''(y) + (1/2 - y)v'(y) - \frac{r}{2\rho}v(y) = 0. \qquad (28)$$

It is straightforward to verify that $u(x) = v\left(\frac{\rho}{\sigma_g^2}x^2\right)$ satisfies equation (26). The general solution of equation (28) is

$$v(y) = \alpha M\left(\frac{r}{2\rho}, \frac{1}{2}, y\right) + \beta U\left(\frac{r}{2\rho}, \frac{1}{2}, y\right).$$

Given a solution u to equation (26) one can construct two solutions v to equation (28), by using the values of the function for $x < 0$ and for $x > 0$. Write the corresponding linear combinations of M and U as $\alpha_1 M + \beta_1 U$ and $\alpha_2 M + \beta_2 U$. If these combinations are constructed from the same u their values and first derivatives must have the same limit as $x \to 0$. To guarantee that $u(x)$ is positive and increasing for $x < 0$, α_1 must be zero. Therefore,

$$u(x) = \beta_1 U\left(\frac{r}{2\rho}, \frac{1}{2}, \frac{\rho}{\sigma_g^2}x^2\right) \quad \text{if } x \le 0.$$

From the definition of the two Kummer functions, one can show that

$$x \to 0-, \; u(x) \to \frac{\beta_1 \pi}{\Gamma(\frac{1}{2}+\frac{r}{2\rho})\Gamma(\frac{1}{2})}, \qquad u'(x) \to \frac{\beta_1 \pi \sqrt{\rho}}{\sigma_g \Gamma(\frac{r}{2\rho})\Gamma(\frac{3}{2})}$$

$$x \to 0+, \; u(x) \to \alpha_2 + \frac{\beta_2 \pi}{\Gamma(\frac{1}{2}+\frac{r}{2\rho})\Gamma(\frac{1}{2})}, \; u'(x) \to -\frac{\beta_2 \pi \sqrt{\rho}}{\sigma_g \Gamma(\frac{r}{2\rho})\Gamma(\frac{3}{2})}$$

By matching the values and first order derivatives of $u(x)$ from the two sides of $x = 0$, we have

$$\beta_2 = -\beta_1, \quad \alpha_2 = \frac{2\beta_1 \pi}{\Gamma\left(\frac{1}{2}+\frac{r}{2\rho}\right)\Gamma\left(\frac{1}{2}\right)}.$$

The function h is a solution to equation (26) that satisfies

$$h(0) = \frac{\pi}{\Gamma\left(\frac{1}{2}+\frac{r}{2\rho}\right)\Gamma\left(\frac{1}{2}\right)} > 0,$$

and $h(-\infty) = 0$. Equation (26) guarantees that at any critical point where $h < 0$, h has a maximum, and at any critical point where $h > 0$ it has a minimum. Hence h is strictly positive and increasing in $(-\infty, 0)$. $\square$

Further properties of the function h are summarized in the following Lemma.

Lemma 1. *For each $x \in R$, $h(x) > 0$, $h'(x) > 0$, $h''(x) > 0$, $h'''(x) > 0$, $\lim_{x\to-\infty} h(x) = 0$, and $\lim_{x\to-\infty} h'(x) = 0$.*

Since q must be positive and increasing in $(-\infty, k^*)$, it follows from Proposition 4 that

$$q(x) = \begin{cases} \beta_1 h(x), & \text{for } x < k^* \\ \frac{x}{r+\lambda} + \beta_1 h(-x) - c, & \text{for } x \ge k^*. \end{cases} \tag{29}$$

Since q is continuous and continuously differentiable at k^*,

$$\beta_1 h(k^*) - \frac{k^*}{r+\lambda} - \beta_1 h(-k^*) + c = 0,$$

$$\beta_1 h'(k^*) + \beta_1 h'(-k^*) - \frac{1}{r+\lambda} = 0.$$

These equations imply that

$$\beta_1 = \frac{1}{(h'(k^*) + h'(-k^*))(r+\lambda)}, \tag{30}$$

and k^* satisfies

$$[k^* - c(r+\lambda)](h'(k^*) + h'(-k^*)) - h(k^*) + h(-k^*) = 0. \tag{31}$$

The next proposition shows that for each c, there exists a unique pair (k^*, β_1) that solves equations (30) and (31). The smooth pasting conditions are sufficient to determine the function q and the "trading point" k^*.

Proposition 5. *For each trading cost $c \geq 0$, there exists a unique k^* that solves (31). If $c = 0$ then $k^* = 0$. If $c > 0$, $k^* > c(r+\lambda)$.*

Proof: Let $l(k) = [k - c(r+\lambda)](h'(k) + h'(-k)) - h(k) + h(-k)$.

If $c = 0$, $l(0) = 0$, and $l'(k) = k[h''(k) - h''(-k)] > 0$, for all $k \neq 0$. Therefore $k^* = 0$ is the only root of $l(k) = 0$.

If $c > 0$, then $l(k) < 0$, for all $k \in [0, c(r+\lambda)]$. Since h'' and h''' are increasing (Lemma 1), for all $k > c(r+\lambda)$

$$l'(k) = [k - c(r+\lambda)][h''(k) - h''(-k)] > 0,$$
$$l''(k) = h''(k) - h''(-k) + [k - c(r+\lambda)][h'''(k) - h'''(-k)] > 0.$$

Therefore $l(k) = 0$ has a unique solution $k^* > c(r+\lambda)$. $\square$

The next proposition establishes that the function q described by equation (29), with β_1 and k^* given by (30) and (31), solves (23). The proof consists of two parts. First, it is established that (24) and (25) hold and that q' is bounded. A standard argument, e.g. Kobila (1993) or Scheinkman and Zariphopoulou (2001), is then used to show that in fact q solves equation (23).

Proposition 6. *The function q constructed above is an equilibrium option value function. The optimal policy consists of exercising immediately if $g^o > k^*$, otherwise wait until the first time in which $g^o \geq k^*$.*

Proof: Let

$$b \equiv q(-k^*) = \frac{1}{(r+\lambda)} \frac{h(-k^*)}{(h'(k^*) + h'(-k^*))}, \tag{32}$$

then equation (29) implies

$$q(-x) = \begin{cases} \frac{b}{h(-k^*)} h(-x) & \text{for } x > -k^* \\ \frac{-x}{r+\lambda} + \frac{b}{h(-k^*)} h(x) - c & \text{for } x \le -k^*. \end{cases}$$

Equation (24) may be rewritten as $U(x) = q(x) - \frac{x}{r+\lambda} - q(-x) + c \ge 0$. A simple calculation shows that

$$U(x) = \begin{cases} 2c & \text{for } x < -k^* \\ \frac{-x}{r+\lambda} + \frac{b}{h(-k^*)} [h(x) - h(-x)] + c & \text{for } -k^* \le x \le k^* \\ 0 & \text{for } x > k^* \end{cases}$$

Thus, $U''(x) = \frac{b}{h(-k^*)} [h''(x) - h''(-x)]$, $-k^* \le x \le k^*$. Lemma 1 guarantees $U''(x) > 0$ for $0 < x < k^*$, and $U''(x) < 0$ for $-k^* < x < 0$. Since $U'(k^*) = 0$, $U'(x) < 0$ for $0 < x < k^*$. On the other hand, $U'(-k^*) = 0$, so $U'(x) < 0$ for $-k^* < x < 0$. Therefore $U(x)$ is monotonically decreasing for $-k^* < x < k^*$. Since $U(-k^*) = 2c > 0$ and $U(k^*) = 0$, $U(x) > 0$ for $-k^* < x < k^*$. Hence equation (24) holds in $(-\infty, k^*)$.

Equation (25) holds by construction in the region $x \le k^*$. Therefore it is sufficient to show that equation (25) is valid for $x \ge k^*$. In this region $q(x) = \frac{x}{r+\lambda} + \frac{b}{h(-k^*)} h(-x) - c$, thus $q'(x) = \frac{1}{r+\lambda} - \frac{b}{h(-k^*)} h'(-x)$ and $q''(x) = \frac{b}{h(-k^*)} h''(-x)$. Hence,

$$\frac{1}{2} \sigma_g^2 q''(x) - \rho x q'(x) - r q(x)$$

$$= \frac{b}{h(-k^*)} \left[\frac{1}{2} \sigma_g^2 h''(-x) + \rho x h'(-x) - r h(-x) \right] - \frac{r+\rho}{r+\lambda} x + rc$$

$$= -\frac{r+\rho}{r+\lambda} x + rc \le -(r+\rho)c + rc = -\rho c < 0$$

where the inequality comes from the fact that $x \ge k^* > (r+\lambda)c$ (see Proposition 5.)

Also q has an increasing derivative in $(-\infty, k^*)$ and has a derivative bounded in absolute value by $\frac{1}{r+\lambda}$ in (k^*, ∞). Hence q' is bounded.

If τ is any stopping time, the version of Ito's lemma for twice differentiable functions with absolutely continuous first derivatives (e.g. Revuz and Yor (1999), Chapter VI) implies that

$$e^{-r\tau} q(g_\tau^o) = q(g_0^o) + \int_0^\tau \left[\frac{1}{2} \sigma_g^2 q''(g_s^o) - \rho g_s^o q'(g_s^o) - r q(g_s^o) \right] ds + \int_0^\tau \sigma_g q'(g_s^o) dW_s$$

Equation (25) states that the first integral is non positive, while the bound on q' guarantees that the second integral is a martingale. Using equation (24) we obtain,

$$\mathbb{E}^o \left\{ e^{-r\tau} \left[\frac{g_\tau^o}{r+\lambda} + q(-g_\tau^o) - c \right] \right\} \leq \mathbb{E}^o \left[e^{-r\tau} q(g_\tau^o) \right] \leq q(g_0^o).$$

This shows that no policy can yield more than $q(x)$.

Now consider the stopping time $\tau = \inf\{t : g_t^o \geq k^*\}$. Such τ is finite with probability one, and g_s^o is in the continuation region for $s < \tau$. Using exactly the same reasoning as above, but recalling that in the continuation region (25) holds with equality we obtain

$$q(g^o) = \mathbb{E}^o \left\{ e^{-r\tau} \left[\frac{g_\tau^o}{r+\lambda} + q(-g_\tau^o) - c \right] \right\}.$$

□

It is a consequence of Proposition 6 that the process g^o will have values in $(-\infty, k^*)$. The value k^* acts as a barrier, and when g^o reaches k^*, a trade occurs, the owner's group switches and the process is restarted at $-k^*$. $q(g^o)$ is the difference between the current owner's demand price and his fundamental valuation and can be legitimately called a "bubble".

The model determines the magnitude of the bubble and the duration between trades. The magnitude of the bubble can be measured by b, as in equation (32), the value of the resale option when a trade occurs.

If we write h in equation (27) as a function of both x and r, Scheinkman and Xiong (2003) show that the expected duration between two trades is given by

$$\mathbb{E}[\tau] = -\frac{\partial}{\partial r} \left[\frac{h(-k^*, r)}{h(k^*, r)} \right] \Bigg|_{r=0}.$$

When the trading cost is zero ($c = 0$), the trading barrier $k^* = 0$, which implies that the expected duration between trades is also zero. This is due to the fact that once a Brownian motion hits a point, it will hits the same point for infinite many times for any given period immediately afterward.

Various comparative statics results are described in Scheinkman and Xiong (2003). As investors become more overconfident (ϕ increases), the volatility parameter of the difference in beliefs (σ_g) increases, resulting in more trades (shorter duration between trades) and a larger price bubble. As the signals become more informative, the mean reverting speed of the difference in beliefs (ρ) becomes larger, resulting in shorter duration between trades and a larger price bubble. As the trading cost c increases from zero, the duration between trades increases and the magnitude of the bubble is reduced, in a continuous manner. In particular the model predicts large trading volume in markets with sufficiently small transaction costs. The effect from an increase in trading cost is most dramatic for the duration between trades but the effects on the bubble are modest. This suggests that while a tax on transactions (Tobin Tax) would have some effect on trading volume, it would have a small effect on the size of a bubble.

In a risk-neutral world, one may consider several assets and analyze the equilibrium in each market independently. In this way the comparative statics properties described in the previous paragraph can be translated into results about correlations among equilibrium variables in the different markets. Thus this model is potentially capable of explaining the observed cross-sectional correlation between market/book ratio and turnover for U.S. stocks in the period of 1996-2000 as documented by Cochrane (2002). It is also able to account for the analogous cross-sectional correlation that has been found by Mei, Scheinkman, and Xiong (2003) between the price ratio of China's A shares to B shares and turnover.

7 Other Related Models

In this section, we discuss other models that have been proposed in the finance literature to study price bubbles and effects of heterogeneous beliefs on trading and asset prices.

7.1 Models on Rational Bubbles

There has been a large literature studying rational bubbles including Blanchard and Watson (1982), Santos and Woodford (1997) and others. In these papers, all agents have identical rational expectations, and the asset prices can be decomposed into two parts, a fundamental component and a bubble component which is expected to grow at a rate equal to the risk free rate. In fact, such a rational bubble component can also be built into the models discussed in Sections 3 and 6. Given a price process p_t that satisfies equation (1) and m_t, a non-negative $\mathcal{F}_t$-martingale, $\hat{p}_t = p_t + \gamma^{-t} m_t$ also satisfies the equation (1). A corresponding remark holds for equation (20) that describes the equilibrium of the model based on overconfidence. We have ruled out such rational bubbles in our previous discussion.

Campbell, Lo, and MacKinlay (1997, pages 258-260) provide a detailed discussion on the properties of rational bubbles. To make this type of bubbles sustainable, the asset must have a potentially infinite maturity. Another property of rational bubbles is that the asset price grows on average without bounds. In addition, the models of rational bubbles provide no explanation for the increase in trading that is often observed during historical bubble episodes.

As we discussed in the previous sections, the resale option provides an alternative way to analyze asset price bubbles, since its value is determined by heterogeneous beliefs among investors which is a variable orthogonal to the fundamental value of the asset. In contrast to rational bubbles, the resale option component does not need to be explosive although its magnitude could be very significant due to its recursive structure. Consequently, in the model exposited in Section 6, variables such as the asset price in equation (22) and its change have stationary distributions. In addition, the size of the bubble generated from the resale option is positively correlated with trading volume, a property that is apparent in several actual episodes of price bubbles.

Finally, the resale option still exists for an asset with a fixed finite maturity, which is not possible for rational bubbles, that depend on a potentially infinite life. It should be apparent from the analysis in Section 6 that one can, in principle, treat an asset with a fixed terminal date. Equations (20) to (21) would apply with the obvious changes to account for the finite horizon. However, the option value q will now depend on the remaining life of the asset, introducing another dimension to the optimal stopping problem. The infinite horizon problem is stationary, greatly reducing the mathematical difficulty.

7.2 Trading Caused by Heterogeneous Beliefs

Several other models have been proposed to analyze asset trading based on heterogeneous beliefs, such as Varian (1989), Harris and Raviv (1993), Kandel and Pearson (1995), Kyle and Lin (2003), and Cao and Ou-Yang (2004).

Harris and Raviv (1993) analyze a model with two groups of risk neutral speculators who trade a risky asset at dates $t = 1, 2, ..., T - 1$. The final liquidation value of the asset at T is random and can be either high (H) or low (L). At each date a public noisy signal is revealed to the two groups of speculators, who assign different probability distributions for the signal and therefore would hold heterogeneous expectations of the final liquidation value of the asset. This mechanism for generating heterogeneous beliefs is similar to what we discuss in Section 5 with investor overconfidence, but with a different random process and noise distribution. The beliefs of the two groups are denoted by $\pi_H^j(t)$, the posterior probability that the final liquidation value will be high ($j = 1, 2$).

To analyze trading between the two groups of speculators resulting from difference in their beliefs, the model also imposes a short-sales constraint, and that one group has sufficient market power to offer a price on a take-it-or-leave-it basis to the other group. As a result, the trading price of the asset will always equal the reservation price of the other group (the "price-taking" group). The existence of such a price-taking group, is not a natural assumption and rules out the presence of a bubble in the observed trading prices, but greatly simplifies the recursive structure in the determination of equilibrium prices that arises in the models of Harrison and Kreps (1978) and Scheinkman and Xiong (2003). Harris and Raviv demonstrate in their model that trade will only occur when the two groups switch side (π_H^1 and π_H^2 flip order), and that there is a positive correlation between trading volume and absolute price changes (but not necessarily the level of prices.)

7.3 Effects of Heterogeneous Beliefs on Risk Premia

Lucas (1978) provides a simple and elegant equilibrium model to analyze the relation between equity premium and aggregate consumption. Consider an economy with a representative agent and an infinite time horizon ($t = 1, 2, 3, ...$). The agent maximizes his lifetime utility from consumption:

$$\mathbb{E}\left[\sum_{t=0}^{\infty} \beta^t u(c_t)\right]$$

where β, ($0 < \beta < 1$), is the agent's subjective discount factor, c_t is the consumption in period t, and the agent's utility from consumption $u(c)$ is often assumed to have a power form: $u(c) = \frac{1}{1-\gamma} c^{1-\gamma}$. There are two assets – one is a risk-free asset and the other is a claim to aggregate endowment in the economy c_t ($t=1, 2, ...$). The risk-free asset is in zero net supply. The agent's marginal rate of consumption provides his pricing kernel for future cashflow:

$$m_{t+1} = \beta \frac{u'(c_{t+1})}{u'(c_t)}.$$

More specifically, a random cashflow of x_{t+1} at $t + 1$ is worth

$$p_t = \mathbb{E}_t\left(m_{t+1} x_{t+1}\right)$$

at period t. Thus, the riskfree rate is given by

$$R^f = 1/\mathbb{E}(m_t),$$

and the return on a risky asset R^i should satisfy

$$\mathbb{E}(m_t R_t^i) = 1.$$

By decomposing the last equation, we obtain $\mathbb{E}(m_t)\mathbb{E}(R_t^i) + cov(m_t, R_t^i) = 1$. Therefore,

$$\mathbb{E}(R_t^i) - R^f = -\frac{cov(m_t, R_t^i)}{\mathbb{E}(m_t)}.$$

This relation implies an upper bound on any risky asset's Sharpe ratio:

$$\left|\frac{\mathbb{E}(R_t^i) - R^f}{\sigma(R_t^i)}\right| = \left|-\frac{cov(m_t, R_t^i)}{\mathbb{E}(m_t)\sigma(R_t^i)}\right| \leq \frac{\sigma(m_t)}{\mathbb{E}(m_t)}. \tag{33}$$

The return from the stock market portfolio provides a way to calibrate the relationship between the equity risk premium and the pricing kernel implied by aggregate consumption. In the case of the market portfolio R^{mv}, the relation in formula (33) holds exactly, and a power utility function would imply

$$\left| \frac{\mathbb{E}(R^{mv}) - R^f}{\sigma(R^{mv})} \right| = \frac{\sigma(m_{t+1})}{\mathbb{E}(m_{t+1})} = \frac{\sigma[(c_{t+1}/c_t)^{-\gamma}]}{\mathbb{E}[(c_{t+1}/c_t)^{-\gamma}]} \approx \gamma\sigma(\Delta \ln c),$$

where R^{mv} is the return on the market portfolio and R^f the risk-free rate. As summarized by Cochrane (2001, page 23), "over the last 50 years in the United States, real stock returns have averaged 9% with a standard deviation of about 16%, while the real return on treasury bills has been about 1%. Thus, the historical annual market Sharpe ratio has been about 0.5. Aggregate nondurable and services consumption growth had a mean and standard deviation of about 1%. We can only reconcile these facts if investors have a risk-aversion coefficient of 50," which is much higher than what economists have usually assumed. This is called an "equity premium puzzle" by Mehra and Prescott (1985).

The equity premium puzzle has motivated a large number of papers since the mid eighties. The objective of these papers is to present modifications of the standard model that justify a Sharpe ratio that is considerably higher than the one implied by the standard model. Part of this literature has used heterogenous beliefs. Williams (1977), Abel (1990) Detemple and Murthy (1994), Zapatero (1998), and Basak (2000), among others, analyze the effects of heterogenous beliefs on the equilibrium risk premium and interest rates.

Detemple and Murthy (1994) consider a continuous time production economy of Cox, Ingersoll and Ross (1985) with a Brownian uncertainty structure. They assume a risky production technology with a return that is invariant to scale and that has an unobservable mean. In addition, there are two groups of risk-averse agents who have heterogeneous prior beliefs on the mean return of the production technology. There are two assets - a claim to the aggregate output and a risk-less asset in zero net supply. Heterogeneous beliefs motivate agents in one group to borrow from agents in the other group using the risk-less asset. In equilibrium, the interest rate and risk premium on risky securities are determined by the wealth distribution across the two groups and each groups' estimates of the production growth rate. Zapatero (1998) considers a similar model in a pure exchange economy of Lucas (1978). He shows that heterogeneous beliefs can lead to a reduced risk-free interest rate in equilibrium from an otherwise identical economy with homogeneous beliefs. This lower interest rate induces a higher excess return. Basak (2000), using a model similar to Detemple and Murthy (1994), further shows that heterogeneous beliefs could add risk to investors' financial investment and therefore may lead to a greater equity premium than an economy with homogeneous beliefs.

8 Survival of Traders with Incorrect Beliefs

In the earlier sections, we discussed various effects that can arise when traders with heterogeneous beliefs interact with each other in an asset market. Some traders may have incorrect beliefs which are generated from incorrect prior beliefs or from incorrect information processing rules, while some others may be smarter and have beliefs that are closer to the objective ones. In such an environment, an important question is whether traders with incorrect beliefs will lose money trading with smarter traders and eventually disappear from the market.

There has been a long debate on this fundamental issue. Friedman (1953) argues that irrational traders who use wrong beliefs cannot survive in a competitive market, since they will eventually lose their wealth to rational traders in the long run. More recently, De Long, Shleifer, Summers and Waldman (1991) suggest that traders with wrong beliefs may survive in the long-run since they may hold a portfolio with excessive risk but also higher expected return and therefore their wealth can eventually outgrow that of rational traders. Several recent studies have been devoted to analyze this issue, *e.g.* Sandroni (2000), Blume and Easley (2001), and Kogan, Ross, Wang and Westerfield (2004). However, the answer is still inconclusive. Depending on the model assumptions, different results have been found in these studies. Here, we discuss a model from Kogan *et al.* (2004) as an example.

Kogan *et al.* consider an economy that has a finite horizon and evolves in continuous time. Uncertainty is described by a one-dimensional, standard Brownian motion Z_t for $0 \leq t \leq T$, which is defined on a complete probability space (Ω, F, P). There is a single share of a risky asset in the economy, the stock, which pays a terminal dividend payment D_T at time T, determined by process

$$dD_t = D_t(\mu dt + \sigma dZ_t)$$

where $D_0 = 1$ and $\sigma > 0$. There is also a zero coupon bond available in zero net supply. Each unit of the bond makes a sure payment of one at time T. We use the risk-free bond as the numeraire and denote the price of the stock at time t by S_t.

There are two types of competitive traders in the economy. Each set corresponds to half of the total number of traders. At time zero, all traders have an equal endowment of the stock (that we normalize as 1/2) and have zero units of the bond. Traders differ in their beliefs of the drift parameters of the dividend process. One set of traders, the rational traders, knows the true probability measure P, while the other set of traders, the irrational traders, believes incorrectly that the probability measure is Q, under which

$$Z_t^Q = Z_t - \int_0^t (\sigma\eta)ds$$

is a standard Brownian motion. Hence, for an irrational trader

$$dD_t = D_t[(\mu + \sigma^2\eta)dt + \sigma dZ_t^Q],$$

where Z_t^Q is a Brownian motion. The constant η parameterizes the irrationality of these traders. When η is positive, the irrational traders are optimistic about the prospects of the evolution of the dividend and overestimates the rate of growth of the dividend. Conversely, a negative η corresponds to a pessimistic view. According to this specification, the two types of traders do not update their beliefs of the drift rate, and one group will always stay as the optimistic one. This structure is different from the one introduced in Section 5, where agents' beliefs fluctuate with the flow of information.

Rational traders and irrational traders consume only at time T, and they share the same constant relative risk aversion. They can trade continuously before time T and would aim to maximize their utility based on their own probability measure. Thus, a rational trader's objective is to maximize

$$\mathbb{E}_0^P \left[\frac{1}{1-\gamma} C_{r,T}^{1-\gamma} \right]$$

where $C_{r,T}$ is the consumption of the typical rational trader at time T. Correspondingly an irrational trader's objective is to maximize

$$\mathbb{E}_0^Q \left[\frac{1}{1-\gamma} C_{n,T}^{1-\gamma} \right]$$

where $C_{n,T}$ is the consumption of an irrational trader at time T. In addition, the model assumes that there is no trading cost and short-sales of shares are allowed.

The probability measure used by irrational traders Q is absolutely continuous with respect to the objective measure P. Thus, the expectation using probability measure Q can be transformed into an expectation using probability measure P through the density (Radon-Nikodym derivative) of the probability measure Q with respect to P:

$$\xi_t \equiv \left. \frac{dQ}{dP} \right|_t = e^{-\frac{1}{2}\eta^2\sigma^2 t + \eta\sigma Z_t}. \tag{34}$$

Hence the irrational trader maximizes

$$\mathbb{E}_0^Q \left[\frac{1}{1-\gamma} C_{n,T}^{1-\gamma} \right] = \mathbb{E}_0^P \left[\xi_t \frac{1}{1-\gamma} C_{n,T}^{1-\gamma} \right].$$

Effectively, an irrational trader is maximizing a state dependent utility function, $\xi_t \frac{1}{1-\gamma} C_{n,T}^{1-\gamma}$, under the true probability measure P.

The competitive equilibrium of the economy described above can be solved analytically. Since there is only one source of uncertainty in the economy and there are no trading cost or short-sales constraints, it is expected that continuous trading on the stock and the bond is sufficient to dynamically complete the markets. Since complete markets yield Pareto efficient allocations, it is natural to first examine the set of Pareto efficient allocations and then show that with the corresponding support prices the two assets actually yield complete markets. This is the route chosen by Kogan *et al.* First they examine the set of allocations that solve:

$$\max \quad \frac{1}{1-\gamma} C_{r,T}^{1-\gamma} + b\xi_t \frac{1}{1-\gamma} C_{n,T}^{1-\gamma}$$
$$\text{s.t.} \quad C_{r,T} + C_{n,T} = D_T$$

where b is the ratio of the utility weights for the two types of traders. The optimal allocations are

$$C_{r,T} = \frac{1}{1 + (b\xi_t)^{1/\gamma}} D_T, \quad C_{n,T} = \frac{(b\xi_t)^{1/\gamma}}{1 + (b\xi_t)^{1/\gamma}} D_T. \tag{35}$$

Based on the traders' marginal utilities, one can derive the supporting state price density given the information at time t as

$$\frac{(1 + (b\xi_t)^{1/\gamma})^\gamma D_T^{-\gamma}}{\mathbb{E}_t[(1 + (b\xi_t)^{1/\gamma})^\gamma D_T^{-\gamma}]}.$$

This state price density allows us to price any contingent claim in the economy, such as the dividend from the stock that is paid at T and the traders' consumption allocations. Kogan et al. show that the equilibrium stock price is given by

$$P_t = \frac{\mathbb{E}_t[(1 + (b\xi_t)^{1/\gamma})^\gamma D_T^{-\gamma} Z_T]}{\mathbb{E}_t[(1 + (b\xi_t)^{1/\gamma})^\gamma D_T^{-\gamma}]}.$$

The utility weights between the two groups of traders, b, is determined so that the budget constraints for the two traders are satisfied. Since the traders start with the same endowments at time $t = 0$, the values of their consumption allocations at time $t = T$ should have the same value at $t = 0$. This gives an equation to identify $b = e^{(\gamma - 1)\eta\sigma^2 T}$. With these calculations, Kogan et al. further show that in fact the stock and the bond dynamically complete markets.

Given the equilibrium consumptions for the two types, Kogan et al. use the asymptotic properties of the two consumptions to discuss the survival of traders. More specifically, they say that irrational traders experience relative extinction in the long-run if

$$\lim_{T \to \infty} \frac{C_{n,T}}{C_{r,T}} = 0 \quad \text{a.s.}$$

The relative extinction of rational traders is defined symmetrically. Kogan et al. use the expression for consumption allocations in equation (35) above, the formula for the Radon-Nykodim derivative of Q with respect to P (equation (34)), and the strong law of large numbers for Brownian motion to establish:

Proposition 7. *Suppose $\eta \neq 0$. Let $\eta^* = 2(\eta - 1)$. For $\gamma > 1$ and $\eta \neq \eta^*$, typically only one type of traders survives. Furthermore:*

Case 1: $\eta < 0$ (pessimistic irrational traders), the rational traders survive.

Case 2: $0 < \eta < \eta^*$ (modestly optimistic irrational traders), the irrational traders survive.

Case 3: $\eta > \eta^*$ (strongly optimistic irrational traders), the rational trader survive.

Interestingly, the irrational trader could survive in the long-run, as in case 2 with a modestly optimistic belief. In such a case, the irrational trader takes more risk and therefore able to outgrow the rational trader. This effect has been pointed out by De Long, Shleifer, Summers and Waldman (1991) using a partial equilibrium model in which irrational traders' trading has no impact on the price. Once their price impact is taken into account as in this model, Kogan *et al.* show that the survival of irrational traders becomes less likely, although still possible. Kogan *et al.* also discuss the price impact of the irrational traders in the long run. They also show that even if the irrational traders do not survive in the long run, irrational traders can still have a persistent impact on the stock price since they are willing to bet strongly on some small probability events when their probability assessment of these events differ greatly from that of rational traders.

9 Some Remaining Problems

Many interesting problems remain in modelling the effects of heterogeneous beliefs on financial markets. Rather than present a long list of unsolved questions we select a few problems that are particularly related to the models discussed above.

The model in Section 5 specifies a normal process for the unobservable fundamental variable. This assumption allows one to use the standard linear filtering technique to analyze the agents' learning process. It also generates a particularly tractable form for the process determining the difference in beliefs. However, a lognormal process would help capture the limited liability feature of many assets such as stocks and bonds. A model of this kind would have to contemplate the difficulties involved in nonlinear filtering and optimal stopping with multiple state variables. Panageas (2003) provides an attempt along this line to analyze the effect of stock market bubbles on firm investment. Multi-dimensional stopping time problems will also result from the introduction of multiple classes of agents.

As discussed above, heterogeneous beliefs should be able to support a bubble even if an asset has a finite life in the context of the model in Section 6. However, in this case, the optimal stopping problem that defines the equilibrium would involve an extra dimension. Tackling this additional complication would allow one to analyze the exact impact of horizon on speculative trading and price bubbles, in addition to establishing rigorously that a bubble generated by heterogeneous beliefs can prevail in assets with a finite life.

The models in Sections 3 and 6 ignore risk aversion of agents by assuming risk neutrality. When agents are risk averse, subtle effects arise in equilibrium. As shown by Hong, Scheinkman and Xiong (2003) in a model with finite periods and myopic risk averse agents, the payoff from the resale option appears to be similar to a standard call option with a strike price determined by the asset float (number of tradable shares) and the risk bearing capacity of investors. The model suggests that asset float could have an important effect on price bubble and trading volume. A more elaborate model remains to be developed to analyze the effects of risk aversion and asset float when agents have heterogeneous beliefs and short-sales of assets are constrained.

In the model in Section 6 the level of overconfidence is a constant. It is perhaps more realistic to assume that the level of investor's overconfidence fluctuates as investors learn of their own ability to forecast. Gervais and Odean (2001) analyze a model in which a trader infers his ability to forecast from his past successes and failures. They show that overconfidence can be generated in this learning process if the trader attributes success to himself and failure to external forces, an attribution bias that has been documented in psychological studies of human behavior. It remains an open problem to analyze the dynamics of heterogeneous beliefs in the presence of endogeneously generated overconfidence and the market equilibrium that would obtain.

References

Abel, Andrew (1990), Asset prices under heterogeneous beliefs: implications for the equity premium, Working paper, Wharton School, University of Pennsylvania.

Abramowitz, Milton and Irene Stegun (1964), *Handbook of Mathematical Functions*, Dover Publications, New York.

Alpert, Marc and Howard Raiffa (1982), A progress report on the training of probability assessors, in Daniel Kahneman, Paul Slovic, and Amos Tversky, ed.: *Judgement under Uncertainty: Heuristics and Biases*, Cambridge University Press, Cambridge.

Barberis, Nicholas and Richard Thaler (2003), A survey of behavioral finance, in George Constantinides, Milton Harris and Rene Stulz (ed.), *Handbook of the Economics of Finance*, North-Holland.

Basak, Suleyman (2000), A model of dynamic equilibrium asset pricing with heterogeneous beliefs and extraneous beliefs, *Journal of Economic Dynamics and Control* 24, 63-95.

Bernardo, Antonio, and Ivo Welch (2001), On the evolution of overconfidence and entrepreneurs, *Journal of Economics and Management Strategy* 10, 301-330.

Biais, Bruno and Peter Bossaerts (1998), Asset prices and trading volume in a beauty contest, *Review of Economic Studies* 65, 307-340.

Blanchard, Olivier and Mark Watson (1982), Bubbles, rational expectations and financial markets, in P. Wachtel (ed.) *Crisis in the Economic and Financial Structure: Bubbles, Bursts, and Shocks,* Lexington Press, Lexington, MA.

Blume, Lawrence and David Easley (2001), If you're so smart, why aren't you rich? Belief selection in complete and incomplete markets, Cowles Foundation Discussion Papers, 1319.

Brav, Alon and J.B. Heaton (2002), Competing Theories of Financial Anomalies, *Review of Financial Studies* 15, 575-606.

Brenner, Lyle, Derek Koehler, Varda Liberman, and Amos Tversky (1996), Over-confidence in probability and frequency judgements, *Organizational Behavioral and Human Decision Processes* 65 (3), 212-219.

Brunnermeier, Markus and Jonathan Parker (2003), Optimal expectations, Working paper, Princeton University.

Camerer, C. (1995), Individual decision making, in John Kagel and Alvin Roth (ed.) *The Handbook of Experimental Economics,* Princeton University Press, Princeton, NJ.

Campbell, John, Andrew Lo, and Craig MacKinlay (1997), *The Econometrics of Financial Markets*, Priceton University Press, Princeton, NJ.

Cao, Henry and Hui Ou-Yang (2004), Differences of opinion and options, Working paper, University of North Carolina and Duke University.

Chen, Joseph, Harrison Hong and Jeremy Stein (2002), Breadth of ownership and stock returns, *Journal of Financial Economics* 66, 171-205.

Cochrane, John (2001), *Asset Pricing*, Princeton University Press, Princeton, NJ.

Cochrane, John (2002), Stocks as money: convenience yield and the tech-stock bubble, NBER Working Paper 8987.

Cox, John, Jonathan Ingersoll and Stephen Ross (1985), An intertemporal general equilibrium model of asset prices, *Econometrica* 53, 363-384.

De Long, Bradford, Andrei Shleifer, Lawrence Summers, and Robert Waldman (1991), The survival of noise traders in financial markets, *Journal of Business* 64, 1-19.

Detemple, Jerome and Shashidhar Murthy (1994), Intertemporal asset pricing with heterogeneous beliefs, *Journal of Economic Theory* 62, 294-320.

Diamond, Douglas and Robert Verrecchia (1987), Constraints on short-selling and asset price adjsutment to private information, *Journal of Financial Econonomics* 18, 277-311.

Friedman, Milton (1953), The case for flexible exchange rates, *Essays in Positive Economics,* University of Chicago Press.

Gallmeyer, Michael and Burton Hollifield (2004), An examination of heterogeneous beliefs with a short sale constraint, Working paper, Carnegie Mellon University.

Gervais, Simon and Terrance Odean (2001), Learning to be overconfident, *Review of Financial Studies* 14, 1-27.

Grossman, Sanford and Joseph Stiglitz (1980), On the impossibility of informationally efficient markets, *American Economic Review* 70, 393-408.

Harris, Milton and Artur Raviv (1993), Differences of opinion make a horse race, *Review of Financial Studies* 6, 473-506.

Harrison, Michael and David Kreps (1978), Speculative investor behavior in a stock market with heterogeneous expectations, *Quarterly Journal of Economics* 92, 323-336.

Hirshleifer, David (2001), Investor psychology and asset pricing, *Journal of Finance* 56, 1533-1597.

Hong, Harrison, José Scheinkman, and Wei Xiong (2003), Asset float and speculative bubbles, Working paper, Princeton University.

Jarrow, Robert (1980), Heterogeneous expectations, restrictions on short-sales, and equilibrium asset prices, *Journal of Finance* 35, 1105-1113.

Kandel, Eugene and Neil Pearson (1995), Differential interpretation of public signals and trade in speculative markets, *Journal of Political Economy* 103, 831-872.

Kogan, Leonid, Stephen Ross, Jiang Wang, and Mark Westerfield (2004), The price impact and survival of irrational traders, Working paper, MIT.

Kyle, Albert (1985), Continuous auctions and insider trading, *Econometrica* 53, 1315-1336.

Kyle, Albert and Tao Lin (2003), Continuous trading with heterogeneous beliefs and no noise trading, Working paper, Duke University.

Kyle, Albert and Albert Wang (1997), Speculation duopoly with agreement to disagree: Can overconfidence survive the market test? *Journal of Finance* 52, 2073-2090.

Lintner, John (1969), The aggregation of investor's diverse judgements and preferences in purely competitive security markets, *Journal of Financial and Quantitative Analysis* 4, 347-400.

Liptser, R. S. and A. N. Shiryayev (1977), *Statistics of Random Processes,* Spring-Verlag, New York.

Lucas, Robert (1978), Asset prices in an exchange economy, *Econometrica* 46, 1429-1446.

Mehra, Rajnish and Edward Prescott (1985), The equity premium puzzle, *Journal of Monetary Economics* 15, 145-161.

Mei, Jianping, Jose Scheinkman and Wei Xiong (2003), Speculative trading and stock prices: An analysis of Chinese A-B share premia, Working paper, Princeton University.

Milgrom, Paul and Nancy Stokey (1982), Information, trade and common knowledge, *Journal of Economic Theory* 12, 112-128.

Miller, Edward (1977), Risk, uncertainty and divergence of opinion, *Journal of Finance* 32, 1151-1168.

Morris, Stephen (1996), Speculative investor behavior and learning, *Quarterly Journal of Economics* 110, 1111-1133.

Odean, Terrance (1998), Volume, volatility, price, and profit when all traders are above average, *Journal of Finance* 53, 1887-1934.

Ofek, Eli and Matthew Richardson (2003), Dotcom mania: The rise and fall of internet stock prices, *Journal of Finance* 58, 1113-1137.

Panageas, Stavros (2003), Speculation, overpricing and investment – theory and empirical evidence, Job market paper, MIT.

Revuz, Daniel and Marc Yor (1999), *Continuous Martingales and Brownian Motion,* Springer, New York.

Rogers, L. C. G. and David Williams (1987), *Diffusions, Markov Processes, and Martingales, Volume 2: Ito Calculus,* John Wiley & Sons, New York.

Sandroni, Alvaro (2000), Do markets favor agents able to make accurate predictions? *Econometrica* 68, 1303-1341.

Santos, Manuel, and Michael Woodford (1997), Rational asset pricing bubbles, *Econometrica* 65, 19-57.

Scheinkman, Jose and Thaleia Zariphopoulou (2001), Optimal environmental management in the presence of irreversibilities, *Journal of Economic Theory* 96, 180-207.

Scheinkman, Jose and Wei Xiong (2003), Overconfidence and Speculative Bubbles, *Journal of Political Economy* 111, 1183-1219.

Tirole, Jean (1982), On the possibility of speculation under rational expecations, *Econometrica* 50, 1163-1181.

Varian, Hal (1989), Differences of opinion in financial markets, in Courtenay Stone (ed.) *Financial Risk: Theory, Evidence and Implications*, Kluwer Academic Publishers, Boston.

Wang, Jiang (1993), A model of intertemporal asset prices under a asymmetric information, *Review of Economic Studies* 60, 249-282.

Williams, Joseph (1977), Capital asset prices with heterogeneous beliefs, *Journal of Financial Economics* 5, 219-239.

Zapatero, Fernando (1998), Effects of financial innovations on market volatility when beliefs are heterogeneous, *Journal of Economic Dynamics and Control* 22, 597-626.

Printing and Binding: Strauss GmbH, Mörlenbach